W9-CRX-627

NATO ASI Series

Advanced Science Institutes Series

A series presenting the results of activities sponsored by the NATO Science Committee, which aims at the dissemination of advanced scientific and technological knowledge, with a view to strengthening links between scientific communities.

The Series is published by an international board of publishers in conjunction with the NATO Scientific Affairs Division

A Life Sciences	Plenum Publishing Corporation
B Physics	London and New York
C Mathematical and Physical Sciences	Kluwer Academic Publishers
D Behavioural and Social Sciences	Dordrecht, Boston and London
E Applied Sciences	
F Computer and Systems Sciences	Springer-Verlag
G Ecological Sciences	Berlin Heidelberg New York
H Cell Biology	London Paris Tokyo Hong Kong
I Global Environmental Change	Barcelona Budapest

PARTNERSHIP SUB-SERIES

1. Disarmament Technologies	Kluwer Academic Publishers
2. Environment	Springer-Verlag/Kluwer Acad. Publishers
3. High Technology	Kluwer Academic Publishers
4. Science and Technology Policy	Kluwer Academic Publishers
5. Computer Networking	Kluwer Academic Publishers

The Partnership Sub-Series incorporates activities undertaken in collaboration with NATO's Cooperation Partners, the countries of the CIS and Central and Eastern Europe, in Priority Areas of concern to those countries.

NATO-PCO DATABASE

The electronic index to the NATO ASI Series provides full bibliographical references (with keywords and/or abstracts) to about 50 000 contributions from international scientists published in all sections of the NATO ASI Series. Access to the NATO-PCO DATABASE compiled by the NATO Publication Coordination Office is possible in two ways:

- via online FILE 128 (NATO-PCO DATABASE) hosted by ESRIN,
  Via Galileo Galilei, I-00044 Frascati, Italy.

- via CD-ROM "NATO Science & Technology Disk" with user-friendly retrieval software in English, French and German (© WTV GmbH and DATAWARE Technologies Inc. 1992).

The CD-ROM can be ordered through any member of the Board of Publishers or through NATO-PCO, Overijse, Belgium.

Series I: Global Environmental Change, Vol. 44

Springer
Berlin
Heidelberg
New York
Barcelona
Budapest
Hong Kong
London
Milan
Paris
Santa Clara
Singapore
Tokyo

Decadal Climate Variability

Dynamics and Predictability

Edited by

David L.T. Anderson
Department of Atmospheric Physics
Clarendon Laboratory
Parks Road, Oxford, OX1 3PU, UK

Jürgen Willebrand
Institut für Meereskunde an der Universität Kiel
Düsternbrooker Weg 20
D-24105 Kiel, Germany

Springer

Published in cooperation with NATO Scientific Affairs Division

Proceedings of the NATO Advanced Study Institute "Decadal Climate Variability: Dynamics and Predictability", held at Les Houches, France, February 13–24, 1995

Library of Congress Cataloging-in-Publication Data

Decadal climate variability : dynamics and predictability / edited by
   David L.T. Anderson, Jürgen Willebrand.
      p.   cm. -- (NATO ASI series. Series I, Global environmental
   change ; vol. 44)
      "Published in cooperation with NATO Scientific Affairs Division."
      "Proceedings of the NATO Advanced Study Institute 'Decadal Climate
   Variability: Dynamics and Predictability', held at Les Houches,
   France, February 13-24, 1995"--Verso t.p.
      Includes bibliographical references and index.
      ISBN 3-540-61459-1 (hard cover)
      1. Climatic changes--Congresses.   I. Anderson, D. L. T. (David L.
   T.) II. Willebrand, J. (Jürgen), 1941-   . III. North Atlantic
   Treaty Organization. Scientific Affairs Division.   IV. NATO
   Advanced Study Institute "Decadal Climate Variability: Dynamics and
   Predictability" (1995 : Les Houches, Haute-Savoie, France)
   V. Series.
   QC981.8.C5D36   1996
   551.6--dc20   96-27903
                                                                  CIP

ISBN 3-540-61459-1 Springer-Verlag Berlin Heidelberg New York

© Springer-Verlag Berlin Heidelberg 1996
Printed in Germany

Typesetting: Camera ready by the editors
Printed on acid-free paper
SPIN: 10475273   31/3137 - 5 4 3 2 1 0

Preface

Decadal climate variability is currently an area of considerable scientific interest whose importance is likely to grow over the next decade. Climate Change on decadal time scales can result, for example, from changes in the concentration of radiatively active gasses but also from natural variations. The causes of natural variability of the climate system on decadal time scales are presently not well known; understanding the mechanisms is however a prerequisite for any climate prediction of changes on these time scales. To this extent a major new experiment, called CLIVAR, has recently been launched by the World Climate Research Programme to determine climate predictability both globally and regionally, and to understand the physical processes giving rise to that predictability. The time scales of interest range from seasons to decades and centuries, and include phenomena such as El Niño Southern Oscillation and the North Atlantic Oscillation, but also long-term changes in the ocean thermohaline circulation.

The purpose of the NATO Advanced Study Institute, which was held in February 1995 in Les Houches, France was to introduce young scientists into the current state of the field. The focus was on modelling, analysing and understanding of ocean-atmospheric interactions, primarily on decadal time scales, although these can not be well separated from both shorter and longer time scale processes. Major topics addressed were observed characteristic patterns of natural variability on decadal time scales in the coupled atmosphere-ocean system, the generation mechanisms which are presently not well understood, the potential predictability of long-term variability, the possibility of rapid climate regime shifts on several timescales, the causes for the large changes taking place at the present time in the northern seas of the Atlantic ocean.

A number of mechanism which are most likely to cause decadal variability have been proposed. Nonlinear interactions within the atmospheric circulation can, in the presence of orography and topography, induce low frequency variability of definite spatial organisation. Interaction with slower components of the climate system (land, ocean, ice) can then produce apparent lower frequency variability. Variability can also arise due to the coupling of atmosphere and ocean. The variability due to El Niño-Southern Ocean (ENSO) is one example of a coupled mode which primarily associated with interannual time scales, but can also have decadal variability.

A coupled mode of decadal time scales, centered over the North Pacific in which the interactions of the atmosphere and the ocean are essential to the mode has been suggested, with the time scale set by ocean planetary wave propagation. Recent model results have shown the possibility that the thermohaline circulation can undergo substantial oscillations on decadal to millenial time scales. Apparently the atmosphere plays no active role but would have to adjust to changes in oceanic heat transport and consequently sea surface temperature. Only those mechanisms involving the ocean are likely to lead to some predictability, but this has yet to be demonstrated.

To provide a coherent structure to the course, principal lecturers developed their topics in depth over a series of lectures. Specific applications were further developed by supporting lecturers. We would like to thank all the lecturers who gave a stimulating series of lectures which are recorded in this volume.

Primary support for the Institute was provided by NATO, through the Special Programme on The Science of Global Environmental Change. Additional support was provided by the European Union and by CNRS, France. This support is gratefully acknowledged. We also thank Shona Anderson, Arne Biastoch and Nils Rix who converted the manuscripts into Latex, and a number of reviewers for helpful comments.

David Anderson
Jürgen Willebrand

April 1996

Contents

OBSERVED CLIMATIC VARIABILITY: TIME DEPENDENCE

JOHN M. WALLACE
University of Washington
Seattle, USA

Contents

1 Introduction

Preconceived notions concerning the nature and causes of climate variability determine the datasets that scientists examine, the analysis tools they employ, and the questions they address in their research. Their choices, in turn, define and limit the range of possible outcomes. If these notions are wrong, the research is likely to get off on the wrong track. If they are lacking altogether, the course of the research may be determined by default, through ad-hoc choices, as in the maxim: "If all you have is a hammer, all you'll see is nails".

This chapter is designed to complement the abundant literature on statistical analysis techniques used in climate research (e.g., Barnett 1981, Joliffe 1986; Richman 1986; Preisendorfer 1988; Vautard and Ghil 1989; Bretherton et al. 1992; von Storch et al.1995). Rather than illustrating

NATO ASI Series, Vol. I 44
Decadal Climate Variability
Dynamics and Predictability
Edited by David L. T. Anderson and Jürgen Willebrand
© Springer-Verlag Berlin Heidelberg 1996

how various techniques might be used, we will attempt to identify some of the opportunities and unresolved issues in climate research that invite further analysis of observed and model generated datasets. Rather than focusing on the solutions we will explore the problems. The choice of topics to be included was largely dictated by the author's preconceived notions concerning the nature of climate variability, which include the following:

1. The statistically robust, richly textured response to the annual cycle in sun-earth geometry is still capable of yielding new insights into the inner workings of the climate system (§2.1)

2. Quasi-periodic phenomena, even if they exist, are of only academic interest from the standpoint of short term climate variability, because they cannot possibly account more than a minute fraction of the total variance of the climatic variables of the greatest interest from the point of view of prediction. By far the most prominent quasi-periodic climate 'signal' is the equatorial stratospheric quasi-biennial oscillation (§2.2, 3), but this phenomenon exhibits only a tenuous connection with tropospheric climate variability. It will be argued that the dominant interannual signal in the climate system: the El Niño Southern Oscillation (ENSO) phenomenon (§2.2) occupies too broad a band of frequencies to be regarded as quasi-periodic.

3. As demonstrated clearly by recent numerical modeling results of Manabe and Stouffer (1996), not all interdecadal variability in climatic time series is caused by processes operative on the interdecadal time scale. Much of it is merely a reflection of inherently unpredictable sampling fluctuations due to the presence of variability on the intraseasonal and interannual time scales (§2.3-4, and 8).

4. Well defined interdecadal to century scale variability is clearly evident in certain climatic time series such as hemispheric- or global-mean surface air temperature, even in their raw, unfiltered form (§2.4).

5. Because of the inherent lack of stationarity of many climatic time series, in the absence of a priori predictions it is extremely difficult to demonstrate the statistical significance of 'regime shifts' or unprecedented events (§2.5).

6. With a few notable exceptions involving quasi-periodic phenomena, lead/lag relationships observed in association with interannual to in-

terdecadal climate variability are not particularly strong. Much of the observed variability can be described as separable functions of space and time (§3).

One-dimensional time series will be considered in the next section and multi-dimensional time series in the following one.

2 Climate time series

The evolution of the climate system can assume a wide variety of forms, but for the purpose of this chapter, it will be convenient to group them into the following categories: periodic and quasi-periodic phenomena, aperiodic and random variability, low-frequency trends, and distinctive temporal signatures such as discrete jumps.

2.1 Periodic phenomena

For the range of time scales emphasized in this volume, by far the most important periodic climate signal is the response to the annual march in the intensity and latitudinal distribution of incoming solar radiation. Even the more subtle features associated with the annual march are comparable to or larger in amplitude than the strongest interdecadal to century scale climatic signals and, because of their regularity, they are much easier to isolate and diagnose in observations and model output. The most dramatic and easily explained features of the annual march are the large summer-winter swings in temperature throughout extratropical latitudes and the related summer monsoon rainfall maxima over the subtropical continents. Let us pass over these and consider a few of the more subtle, less easily explained features of the annual march, which lend insight into the inner workings of the climate system.

The annual march of zonally averaged temperature at the mesopause level, near 80 km is exactly the opposite of what one would expect on the basis of radiative considerations: the summer hemisphere is cold and the winter hemisphere is warm. This anomalous behaviour is a consequence of a seasonally reversing pole-to-pole mean meridional circulation cell, characterized by ascent and adiabatic expansion over the summer pole and subsidence and adiabatic compression over the winter pole which drives the temperatures away from local radiative equilibrium. Throughout most of the layer that it occupies, this circulation cell is thermally direct, with warm air rising and cold air sinking, but at levels above 65 km it is thermally indirect (Leovy 1964).

In a similar manner, the tropical (20N-20S) tropopause is remarkably cold year round because of the upwelling associated with the Hadley Circulation, and the annual march of temperature at this level bears little direct relation to the variations in radiative heating associated with the earth's orbital geometry. Throughout this belt temperatures are higher in July, when the earth is farthest from the sun, than they are in January. Over the equator, where one would expect them to be zero, the differences range as high as 10 K (Reed 1963, Reed and Vlcek 1968). Furthermore, there is little indication of a semiannual cycle in response to the seasonal variations in solar declination angle.

Yulaeva et al. (1994) have argued that this anomalous annual cycle is a consequence of the fact that the broad mountain ranges in the Northern Hemisphere (the Rockies, in particular) are much more effective in generating vertically propagating planetary-waves than their narrower counterpart, the Andes, in the Southern Hemisphere. The planetary waves in the winter hemisphere disturb the stratospheric polar vortex, transferring heat from low latitudes to higher latitudes. The same planetary-waves induce a time-mean poleward Lagrangian drift of air parcels in the winter hemisphere, which is fed by ascending motion throughout the tropics. The adiabatic cooling induced by this wave-driven ascent contributes to the remarkable coldness of the tropical tropopause. The more intense Northern Hemisphere wintertime planetary-waves, which attain their peak amplitude in January induce a stronger high latitude warming, accompanied by a stronger cooling of the tropical tropopause, than their Southern counterparts, which attain their peak amplitude in July. The almost perfect compensation between the tropical and extratropical annual cycles, as illustrated in Fig. 1 reflects the constraint that the wave-driven Lagrangian circulation cannot produce any net warming or cooling at any given level: it can only move heat poleward (e.g., see Andrews et al. 1987).

Figure 2 shows an extended time series of total column ozone at Arosa, Switzerland, as inferred from ground based measurements. During most years the column ozone exhibits a pronounced springtime maximum, believed to be a consequence of the same wave-induced poleward Lagrangian circulation mentioned in connection with Fig. 1, which carries ozone from its photochemical source region at the 25-km level over the tropics, pole-

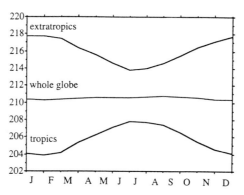

Figure 1: *The climatological mean annual march of lower stratospheric temperature over the tropics (30°N-30°S), the extratropics (poleward of 30° in both hemispheres) and the entire globe, based on data from channel 4 of the the microwave sounding unit for the period 1979 through 1991. The weighting function for this channel is centered near the 70-mb (18 km) level. From Yulaeva et al. (1994).*

ward and downward into a high-latitude reservoir centered near the 15-km level. The distinctive annual march of ozone concentrations at Arosa and other middle and high latitude stations reflects the cumulative effect of the wintertime transports.

Ozone is not the only climatic variable that exhibits a pronounced maximum in the March-April time frame. Sea surface temperatures (SST) over the equatorial Atlantic and tropical eastern Pacific, shown in Fig. 3 exhibit an annual cycle with March-April maxima. The year-to-year variability is larger in the Pacific than in the Atlantic because of the more pervasive influence of the El Niño/Southern Oscillation (ENSO) phenomenon. For example, the row of outlier points along the upper margin of the plot corresponds to the record breaking 1982-83 warm episode. The annual march in the Atlantic is not a pure sine wave: SST cools rather abruptly from May through July and warms more gradually throughout the remainder of the year. It is not at all obvious why SST along the equator should exhibit such a pronounced annual cycle. For discussions of this issue the reader is referred to the observational study of Mitchell and Wallace (1992), the coupled modelling study by Giese and Carton (1994), and the ocean modelling study of Köberle and Philander (1994). The annual march of SST in the equatorial eastern Pacific exerts a profound influence upon the rainfall in adjacent coastal regions of South America. Significant stream-flow in the Piura river in northern Peru (Fig. 4) is largely confined to the warm

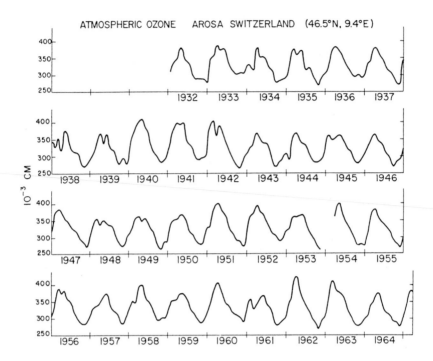

Figure 2: *Monthly mean column ozone amounts at Arosa, Switzerland, 1932-65, as inferred from ground based measurements. Units 10^{-3} cm at STP.*

season for offshore SST (January - May). Stream flow during the rainy season ranges from near zero during years with colder than normal SST to flood conditions during El Niño years.

The phenomena documented in Figs. 1-4 serve to illustrate the complexity of the response of the climate system to a simple periodic thermal forcing. Dynamical and physical linkages between the various components of the system give rise to a wide variety of phases, and amplitudes can be quite substantial, even in regions not subject to strong direct solar forcing.

For perfectly periodic phenomena such as the annual cycle, harmonic analysis is the best suited analysis tool. When performed on a suitably chosen period of record, it yields a "line spectrum" in which the lines are integral multiples of the fundamental frequency. In contrast, quasi-periodic phenomena correspond to peaks in a continuous spectrum, as defined by the methods of power spectrum analysis. Brier and Bradley's (1964) convincing demonstration that the lunar synodic cycle is evident in precipitation frequencies over the United States (Fig. 5) is a tribute to the

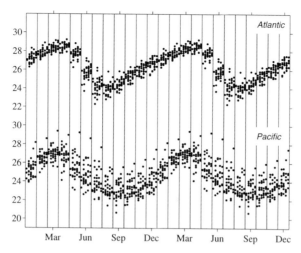

Figure 3: *Scatter plot of monthly mean sea surface temperature (°C) in the equatorial cold tongue regions of the Atlantic and Pacific for individual year/months, grouped by calendar month. The dots for each calendar month have been scattered along the x axis to make them more visible and the calendar year is repeated. Based on data from the Comprehensive Ocean Atmosphere Data Set (COADS) for 1946-85. From Mitchell and Wallace (1992).*

power of harmonic analysis for detecting weak periodic signals.

2.2 Quasi-periodic phenomena

Because climate research tends to be driven, in large part, by the quest for prediction, quasi-periodic phenomena have attracted a disproportionate amount of attention per unit explained variance in the climate record. The literature of the first half of this century abounds with claims of periodicities and quasi-periodicities in the climate record; many of them related to the 11-year cycle in solar activity and its harmonics and subharmonics. Few, if any of these claims have withstood the test of time. Decidedly fewer quasi-periodicities have been reported since the advent of power spectrum analysis, which imposes more rigorous standards for assessing statistical significance.

By far the most notable, and perhaps the only universally recognized quasi-periodic phenomenon in the climate system on the interannual to interdecadal time scale is the regular downward progression of alternating, zonally symmetric easterly and westerly wind regimes in the equatorial

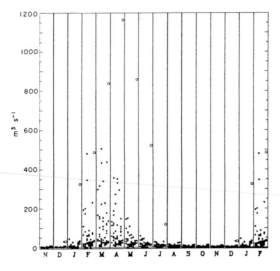

Figure 4: *Monthly-median discharge from the Piura River, which flows westward from the Andes to the Pacific in northern Peru. Each dot represents a particular year month. The dots for a particular calendar are scattered along the x axis to make them more visible and the calendar year is repeated. The circles represent months during the 1982-83 El Niño. From Deser and Wallace (1987).*

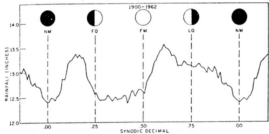

Figure 5: *The total rainfall for a network of more than 100 stations in the United States summarized according to the lunar synodic (29.53 day) month for the period of record 1900-62. From Brier and Bradley (1964).*

stratosphere, first pointed out by Reed (1960) and Ebdon (1961). The time required for the wind to execute a full easterly/westerly cycle ranges from slightly less than two years to about three years: hence the term quasi-biennial oscillation (QBO). This phenomenon dominates time series of zonal wind at levels between 10 and 100 hPa (17 and 30 km) at stations within 10 degrees of the equator, as illustrated in Figs. 6 and 7.

In comparison to the equatorial stratospheric QBO, time series related to ENSO such as the one shown in Fig. 8, exhibit a less regular behaviour, as evidenced by the greater breadth of its spectral peak in Fig. 9. The

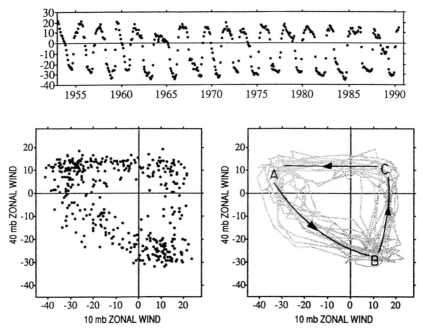

Figure 6: *Top panel: monthly mean zonal wind at the 30-mb (24 km) level over the equator in units of m/s. Lower left: scatterplot of 5-month running mean zonal wind anomalies over the equator at the 10 versus 40-mb (30 vs. 22 km) levels. Lower right: gray lines represent the trajectories in the same two dimensional "phase space", constructed by connecting the points in the bottom left panel in chronological order. Arrows indicate idealized orbits showing the sense of the motion along the trajectories. From Wallace et al. (1993). Data courtesy of Barbara Naujokat, Free University of Berlin.*

power spectrum of the QBO is dominated by a spectral peak centered near a period of 28 months, while the ENSO time series exhibits a much broader band of enhanced power extending all the way from 2 to almost 10 year periods. A measure of the breadth of the peaks is the ratio of the lower to the higher frequencies that bound them: it is of order 0.6 for the QBO compared to 0.2 for the ENSO time series. It is evident from a comparison of Figs. 6 and 8 that, whilst several cycles of the QBO are sufficient to illustrate how it behaves, many more realizations of the ENSO cycle are required in order to sample its full range of variability. One might even question whether even the 120 year record in Fig. 8 is long enough.

Most climatic time series from which the climatological mean annual march has been removed lack robust, well-defined peaks in their power spectra. Rather than focusing one's attention on marginally significant, narrow spectral bumps and dips, which account for only very small frac-

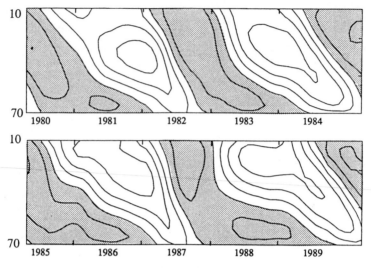

Figure 7: *Time-height section of 5-month zonal wind anomalies over the equator. Contour interval 10 m/s. Westerlies are shaded. The ordinate ranges from the 70-mb pressure level near the tropical tropopause at 17 km to the 10-mb level near 31 km and the scale is logarithmic in pressure and approximately linear in height. Labels on the abscissa denote the midpoint of the calendar year. From Wallace et al. (1993)*

tions of the total variance of the time series, is usually more informative to regard the time series as aperiodic (i.e., without periodicities or quasi-periodicities) and to focus on the gross features of the spectra and the relationships between variables that prevail across wide ranges of frequencies and account for substantial fractions of the total variance.

2.3   Aperiodic and random variability

Whether the variability associated with a particular aperiodic phenomenon should be viewed as deterministic or random depends upon the range of frequencies under consideration. For example, baroclinic waves and their attendant cyclones and fronts are deterministic and quite predictable out to several days, yet they have to be treated stochastically in seasonal-to-interannual climate prediction. The "stormtracks" (the envelopes in which the most vigorous baroclinic wave activity resides) might evolve in a deterministic manner on the seasonal to interannual time scale, but not the individual cyclones within them. In a similar manner, ENSO is deterministic and to some extent predictable on the seasonal to interannual time scale, but it would have to be treated stochastically in dealing with decade-to-century scale variations.

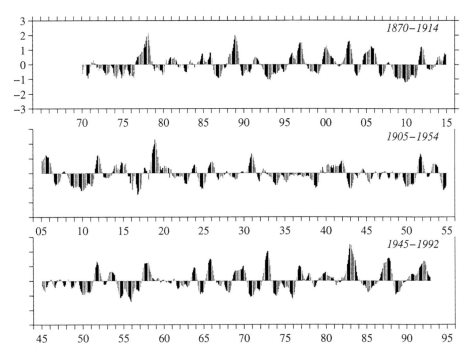

Figure 8: *Sea surface temperature (SST) averaged over the region 6°N-6°S, 90°- 180°W depicted by the shading in Fig. 18 of chapter 2. Gray bars denote warm season (January through May) values and black bars denote cold season (July through November) values. Based on data from the Comprehensive Ocean-Atmosphere Data Set (COADS). The time series has been smoothed with a 5-month running mean filter. To compensate for changes in instrumentation discussed in section 2.3 of chapter 2 a constant value 0.3 K has been added to monthly values prior to December 1941.*

The distinguishing characteristic of random variability, in the sense that the term will be used in this chapter, is the lack of autocorrelation at all lags other than zero. Successive values in the time series are independent of one another: the next data point cannot be predicted with any degree of skill whatsoever on the basis of a knowledge of the previous time history. The Fourier transform of such a lag correlation function, which defines the power spectrum of the associated time series exhibits equal power at all frequencies. It is commonly referred to as "white noise" in analogy with the electromagnetic signature of white light, whose electromagnetic spectrum

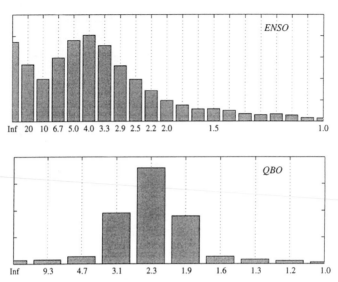

Figure 9: *Power spectra of (top panel) equatorial Pacific SST (the time series in Fig. 8) based on twice-yearly data for 1882-1992 and (bottom panel) zonal wind at the 30-mb level over the equator (the time series in the top panel of Fig. 6). The abscissas of both panels are linear in frequency, but are labeled in terms of period, in years.*

is independent of frequency. It follows from elementary sampling theory that the expected value of the variance of such a time series is independent of the length of the time series and the expected value of the variance of time means of consecutive segments of the series, each consisting of N successive (independent) data points, decreases in inverse proportion to N.

In contrast, the spectra of aperiodic time series generated by deterministic processes tend to be "red": i.e., their spectra tend to drop off with increasing frequency. Such time series are sometimes modeled stochastically in terms of a first order autoregressive process, defined by the relation

$$x_{n+1} = ax_n + (1 - a)\varepsilon_n$$

where the subscripts n and $n+1$ refer to successive values of the series x, ε is a "white noise" time series of unit variance, and a is the linear correlation between successive data points in the time series, commonly referred to as the "lag 1 autocorrelation", a measure of the "redness" of the time series. When a random time series is being generated to be used as a proxy for an observed time series, a is chosen to match the observed time series. The autocorrelation at lag N is a^N and the characteristic "decorrelation time" τ (the time required for the autocorrelation between lagged data points in the time series to drop off by a factor of e) is given by $\delta t/(-\ln a)$, where δt is the time increment between successive data points. For purposes of significance testing, the effectively time between independent data samples in the time series is 2τ (Leith 1973). The expected value of the variance of a such a "red noise" time series increases with sampling interval, but the rate of increase slows as the sampling interval becomes much longer than the decorrelation time. In the limiting case $a = 1$, the time series can be modelled as a 'random walk' process whose expected variance increases linearly with sampling interval.

The expected variance of means of N successive data points in a "red" time series decreases with increasing N at a rate more gradual than $1/N$, but approaches $1/N$ as the averaging interval becomes much longer than the decorrelation time. In other words, for averaging intervals much longer than τ, non-overlapping means of a "red" time series are linearly independent: their variability must be regarded as random and inherently unpredictable sampling fluctuations associated with whatever physical processes happen to be operative at higher frequencies. In a similar manner, the power in the spectrum of a "red" time series increases with decreasing frequency, but the rate of increase slows as frequency becomes smaller than the inverse of the decorrelation time and it eventually levels off. Hence, at sufficiently low frequencies, even 'red' time series exhibit 'white' spectra.

Madden (1976) expressed concern that the month-to-month and winter-to-winter variability inherent in the climate record might be nothing more than sampling variability associated with the presence of higher frequency phenomena such as baroclinic waves with characteristic time scales of days and sporadic blocking episodes that might last as long as a week or two. To illustrate the validity of his concern, we will make use of a synthetic "climate", whose low frequency behaviour is known with much greater precision than that of the real atmosphere. The results presented in this subsection are based on a 100,000 day, 'perpetual January' simulation with a low resolution (rhomboidal 15 truncation) GFDL general circulation model (GCM) run with fixed climatological mean SST. If a single winter is regarded as being 100 days in length, this run provides a sample size equivalent to 1000 winters: roughly 20 times as many as in the observational record.

The statistics presented in Fig. 10 are based on the simulated 500-hPa height field poleward of 20°N. The calculations described below are roughly equivalent to what would be obtained if they were performed on the time series for each individual gridpoint and then area averaged over the entire domain to form a single plot for each type of calculation. First a set of 'aggregated' time series was generated from each of the daily time series derived from the GCM run: the first N daily values were averaged to form the first data point of the aggregate time series, daily values from $N + 1$ to $2N$ were averaged to form the second data point, etc. This procedure is designed to emulate the conventional formatting of climate time series in terms of non-overlapping monthly means, annual means, decadal means etc. The autocorrelation was then computed at a lag of one data point for the daily time series and the entire set of aggregated time series generated from them. The results are shown in the upper left panel. Lag on the abscissa refers to the averaging interval N used in forming the aggregated time series. The figure shows that for averaging intervals less than about 100 days, successive values of the aggregated time series are positively correlated (i.e., the series are "red"), whereas for longer intervals they are uncorrelated (i.e., the series resemble white noise).

The upper right panel shows how the variance of the 500-hPa height field in the GCM drops off with increasing averaging interval N. For averaging intervals longer than 100 days the curve becomes tangent to the dotted straight line, whose slope is indicative of an inverse proportionality between variance and averaging interval, the characteristic signature of white noise. The lower left panel shows the variance based on subsamples of the record N days in length, averaged over all such samples obtained from the 100,000 day record. Variance increases with the length of the subsamples for subsample lengths up to around 100 days, beyond which only a very small further increase is evident.

The lower right panel shows the corresponding frequency spectra based on the Tukey lag-correlation method, using a maximum lag of 1000 days (1% of the length of the record), the longest that was found to yield an acceptably smooth spectrum. The spectral resolution (which is also the lowest frequency that can be resolved) corresponds to one cycle per two lag intervals (2000 days) and the curve in Fig. 10 begins with the second spectral estimate (1 cycle per 500 days): the lowest frequency that is not affected by the detrending of the time series. The transition from white to red, where the variance begins to drop off with increasing frequency, is seen to occur around a period of a few hundred days, which corresponds to the around the 5th spectral estimate. The white (flat) segment of the spectrum is confined to the region so close to zero frequency that it is not

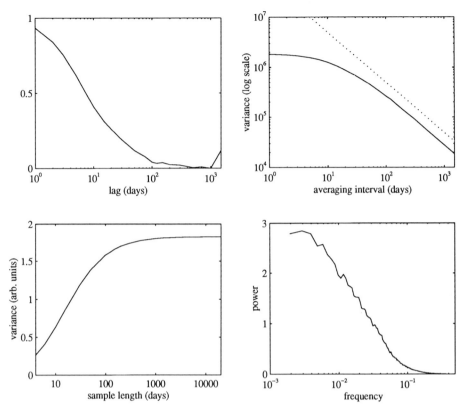

Figure 10: *Hemispheric-mean 500-hPa height statistics derived from a 100,000 day per-petual January integration of the NOAA/GFDL R15 truncation general circulation model run with fixed climatological mean SST. Upper left: lag-1 autocorrelation for the aggre-gated time series as a function of lag between successive data points. Upper right: variance of the aggregated time series as a function of averaging interval, plotted on a log scale in arbitrary units. The slope of the dotted line corresponds to an inverse relation between variance and averaging interval. Lower left: average variance of consecutive segments of the time series as a function of sample length. Lower right: log/log plot of the power spectral density (variance per unit frequency interval) as a function of frequency in cycles per day. Note that variance is not proportional to area under the curve. All statistics represent approximations based on the 10 leading EOF's of the simulated 500-hPa height field.*

clearly visible when this spectrum is plotted on a linear frequency scale (not shown).

Although each involves a different way of analyzing the time series, the four displays in Fig. 10 contain essentially equivalent information about the transition between aperiodic and random behaviour. Such redundancy is to be expected, since the power spectrum is the Fourier transform of the autocorrelation function, and for each value of N, the sum of the mean variance of samples of length N days plus the average variance of means over N successive days (i.e., the sum of the variances plotted in the upper right and lower left panels) is equal to the total variance of the daily data, based on the full length of the record. Regardless of which way the data are presented, the transition between deterministic, aperiodic behaviour at the higher frequencies and white noise at the lower frequencies is quite gradual: it is impossible to say at precisely what frequency or on what time scale it occurs. The 'transition zone' in the frequency domain, in which the behaviour is neither purely deterministic, nor completely random is particularly broad in this example because the GCM mimics several types of aperiodic variability in the real atmosphere, ranging from baroclinic instability, with a characteristic time scale of a day or two, to a zonally symmetric mode that varies on a time scale of a few weeks. The transition zone in Fig. 10 may be viewed as representing a superposition of a number of somewhat narrower zones, each bracketing the 'low-frequency cutoff' of a particular mode of deterministic aperiodic variability. In the coupled climate system, deterministic aperiodic variability extends out to much longer time scales where it coexists with random sampling fluctuations associated with the atmosphere's own internal variability.

Even in the case of this extremely long time series, the "white" segment of the spectrum in Fig. 10 is barely resolved. It would be possible, in principle, to extend the spectrum to lower frequencies, but only at the expense of reducing the number of degrees of freedom of the spectral estimates, rendering them even more noisy than those in Fig. 10. If the time series had been, say, 10,000 days long instead of 100,000 days, and/or if lower frequency dynamical processes such as those associated with the ENSO cycle had been present in the model, as in the real world, the transition zone would have extended to the lowest resolvable frequencies, as it typically does in the analysis of observational data.

Even though the power spectrum of a "red noise" time series is featureless, such a time series can yield what could mistakenly be interpreted as intriguing looking results when subjected to certain analysis techniques. When a large ensemble of 'red noise' time series of length N is subjected to EOF analysis in the time domain, one obtains a set of modes that resemble the results of harmonic analysis except that the frequencies of the family of waves are not constrained to be integral multiples of $2\pi/N$.

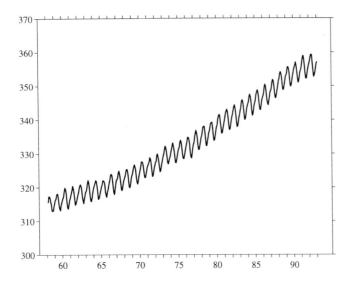

Figure 11: *Monthly averaged CO2 concentrations in parts per million of dry air at the Mauna Loa Observatory, Hawaii.*

Hence, the recovery of periodic or quasi-periodic modes in singular spectrum analysis, which are the EOF's derived from an ensemble of segments of an extended time series, does not necessarily imply that periodicities or quasi-periodicities are present. It might merely be an indication that the time series under investigation resembles 'red noise'.

2.4 Trends

Decadal-to-century scale trends in the climate record constitute a particularly important class of variability. Figure 11 shows a familiar example: the upward trend in carbon dioxide concentrations at Mauna Loa. Also evident in this record is a robust annual cycle believed to be associated with the uptake of carbon dioxide by the terrestrial biosphere during the Northern Hemisphere growing season. What makes the trend so obvious in this particular example is the clear indication of an enormous 'spectral gap' between the annual cycle and century scale variability.

The record of Northern Hemispheric-mean surface air temperature anomalies shown in the bottom panel Fig. 2 of chapter 2 , exhibits distinctive variability on the interdecadal-to-century time scale. The spectral gap is not as obvious as in the previous example, but the evidence of warming in the 1920's and '30's, cooling in the 50's and '60's and warming from 1976-90 is nonetheless quite compelling, even in this unfiltered version of the record. For example it is evident, even to the naked eye, that a large fraction of the monthly mean temperatures observed during the decade of the 1980's were above the normals for their respective calendar months based on averages over the previous three decades. If the data were smoothed, say with a one year running mean filter, the interdecadal to century trend would be no less apparent, but visual information on the 'signal to noise ratio', as manifested in the amplitude of the long term changes relative to the width of the envelope within which the individual data points tend to be clustered, would be substantially degraded. Formal statistical measures of the signal to noise ratio could, of course, be calculated, but it is not clear that they would be any more discerning or informative than the subjective impression that emerges from the image processing that goes on inside the human brain when it is provided with the raw data in the analogue format of Fig. 2 of chapter 2.

The warm season (May-October) data points, indicated by the darker dots in the top panel of Fig. 16 of chapter 2 , exhibit substantially less month-to-month variability than their cold season counterparts, rendering the interdecadal to century scale "signal" more prominent. A distinction is also apparent in the corresponding 5-year running mean time series shown in the top panel of Fig. 17 of chapter 2. The cold season curve exhibits a richer spectrum of variability in the range of periods from ten years, which corresponds to the high-frequency cutoff of the filter, up to several decades, whereas the variability in the warm season time series is more concentrated at the interdecadal-to-century time scale. In view of the much larger month-to-month variability in the cold season time series, it is conceivable that much of the excess variability at periods ranging from 10 to 40 years could merely be sampling fluctuations, as discussed in the previous section. If one only had access to the smoothed data in Fig. 17 of chapter 2, one might be inclined to accept the differences between the warm and cold season time series at face value, without questioning whether they have any physical significance. It will be shown in the next chapter that most of the excess variability of hemispheric mean temperature during the cold season is associated with a particular spatial pattern of hemispheric circulation anomalies, whose month-to-month variations are almost, but not quite random.

There is no guarantee that the noteworthy interdecadal-to-century scale variability within a particular segment of a climate record will happen to assume the form of a linear trend within a particular period of record like a century, as is sometimes presumed in studies of global change. For example, in Figs. 16 and 17 of chapter 2 there is little indication of a linear trend in Northern Hemisphere surface air temperature during the period 1945-90, yet there is a distinct 'U shaped' signature of interdecadal variability during this period.

2.5 Regime shifts and unprecedented events

A discussion of conceptual models of temporal climate variability would not be complete without mention of special temporal signatures, which are not likely to be emphasized in studies based on the conventional tools used in time series analysis. For example, the concept of an abrupt 'regime shift' has direct relevance to the interdecadal time scale, as discussed below, and it figures prominently in the influential work of Hansen and Sutera (1986) who have argued that transitions of the Northern Hemisphere winter circulation between high and low wave amplitude regimes account for a significant fraction of the low-frequency variability.

The existence of regime shifts could be manifested in a number of different ways. If the individual regimes can be characterized in terms of distinctive ranges of values of some index, defined as a continuous function of time, the probability density function (PDF) of that index should be bi-or multi-modal. Another manifestation could be the existence of persistent flow regimes, marked by rapid onset and termination, as discussed by Horel (1985).

An oft cited example of a 'regime shift' on the interdecadal time scale is the warming of the equatorial Pacific, accompanied by a drop in the 'Southern Oscillation Index' (SOI, the Darwin minus Tahiti sea-level pressure difference) that took place in 1976. These changes corresponded to the onset of a warm episode of the ENSO cycle, but the SOI has remained depressed ever since, apart from brief positive excursions in 1984-85 and 1988-89, as documented in Fig. 22 of chapter 2. Trenberth (1990) has suggested that the same regime shift was accompanied by a change in the planetary-scale flow pattern over the extratropical North Pacific, marked by a pronounced drop in sea-level pressure over this region. We will discuss these and other changes in the climate system that took place around the same time in §2.7.

The testing of the statistical significance of reported regime shifts is problematical if they were not predicted a priori; especially if the record is not much longer than the duration of the regimes, as is often the case when one is dealing with interdecadal variability in the historical record. Even in random time series with red spectra, chance superposition of the various Fourier harmonics can sometimes create structures that could be interpreted as regime shifts by an analyst intent on finding them. Impressive looking regime shifts can be created artificially by selecting *a posteriori* (i.e., after inspecting the data) a number of 'key dates' characterized by large positive or negative time derivatives, and compositing segments of the time series relative to these key dates. One can be assured that the larger the number of events included in the composite, the more clearly the 'regime shift', centered on the key date, will stand out above the temporal variability within the artificial 'regimes' that precede and follow it.

The physical and statistical significance of what I will refer to as "unprecedented events" (e.g., record high or low values, the longest uninterrupted sequence with anomalies of the same polarity, the highest frequency of record high or low values within an interval of prescribed length, the strongest linear trend within an interval of prescribed length, the longest interval between events of some prescribed type) in climatic time series of finite length is also bound to be controversial. Even random time series exhibit record high and low values from time to time, and extrema of like sign are likely to be clustered in time if the time series is even slightly red. The pervasive nonstationarity inherent in many climatic time series increases the likelihood of encountering unprecedented events. In particular, the time series associated with chaotic nonlinear systems are full of surprises which may involve changes in time-mean state and/or changes in the character of the variability about the time-mean state; e.g., the amplitude, the dominant frequency, the tendency for periodic or aperiodic behaviour. In a time series of finite length one is never sure that one has sampled the full range of possible behaviour of the natural variability. Just because some new kind of behaviour is noted doesn't necessarily imply that the earth has entered into a new climatic regime or that anthropogenic influences must be responsible for it.

The ENSO time series in Fig. 8 is a case in point. The mean state and the character of the variability about the mean state have varied substantially from one decade to the next. For example, the 1930's and 40's were notable for the absence of a strong ENSO cycle; the '60's and early 70's for the regularity of the cycle with a tendency for 'phase locking' with the annual cycle (as evidenced by the fact that most warm and cold episodes reached peak amplitude during the season July-November); the late '70's and '80's for the high mean SST relative to the previous decades. Whether these interdecadal changes are statistically significant is difficult to judge. One could perform Monte Carlo tests on randomly generated time series to determine the probability that the observed changes could have occurred by chance, but in order to do so one would need to justify the choice of an algorithm for generating the random time series and to quantitatively define the features in the time series that were observed to change on the interdecadal time scale. If these definitions were made on an *a posteriori* basis, in such a way that they tended to accentuate the importance of the apparent regime shifts or unprecedented events in the record, the integrity of the significance test would be compromised.

3 The view from "phase-space"

Although it may be conceptually useful to envision the entire climate system as residing in a multi-dimensional "phase space", in which each of the coordinate axes represents an independent "degree of freedom", this concept is too abstract to apply to real observations. Any serious attempt at a comprehensive representation of the evolution of the climate system over the past few decades or centuries would require a phase space with so many dimensions that the observational record would literally be lost in it (i.e., most of the space would be empty) and the description of the behaviour would involve keeping track of so many variables that it would be very difficult to digest it and distill any useful information from the exercise. Even the task of determining what coordinate axes to use to define the space would be a daunting one. In practice, it is feasible to perform analyses only in a very small subspace selected to illuminate some particular aspect of climate variability. As implied by the first paragraph of this chapter, the selection of that subspace may prove to be the most important step in the analysis.

Periodic and quasi-periodic variations lend themselves to a representation in a one- or a two-dimensional phase space, depending upon whether they can be described as "standing oscillations" or "progressive oscillations". In the former, all time series fluctuate either in-phase or one half cycle out of phase, apart from sampling variability, whereas in the latter they exhibit a continuum of phases. Standing oscillations can be represented just as well by individual time series or by a single linear combination of time series, whereas progressive oscillations can be conveniently represented by a pair of time series which oscillate in quadrature with one another, each of which may consist of a linear combination of the original time series included in the analysis. A phase-space representation offers 'value added' only for progressive oscillations.

3.1 Progressive oscillations in two-dimensional phase-space

An individual time series $x(t)$ can be viewed as a trajectory in a one-dimensional phase-space and a pair of time series $x(t)$ and $y(t)$ define a two-dimensional phase-space, for which the lower panels of Fig. 6 serve as a convenient example. The x and y axes in these panels represent the zonal wind component at two different levels that experience the equatorial stratospheric QBO. Referring to the time-height section in Fig. 7, it is readily verified that an individual cycle of the QBO is represented by a clockwise or counterclockwise loop in the diagram, depending upon whether the upper or lower level zonal wind value is plotted on the x-axis. In this particular example, the trajectory in phase-space executes a series of counterclockwise loops.

The appearance of the loops in such a plot obviously depends upon the vertical separation between the levels selected for the phase-space representation. If they are close together (e.g., zonal wind at the 30-mb plotted versus zonal wind at the 40-mb level) the two time series will be positively correlated and the loops will tend to be elongated along a line with a positive slope in the (x, y) plane. In the limit of no separation between levels, the trajectories would collapse to that line. If the levels are chosen such that one lies near the top of the domain of Fig. 7 (say, 10-mb) and the other near the bottom (say, 50- or 70-mb) the negative correlation between the time series will be reflected in an elongation of the loops along a straight line with a negative slope in the (x, y) plane. The evolution of the QBO is most clearly revealed in plots whose axes correspond to levels separated by about 5 km, which corresponds to roughly one quarter of the effective vertical wavelength of the QBO. The zonal wind fluctuations at such pairs of levels tend to occur in quadrature; i.e., with one lagging the other by 1/4 period. They are neither positively nor negatively correlated with one another: i.e., they are mutually orthogonal.

The loops formed by the evolution of the QBO are represented in the lower left panel of Fig. 6 by a series of dots which represent sequential 1-month means of zonal winds at two different levels in the equatorial stratosphere. The dots tend to be clustered in preferred regions of this particular phase space; i.e., near A and B. As the QBO executes a loop, the trajectory tends to move relatively quickly between clusters. These transitions take place when the layers of rather intense vertical shear in Fig. 7 propagate downward through one or the other of the levels represented in Fig. 6. Hence the characteristic "square wave" shape of the time series in the top panel (i.e., the presence of embedded higher harmonics) is responsible for the irregularity of the loops in the bottom panels. Were it not for these higher harmonics, the loops would be circular or elliptical and the density of points would be uniform along the loops.

A simpler and, in some sense, more informative phase space representation of the QBO can be constructed by expanding the time series of zonal wind at the seven levels represented in Fig. 7 into EOFs and using the time series of the expansion coefficients (often referred to as "principal components" or PCs) of the two leading EOFs as the axes in the phase space plot. In this particular example, the two leading EOFs account for 90% of the variance of the monthly mean time series and over 95% of the 5-month running mean time series of zonal wind at the seven levels. Hence, their PC time series contain the information required to reconstruct the more robust features of Fig. 7 (Wallace et al., 1993). The EOFs themselves, shown in Fig. 12, resemble sinusoidal functions of height, in quadrature with one another, with a vertical half-wavelength $\sim$ 10km.

When represented in the phase space defined by the two leading PCs (Fig. 13), the QBO appears simpler and more coherent than in Fig. 6. The tendency for clustering is gone and the trajectories are much more circular[1]. The inherent predictability of the QBO is evident in the regularity of the circular orbit.

In perfectly periodic phenomena such as the annual march, the orbits in phase space are perfectly reproducible from one cycle to the next. In the absence of embedded higher harmonics, they can be fully represented in a two-dimensional phase space in which the x and y axes can be iden-

---

[1]Upon careful inspection it is evident that the circular loops in Fig. 13 are centered, not on the origin, but on a point in the lower right quadrant and the density of points tends to be highest in that quadrant. This asymmetry reflects the tendency for westerly regimes to propagate downward more rapidly than easterly regimes in Fig. 7. The lower right quadrant in Fig. 13 corresponds to the phase of the QBO in which easterlies overlie westerlies, and the phase progression is relatively slow.

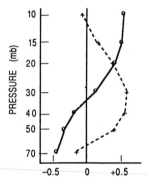

Figure 12: *The two leading EOFs of the temporal correlation matrix constructed from the time series of 5-month running mean zonal wind anomalies at the 70-, 50-, 40-, 30-, 10-, 15-, and 10-mb levels over the equator based on data for the period May 1956 through April 1990. EOF1 is solid and EOF2 is dashed. From Wallace et al. (1993).*

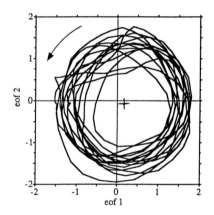

Figure 13: *As in the lower left panel of Fig. 6, but based on the expansion coefficient (or principal component) time series of the two leading EOF's shown in Fig. 12. From Wallace et al. (1993)*

tified with spatial patterns that appear in quadrature with one another in the time domain. For example, in representing the annual cycle, the x axis might represent the July minus January pattern and the y axis the October minus April pattern. The angle relative to the x axis represents the phase of the oscillation (e.g., calendar month in the annual cycle relative to an arbitrary reference time such as 1 January). The 'climate state', which can be reconstructed based on a knowledge of PC1 and PC2, progresses around the circular orbit at a uniform rate: i.e., the spatial pattern associated with the positive polarity of EOF2 appears exactly 1/4 cycle after the spatial pattern associated with the positive polarity of EOF1; the spatial pattern associated with the negative polarity of EOF1 appears exactly 1/4 cycle after the pattern associated with the positive polarity of EOF2, etc. In abbreviated notation: +EOF1 →+EOF2 →-EOF1 →-EOF2 →+EOF1 .... This kind of sequence may be viewed as the canonical form a progressive oscillation.

Although their time dependence is not perfectly periodic, normal mode solutions exhibit a behaviour similar, in some respects, to the progressive oscillation described in the previous paragraph. The general form $\Psi = \exp(\sigma t)[A\cos\omega t + B\sin\omega t]$ may be represented either as an expanding or decaying spiral pattern depending upon whether ω is positive or negative. (For example, if $A = \cos kx$ and $B = \sin kx$, and $\sigma = 0$, Ψ is simply an eastward propagating wave.) If such modal behaviour is evident in observational data or model output, it should be possible to represent it in a two-dimensional phase space. Statistical analysis methods such as complex EOF analysis (Barnett 1981), Principal Oscillation Pattern (POP) analysis; (Hasselmann 1988), and Multi-channel Singular Spectrum Analysis (MCSSA; Vautard and Ghil 1989) are well-suited for detecting such oscillations, if they exist, though if not used in conjunction with stringent tests of statistical significance, they are also capable of detecting spurious oscillations.

Some normal mode solutions, and some quasi-periodic phenomena lack a statistically significant "imaginary component" (i.e., a second EOF whose expansion coefficient time series oscillates in quadrature with that of the leading EOF). Periodic or quasi-periodic phenomena may possess progressive wave signatures when viewed in the appropriate phase space, but they may be detectable only as standing oscillations if the phase space chosen for the analysis does not include the variables or spatial patterns that exhibit a sufficient range of phase lags relative to the leading EOF. In the presence of sampling variability, standing oscillations can easily be mistaken for progressive oscillations if careful significance testing is not performed.

In general, the more periodic the phenomenon, the more its evolution is constrained to resemble a progressive or standing oscillation in some appropriately chosen two-dimensional phase space. Strong bandpass filtering or methods of analysis that have a predilection for quasi-periodic modes (i.e., POP analysis and MCSSA) will inevitably tend to focus the investigator's attention on such phenomena, regardless of whether they have any physical significance. The fact that even red noise can be formally decomposed into quasi-periodic modes underscores the need for rigorous statistical significance testing in conjunction with the use of these methods.

3.2   Random variability in phase space

The distinctions between red and white noise discussed in section 2.3 are easily extended to a multi-dimensional phase space. The autocorrelation between successive points in an individual time series is analogous to the coherence of the points along a trajectory in phase space. If the correspondence between successive points along the trajectory is so weak that it is impossible to identify which point follows which, then that trajectory is more appropriately viewed as 'white noise'. The decrease in temporal variance with lengthening averaging time is analogous to the collapse of the "cloud" of time averaged realizations in phase space toward the origin. The increase in temporal variance with sampling interval is analogous to the growth of the cloud of data points in phase-space.

   The transitional frequency between red and white noise, if it exists, need not be the same along all axes. If some variables (or PC's) are red out to lower frequencies than other variables, increasingly strong time averaging will tend, in addition to collapsing the cloud in all dimensions, to cause what remains of the cloud to become increasingly elongated along the axes of the "reddest" variables. In this situation time averaging has the effect of reducing the dimensionality of the phase space.

4   Concluding Remarks

In view of the complexity of the climate system, with its large number of degrees of freedom in the space and parameter domains, and its multiplicity of feedbacks, highly ordered structures and modes of evolution are likely to be the exception, rather than the rule. It follows that the more conservative or 'coarse' ways of conceptualizing and describing the climate variability (i.e., those that require calculating less detailed information on the structure and evolution of climate anomalies) are likely to be the more robust and informative. In those relatively rare instances in which a more detailed, refined statistical description is warranted, the more coarse description of the observations or model output can serve as a properly elaborated "null hypothesis" for use in establishing the strength and statistical significance of the more specialized "signal" in relation to the background space-time variability. With certain notable exceptions such as the annual march, the more specialized the conceptual model (e.g., the more periodic or the more step-like), the smaller the fraction of the variance of the climate record that it is likely to be able to account for and the more difficult the task of establishing the statistical significance of the products of the analysis. For example, it is easier to determine the variance or skewness of a time series than to determine whether its PDF is unimodal or multi-modal; it is easier to determine the degree of redness of a time series than to determine in detail the shape of its power spectrum. Unless the period of record can be made arbitrarily long, the investigator is bound to be faced with a tradeoff between quantity (or degree of refinement) and quality (i.e., statistical significance) of the statistical information that can be derived from the analysis.

Regardless of the phenomenon under investigation, the simpler and more straightforward the analysis scheme used in presenting the results, the more convincing the presentation and the broader the audience it can reach. I am convinced that the climate 'signals' of primary importance for prediction, detection of global climate change, and elucidation of how the climate system works should be visible to the naked eye in time series, time sections, and/or animations of data that have been subjected to only a minimal amount of processing. The more sophisticated analysis tools can sometimes provide useful guidance as to what parameters or combinations of parameters reveal a particular climate signal most clearly, but once that determination has been made, it should be possible to revert to simpler and more widely used tools to communicate the new results to the scientific public.

In exploratory climate diagnostics, the assessment of statistical significance is often a more formidable task than it might appear, because of the *a posteriori* character of the results (i.e., the fact that the form of whatever space-time structures emerge from the analysis was not predicted beforehand). For example, suppose that a single narrow spectral peak is identified in an exploratory POP's or singular spectrum analysis, whose frequency was not predicted beforehand. In this case it is necessary to assess the probability that a peak of the observed amplitude could have occurred by chance, not at some specified frequency, as in the conventional 'cookbook' formula, but in *any* frequency band of comparable width within the entire spectrum (e.g., see Madden and Julian 1972). If the newly discovered peak occupies only a few percent of the range of frequencies included within the spectrum, the requirements for establishing the statistical significance of the peak at a given confidence level will prove to be far more stringent than in the conventional test. To cite another example, suppose that a bi-modal probability density function is observed for some prescribed circulation index. If the investigator has explored a number of different ways of computing the index, and has singled out one variant of the index for further investigation because it yielded the most convincing evidence of bi-modality, the significance testing needs to appropriately account for the manner in which this exploration was conducted, and the sensitivity of the results to the various permutations of the index that were tried. For example, if the frequency distributions proved to be highly sensitive to the exact definition of the index, and "significant" bi-modality, say at the 5% confidence level in the conventional *a priori* sense, were found for only one of five variants of the index that were explored, then the probability that the reported bi-modality could have occurred by chance would actually be closer to 25%. Even the most sophisticated formal significance tests are bound to overestimate the significance of the results unless they are conducted in the proper context, with due regard for what was, as well as what was not predicted beforehand, as well as any voluntary choices in the analysis procedure that might have served to enhance the apparent significance of the results (regardless of how those choices might be justified after the fact). The assessment of statistical significance of regime shifts and unprecedented events in a time series of limited duration is particularly problematical in view of the inherent nonstationarity of most climatic time series and the *a posteriori* flavour of any statistic that involves a prescribed sampling interval, type of event, or specially designed index.

Access to and familiarity with a wide range of climate data sets is essential if an investigator is to have any hope of exploring the multi-dimensional phase space in which the more important climate signals reside. Among the most important functions of national and international climate programs is the dissemination of the more fundamental climate datasets in easily accessible formats.

In this chapter we have managed to avoid the issue of spatial structure by confining our attention to phenomena that can be represented by just a few time series or, in the case of Fig. 10, averages over an ensemble of time series. The next chapter is largely devoted to this topic.

5 Acknowledgments

Figures 3, 8, 9 and 11 were kindly provided by Todd P. Mitchell and 12 and 13 by Yuan Zhang. Fig. 10 is based on unpublished work of Gregor Nitsche, based on a data set provided by Ngar-Cheung Lau and Mary Jo Nath of NOAA/GFDL. Imke Durre proofread the manuscript and offered helpful suggestions.

References

ANDREWS, D.G., J.R. HOLTON AND C.B. LEOVY, 1987, Middle Atmospheric Dynamics. *Academic Press, Orlando, 489pp.*

BARNETT, T.P., 1981, Interactions of the monsoon and Pacific trade wind system at inter-annual time scales. Part I: The equatorial zone. *Mon. Wea. Rev.*, **111**, 756-773.

BRETHERTON, C. S., C. SMITH AND J. M.WALLACE, 1992, An intercomparison of methods for finding coupled patterns in climate data. *J. Climate*, **5**, 541-560.

BRIER, G.W. AND D.A. BRADLEY, 1964, The lunar synodical period and precipitation in the United States. *J. Atmos Sci.*, **21**, 386-395.

DESER, C. AND J.M. WALLACE, 1987, El Niño events and their relation to the Southern Oscillation: 1925-1986. *J. Geophys. Res.*, **92**, 14189-14196.

EBDON, R.A., 1960, Notes on the wind flow at 50 mb in tropical and sub-tropical regions in January 1957 and 1958. *Quart. J. Roy. Meteorol. Soc.*, **86**, 540-543.

GIESE, B.S. AND J.A. CARTON, 1994, The seasonal cycle in a coupled ocean-atmosphere model. *J. Climate*, **7**, 1208-1217.

HASSELMANN, K., 1988, PIPs and POPs: the rediction of complex dynamical systems using Principal Interaction and Oscillation Patterns. *J. Geophys. Res.*, **93**, 11015-11021.

HANSEN, A.R. AND A SUTERA, 1986, On the probability density distribution of planetary scale atmospheric wave amplitude. *J. Atmos Sci.*, **43**, 3250-3265.

HOREL, J.D., 1985, Persistence of the 500 mb height field during the northern winter. *Mon. Wea. Rev.*, **113**, 2030-2041.

JOLIFFE, I.T., 1986, Principal Component Analysis. *Springer-Verlag. New York, 271pp.*

JONES, P.D., S.C.B. RAPER, B.D. SANTER, B.S.G. CHERRY, C.M. GOODESS, P.M. KELLY, T.M.L. WIGLEY, R.S. BRADLEY AND H.F. DIAZ, 1985, A Grid Point Surface Air Temperature Data Set for the Northern Hemisphere. *U.S. Dep't. of Energy, Carbon Dioxide Research Division, Technical Report TR022, 251pp.*

KÖBERLE, C. AND S.G.H. PHILANDER, 1994, On the processes that control seasonal varia-tions of sea surface temperatures in the tropical Pacific Ocean. *Tellus*, **46A**, 481-496.

LEITH, C.E., 1973, The standard error of time averaged estimates of climatic means. *J. Appl. Meteor.*, **12**, 1066-1069.

LEOVY, C.B., 1964, Simple models of the thermally driven mesospheric circulation. *J. Atmos. Sci.*, **21**, 327-341.

MADDEN, R.A., 1976, Estimates of the natural variability in time averaged sea-level pressure. *Mon. Wea. Rev.*, **104**, 942-952.

MADDEN, R.A. AND P.R. JULIAN, 1972, Description of global scale circulation cells in the tropics with a 40-50 day period. *J. Atmos. Sci.*, **29**, 1109-1123.

MANABE, S. AND R.J. STOUFFER, 1996, Low-frequency variability of surface air temperature in a 1,000 year integration of a coupled ocean-atmosphere model. *J. Climate, submitted.*

MITCHELL, T.P. AND J.M. WALLACE, 1992, The annual cycle in equatorial convection and sea surface temperature. *J. Climate*, **5**, 1140-1156.

PREISENDORFER, R.W., 1988, Principal Component Analysis in Meteorology and Oceanography. *C. Mobley, ed., Elsevier, 418pp.*

REED, R.J., 1960, The Circulation of the Stratosphere. *Paper presented at the 40th Anniversary Meeting of the Amer. Meteorol. Soc., Boston, MA.*

REED, R.J. AND C.L. VLCEK, 1969, The annual temperature variation in the tropical lower stratosphere. *J. Atmos. Sci.*, **26**, 163-167.

RICHMAN, M.B., 1986, Rotation of principal components. *J. Climatol.*, **6**, 293-335.

TRENBERTH, K. E., 1990, Recent observed interdecadal climate changes in the Northern Hemisphere. *Bull. Amer. Meteor. Soc.*, **71**, 988-993.

VAUTARD R. AND M. GHIL, 1989, Singular spectrum analysis in nonlinear dynamics, with applications to paleoclimatic time series. *Physica D*, **35**, 395-424.

VON STORCH, H., G. BURGER, R. SCHNUR, AND J.-S. VON STORCH, 1995, Principal Oscillation Patterns: A review. *J. Climate*, **8**, 377-400.

WALLACE, J.M., R.L. PANETTA AND J. ESTBERG, 1993, Representation of the equatorial stratospheric quasi-biennial oscillation in EOF phase space. *J. Atmos. Sci.*, **50**, 1751-1762.

YULAEVA, E., J.R. HOLTON AND J.M. WALLACE, 1994, On the cause of the annual cycle in tropical lower stratospheric temperatures. *J. Atmos Sci.*, **51**, 169-174.

OBSERVED CLIMATIC VARIABILITY: SPATIAL STRUCTURE

JOHN M. WALLACE
University of Washington
Seattle, USA

Contents

1 Introduction

Atmospheric dynamicists and statisticians have been working more or less independently, in an effort to make sense out of the structure and behaviour of the climate system. Dynamicists have tended to focus on variables such as geopotential height and wind, whereas statisticians have been primarily concerned with long term trends in regionally, hemispherically or globally averaged quantities such as surface air temperature. The dynamicists have confined their attention almost exclusively to wintertime phenomena, whilst those working to document global temperature trends

NATO ASI Series, Vol. I 44
Decadal Climate Variability
Dynamics and Predictability
Edited by David L. T. Anderson and Jürgen Willebrand
© Springer-Verlag Berlin Heidelberg 1996

have considered all seasons. The differing research priorities are rooted in differing paradigms concerning the spatial structure and seasonality of climate variability. In this chapter we will examine these contrasting ways of conceptualizing and analyzing climate variability and attempt to reconcile them.

2   Background

2.1   Some dynamical considerations

The sea-level pressure field is unique among climatic variables in the sense that in the global average, the anomalies at any given time must be very close to zero: i.e., there must be almost perfect compensation between positive and negative anomalies. This constraint is a consequence of the fact that the mass of the atmosphere does not change appreciably on the interdecadal time scale. Strictly speaking, it applies to surface pressure as opposed to sea-level pressure, but the two are very closely related over most of the globe (Trenberth and Christy, 1985). Because of this constraint, the anomalies in the sea-level pressure field tend to assume the form of dipole structures or wavelike patterns. In contrast, there is no reason why positive temperature anomalies or trends in one region need to be compensated by negative anomalies or trends somewhere else. This structural distinction between the sea-level pressure and temperature fields is clearly reflected in the statistical correlation patterns for the Southern Oscillation, as defined by Walker and Bliss (1932), in which the pressure pattern is characterized as an east-west 'seesaw' between the eastern and western sides of the tropical South Pacific, whereas the corresponding temperature pattern is characterized by fluctuations of the same polarity throughout virtually the entire tropics.

Even though the anomaly patterns in the temperature and pressure fields are capable of assuming quite different forms, their horizontal structures are related through the hypsometric equation. Hence, temperature anomalies are bound to be reflected in the pressure (or geopotential height) field. For example, a spatially uniform warming would be reflected in rising heights aloft, whereas a localized warming of the lower troposphere might be reflected in a local drop in sea-level pressure. However, it should be noted that pressure changes can occur in the absence of temperature changes, provided that they are purely barotropic. In a similar manner, because of the constraint of geostrophy, pressure changes are invariably accompanied by wind changes somewhere in the atmosphere, but a horizontally uniform warming or cooling could occur without affecting the wind field.

Because enhanced ascent in one region must be accompanied by enhanced subsidence over the remainder of the globe, the field of precipitation anomalies exhibits some tendency for compensation. However, since precipitation rates depend not only upon vertical motion, but also upon the available moisture, the compensation is not as complete as in the case of sea-level pressure. Because vertical velocity is closely related to horizontal divergence which, in turn, is related to the Laplacian of the pressure field, rainfall anomalies tend to occur on smaller horizontal scales than the concomitant pressure or geopotential height anomalies.

2.2 Geometrical considerations

In contrast to the categorization of temporal variability, which is inherently one-dimensional, the description of spatial variability involves consideration of two or three dimensions, though in practice, the task is usually simplified by reducing the dimensionality of the space domain. Perhaps the most common simplification is the use of zonal averaging to eliminate the longitude dimension so that fields such as surface temperature can be displayed as functions of latitude only and three dimensional fields can be displayed in the form of latitude-height sections. This approach is deeply ingrained in the general circulation literature. If the phenomenon of interest is well represented by zonal-mean data at a particular latitude, it may be possible to collapse such displays to one spatial dimension, as in Fig. 7 of chapter 1.

Since extratropical low frequency circulation anomalies tend to be equivalent barotropic (i.e., similar in shape throughout the depth of the troposphere, with amplitude increasing with height), they are often displayed by patterns at a single level such as 500-mb, which is viewed as representative of vertically averaged conditions. The more significant tropical circulation anomalies, on the other hand, project more strongly onto the first baroclinic mode, which is characterized by wind and pressure anomalies of opposing sign in upper and lower troposphere, with large temperature anomalies in the intervening layer (e.g., see Gill 1980). Hence, display of any one of these variables reveals a considerable amount of information about the others. In contrast to the time domain, the space domain is bounded in all three dimensions. The longitude domain is periodic, which enables it to be represented by Fourier harmonics (zonal wave numbers 0, 1, 2..). Latitudinal variability can be represented in terms of Legendre polynomials. Global or hemispheric patterns on a horizontal surface can be represented in terms of spherical harmonics which are products of sine and cosine functions in longitude and Legendre polynomials in latitude (e.g., see Washington and Parkinson 1986). Just as power spectra can be used to describe and categorize temporal variability, one dimensional spectra on latitude circles or two dimensional spectra based on spherical harmonics can be used to describe and categorize spatial variability. Fourier decomposition of wind and pressure patterns on latitude circles is implicit in many theoretical treatments of baroclinic waves, stratospheric planetary waves, and equatorially trapped waves. The disturbances are represented in terms of sinusoidal variations in longitude modulated by a latitude- height profile of amplitude and phase. Observations are often analyzed in terms of the same conceptual framework.

If a continuous field defined on a spherical surface were characterized by isotropic red noise, where 'red', in this context, refers to the space domain, the EOF's would assume the form of spherical harmonics (North, 1975). In analogy with the red noise time series considered in section 2.3 of chapter 1, the degree of separation between successive eigenvalues would be determined by the degree of redness in the space domain (i.e., the strength of the temporal autocorrelation between the fluctuations at adjacent gridpoints). In the absence of longitudinal asymmetries in the bottom boundary conditions, the circulation would be dominated by such structures. The leading EOF's of the wintertime 50-mb height field (located in the lower stratosphere at an altitude near 20 km), exhibit the kind of geometrical symmetry that is characteristic of spherical harmonics (Fig. 1).

2.3 Considerations of data availability

The types of analyses that can be performed to investigate the structure and evolution of climate variability are constrained by the availability of suitable data sets. In general, the longer the temporal scope of the investigation, the more limited the spatial sampling and the greater the concerns about the homogeneity of the record in the time domain. This section

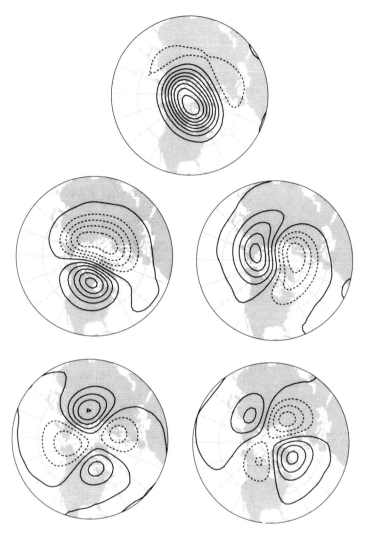

Figure 1: *The five leading EOF's of wintertime-mean (December through February) 50-mb height, computed from the temporal covariance matrix, based on NMC operational analyses for the period 1964 through February 1990. Contour interval 50 m.*

provides a brief, selective chronology of some of the primary observational data sets used in climate research.

The year 1659 marks the beginning of the longest instrumental record of surface air temperature: the one for central England temperature. By 1800 a number of European stations and one U.S. station (New Haven) were recording surface air temperature. The network of roughly 50 stations that were operating in 1856 was extensive enough to enable Jones et al. (1985)

to represent, with a high degree of fidelity, the interdecadal variability of surface air temperature over the Northern Hemisphere continents from that time onward. The gridded monthly surface air temperature data set described in Jones et al. (1985) and Jones (1988) extends back to that date. Marine surface observations of sea surface temperature (SST), surface air temperature, and other meteorological variables, derived from the log books of merchant ships, are also available in a gridded format, extending backward in time to about 1870 for SST and wind (Woodruff et al. 1987, Bottomley et al. 1993, Parker et al. 1995).

A thin global network of stations with barometers was in place by 1900, which marks the beginning of a series of subjectively analyzed Northern Hemisphere daily surface weather maps. The U.S. National Meteorological Center (NMC) began to produce gridded upper air maps in 1946. Manual analysis of these charts was supplanted by a primitive form of objective analysis by the early 60's. The record of visible satellite imagery dates back to 1967 and more quantitative outgoing longwave radiation measurements began in 1974 and have been continuous since that time, apart from a break in 1978. The newly available satellite observations, together with increasingly sophisticated multivariate data assimilation made it possible to extend operational daily surface and upper air analyses to include the tropics and the Southern Hemisphere. The year 1979, which corresponds to the Global Weather Experiment and the first year of operation of the European Centre for Medium-Range Weather Forecasting (ECMWF) marks the effective start date for reliable global upper air data. Coincidentally, it also marks the beginning of the global temperature and oceanic precipitation records derived from the microwave sounding unit (MSU) carried aboard the TIROS satellites (Spencer and Christy 1990, 1992; Spencer 1993).

A few of the more prominent discontinuities in the records deserve mention. Relatively few marine surface observations were archived during World Wars I and II. Prior World War II, the dominant method of inferring sea surface temperature was to measure the temperature of a recently extracted bucket of sea water. Such 'bucket measurements' were supplanted by direct measurements of thermometers housed in condenser intakes, located in hot engine rooms. This change resulted in a spurious upward jump in reported temperatures centered around 1941/42 (Folland and Parker 1995). In recognition of this problem, nighttime marine air temperature records are sometimes used to substantiate the existence of long term trends in SST time series (e.g., Parker and Folland 1991). During the postwar period, anemometer measurements have gradually supplanted observations of sea state (interpreted in terms of the Beaufort Scale) for inference of wind speed. This change in observing practice, together with the increasing height of the bridge decks of ships over the years, has resulted in a spurious increase in reported wind speeds (Cardone et al., 1990). Changes in radiosonde instrumentation and/or analysis procedures in the 1950's and early 1960's raise questions concerning the homogeneity of upper air temperature measurements extending back beyond 1963 (Lambert 1990). An extended record of substantially improved upper air analyses is expected to be available by the end of 1996, as a result of reanalysis projects currently under-way at NMC and ECMWF (Bengtsson and Shukla 1988).

3   How complex are the spatial patterns associated with climate variability?

The climate system is widely regarded as having many spatial degrees of freedom, yet its time dependent behaviour has often been characterized in terms of, or purported to be inferred from the variations in a single time series. This apparent paradox reflects the differing attitudes of researchers with backgrounds in seasonal to interannual climate prediction versus those with backgrounds in paleoclimate. The former are accustomed to working with spatially continuous, gridded fields based on data from a sophisticated observing system. They are inclined to focus on dynamical variables such as geopotential height, which exhibit richly textured spatial patterns, with strong compensation between regional anomalies of contrasting polarity. In contrast, paleoclimatologists are accustomed to working with proxy time series from widely separated sites which are often dominated by similar features with the same polarity. As the two cultures commingle in the melting pot of interdecadal variability, the lack of a commonly accepted paradigm tends to inhibit communication and generate confusion.

Based on dynamical considerations mentioned in section 2.1, it is conceivable that the surface air temperature field could include a substantial hemispheric or even global mean component that might be evident in suitably smoothed time series from local sites. The hemispheric and global mean time series published in the IPCC reports suggests (Folland et al. 1990, 1992) exhibit a distinctive pattern of interdecadal variability that is remarkably well captured in time-filtered data for China alone, as illustrated in Bradley et al. (1987). Whether data for individual stations are capable of revealing interdecadal fluctuations in hemispheric mean temperature is more problematical. On the basis of an examination of twelve seasonal and annual mean surface air temperature records 150 years or more in length, together with results of a more comprehensive study by Jones and Kelly (1983), Jones and Bradley (1992) concluded: "Although there appears to be some agreement between warm and cold decades over the major northern continents, ...no one region can be said to be representative of hemispheric-wide conditions. ...the only way of producing a truly representative time series for the Northern Hemisphere is to include station data from as many regions as possible. There is no short-cut method using only a few stations..." However, they went on to acknowledge that all the sites that they examined showed warming over the period 1851-1980 and that the 10-year lowpass filtered temperature records for 8 of the 12 stations that they examined were quite well correlated with fluctuations in hemispheric mean temperature, with correlation coefficients ranging from 0.64 to 0.76.

The difficulties inherent in relating local temperature fluctuations to hemispheric or global fluctuations are illustrated by Fig. 2. The top panel shows monthly mean surface air temperature anomalies for a localized region of the midwestern United States centered in Iowa. The large month-to-month variability, with a standard deviation of several degrees, is an order of magnitude larger than the observed fluctuations in hemispheric mean surface air temperature on the interdecadal time scale. Even if these month-to-month variations exhibited no persistence, the sampling variability of 10-year running means would be as large as the observed decade-to-decade variability of hemispheric mean temperature. It is only the stronger trends that are sustained over several decades that might, at least in principle, be detectable in individual station records. Monthly mean surface air temperature averaged over the entire United States (exclusive of Alaska and Hawaii), shown in the middle panel of Fig. 2, exhibits considerably less scatter because of the compensation between areas of positive and negative anomalies within this region. Upon close inspection, the distinctive interdecadal signature of hemispheric mean temperature is evident in these data, even without the temporal smoothing. When the spatial averaging is extended to include the entire Northern Hemisphere continents (bottom panel) the scatter is further reduced and the hemispheric interdecadal signal is much more clearly apparent. The similarity between regional and

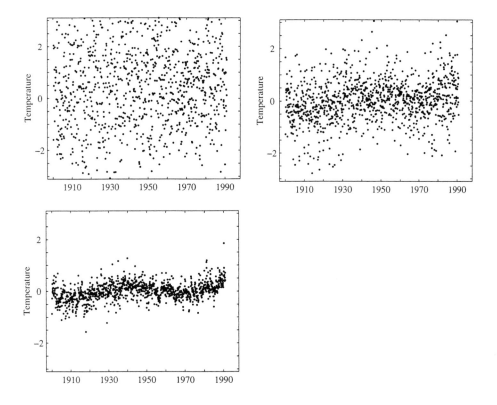

Figure 2: *Time series of monthly-mean surface air temperature anomalies based on the gridded data set compiled by Jones et al. (1985) under the sponsorship of the United States Department of Energy, obtained from the Carbon Dioxide Information Analysis Center (CDIAC). The reference period for computing the anomalies is 1951-80. Upper left panel: the 5 degree latitude x 5 degree longitude grid box centered at (92.5 N, 42.5 W) in the central United States. Many of the individual monthly values lie outside the range of the diagram. Upper right panel: the corresponding time series for the area bounded by (25 and 55 N, 70 and 120 W), which corresponds roughly to the contiguous United States. Lower left panel: the Northern Hemispheric-mean based on all available grid points. The scale on the vertical axis is the same in all three panels.*

hemispheric-mean temperature anomalies is more clearly apparent in the 5-year running mean time series shown in Fig. 3.

It is notable that most of the remaining month-to-month scatter in the lower panel of Fig. 2 is contributed by the cold season months, November through April, as shown in Fig 4. It is also evident from the figure that

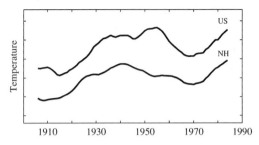

Figure 3: *As in the lower two panels of Fig. 2 but the data have been smoothed with a 5-year running mean filter. The spacing between the curves is arbitrary. The interval between tick marks on the vertical axis is 0.25 K.*

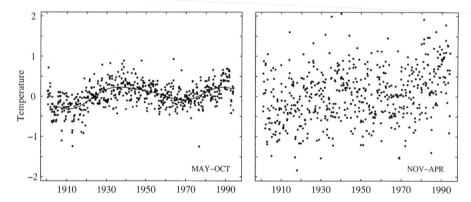

Figure 4: *Northern Hemispheric-mean surface air temperature anomalies based on gridded station data, obtained from the Hadley Centre for Climate Prediction and Research. The data are partitioned into Northern Hemisphere warm and cold seasons, as indicated. From Wallace et al. (1995a)*

warm and cold seasons have exhibited similar patterns of interdecadal to century scale variability. The same is true of surface air temperature over the Southern Hemisphere continents (Wallace, 1995).

From the foregoing, it is apparent that the structure of climate variability on the interannual to interdecadal time scale is quite complex, but the evidence presented here does not contradict the prevailing view that it may be much simpler on the time scale of centuries or longer. In order to understand the nature of the month-to-month variability in local, regional and hemispheric mean surface air temperature and to explain why it is so much larger during the cold season, it will be necessary to learn something about the horizontal structure of the geopotential height and temperature patterns in the troposphere. Since the processes that generate tropical and extratropical variability are somewhat different, these regions will be considered in separate sections. The discussion in these sections will also provide some insights into the nature of regional climate variability on the interdecadal time scale and the extent to which it is coupled to variations in hemispheric- or global-mean temperature.

4 Horizontal structure of extratropical climate variability

The structure of the low-frequency (10-day period and longer) variability of the extratropical circulation is strongly seasonally dependent. Therefore it will be convenient to consider the winter and summer seasons separately. We will begin with winter, whose climatological mean circulation is stronger because of the stronger meridional gradient of diabatic heating that drives it.

The Northern Hemisphere climatological mean wintertime circulation, as reflected in the 500-mb height field, shown in Fig. 5, exhibits strong zonal asymmetries forced primarily by the continent-ocean heating contrasts and the presence of the Rockies and the Himalayas. These asymmetries give rise to a strong longitudinal dependence in the amplitude and structure of the observed low-frequency variability. Figure 6 shows the geographical distribution of the temporal variance of the 500-mb height field based on unfiltered daily data, 10-day low-pass filtered daily data, monthly means and seasonal (December through February) means. All four patterns exhibit pronounced variance maxima over the northern oceans. Variance decreases with the strength of the temporal smoothing that is applied to the data. The four patterns are similar, but upon careful inspection it is evident that the patterns based on monthly and seasonal means exhibit relatively stronger longitudinal contrasts, in a relative sense, than the less heavily smoothed patterns. Although it doesn't show up clearly in the figure, the variance maximum over the North Pacific stands out particularly strongly (in a relative sense) in the seasonal-mean data.

Figure 7 shows a series of one-point correlation maps for the gridpoint (45°N, 165°W) which is centered near the region of largest variance in the monthly and seasonal panels of the previous figure. These maps are constructed by correlating the 500-mb height time series at the 'reference gridpoint' with the corresponding time series at all gridpoints. The one in

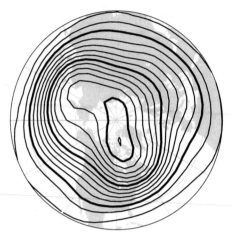

Figure 5: *The wintertime mean 500-mb height field, based on NMC operational analyses for the years 1946-89. Contour interval 60 m. The 5100, 5400 and 5700 m contours are thickened.*

the first panel, which is based on unfiltered data, is dominated by a nearly circular "bull's eye" centered on the reference gridpoint. The fact that it is nearly circular indicates that the correlation coefficient between the time series of 500-mb height anomalies at the reference gridpoint and the corresponding time series at other gridpoints decreases with distance from the reference gridpoint at approximately the same rate, irrespective of direction. Bearing in mind that winds in the atmosphere are quasi-geostrophic, and that the geostrophic wind is proportional to the geopotential height gradient in the transverse direction, the circular correlation pattern implies that the unfiltered wind field in the vicinity of the reference gridpoint is more or less isotropic (i.e., in a statistical sense, the root-mean-squared amplitude of the u and v wind components is the same and is independent of the orientation of the x and y axes). In the subsequent panels based on 10-day lowpass filtered and monthly mean fields, the closed correlation contours surrounding the reference gridpoint is more elliptical in shape, elongated in the zonal direction. Ellipticity implies anisotropy: meridional gradients of 500-mb height tend to be stronger than zonal gradients: hence the zonal component of the geostrophic wind tends to be stronger than the meridional component [1]

---

[1]The corresponding correlation pattern for 500-mb height data that have been highpass filtered to emphasize fluctuations with periods shorter than 10 days (not shown) are anisotropic in the opposite

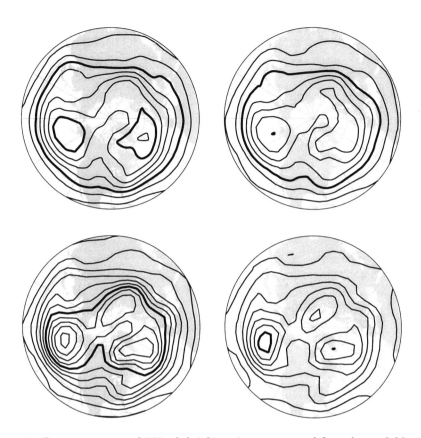

Figure 6: *Root-mean-squared 500-mb height variance computed from (upper left) unfiltered, (upper right) 10-day lowpass filtered, (lower left) 30-day mean and (lower right) wintertime seasonal-mean values, based on NMC operational analyses for the winter months (December through February) of the years 1946-89. Contour interval 20 m.*

Another distinction between the correlation patterns based on filtered and unfiltered data is the greater strength of the features that lie outside the bull's eye pattern; for example, the regions of negative correlation near Hawaii and over western Canada become more prominent as the temporal smoothing becomes stronger. Hence, month-to-month and interannual variability of the 500-mb height field and, by implication, of the wind field as well, are characterized by strong correlations between remote regions, or "teleconnections".

The patterns in Fig. 7 contain elements of the leading EOF of the 10-

sense; i.e., the features are meridionally elongated and assume the form of zonally oriented wavetrains that exhibit many of the properties of baroclinic waves (Blackmon et al. 1984). Wallace and Lau (1985) offer a dynamical interpretation of the contrasting anisotropy between high- and low-pass filtered fields.

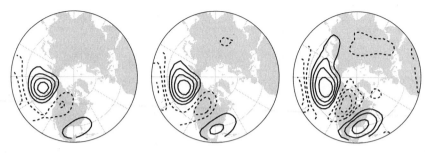

Figure 7: *One-point correlation maps based on the grid-point (45 N, 165 W) which corresponds to the primary centre of action of the PNA pattern. Based on (left to right) unfiltered, 10-day lowpass filtered, and monthly-mean NMC operational analyses for the winter months (December through February) of the years 1946-93. Contour interval 0.2.*

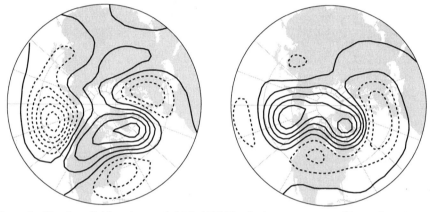

Figure 8: *The first (left) and second (right) EOF's of wintertime-mean (December through February) 500-mb height, computed from the temporal covariance matrix, based on 10-day lowpass filtered NMC operational analyses for the period 1946-89. Contour interval 20 m.*

day lowpass filtered 500-mb height field, shown in Fig. 8. Since EOF's, by their very definition, are structured so as to explain as much as possible of the variance, it is to be expected that the primary "center of action" of the leading EOF should coincide with the region of strongest variability.

Are the correlation patterns in Fig. 7 typical of correlation patterns for gridpoints throughout the hemisphere, or are they particularly strong ones? The most straightforward way to answer this question is to systematically examine the complete set of one-point correlation maps analogous to the ones in Fig. 7, but for all grid points. Such a survey has, in fact been

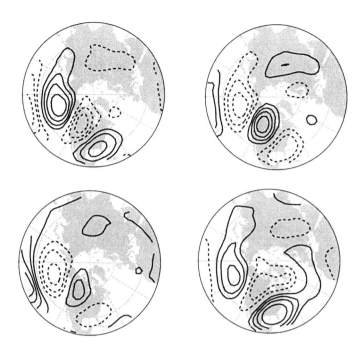

Figure 9: *One-point correlation maps based on four centres of action of the PNA pattern, based on monthly-mean NMC operational analyses for the winter months (December through February) of the years 1946-93. Contour interval 0.2.*

carried out, and it was found that there is only one other map that exhibits a negative correlation as strong as the negative correlations over western Canada in Fig. 7, and that map, shown in Fig. 9c, exhibits a very similar pattern, apart from a sign reversal. Note that the reference gridpoint for Fig. 9c corresponds to the gridpoint with the strongest negative correlation in Fig. 9b and vice versa. An analogous reciprocal relationship is evident for the gridpoints near Hawaii and Florida, whose correlation maps are shown in Fig. 9a,d. Evidently the strong centers of action in Figs. 7 and 9 are associated with the same "teleconnection pattern", which has been labeled the Pacific/North America (PNA) pattern (Wallace and Gutzler, 1981).

It remains to be determined whether the pattern in Fig. 9 is really outstanding among the patterns that appear on one-point correlation maps. Suppose, for example, that one examined such maps for all gridpoints along the "great circle route" connecting the reference gridpoints in Fig. 9. Would one observe equally strong patterns on all maps, or would the correlation maps for the gridpoints located at the nodal lines of the PNA pattern be featureless, apart from the inevitable bull's eye surrounding the reference gridpoint? This question has been addressed by Wallace and Gutzler (1981) and, in a somewhat different context, by Kushnir and Wallace (1987), who showed that anisotropic patterns with multiple centers of action far removed from the reference gridpoint are, in fact, observed on virtually all such maps. The zonal and meridional scales of the features on different maps tend to be similar, though the intensity and the position of the features relative to the reference gridpoint varies from map to map. Hence, the interannual variability of the wintertime circulation is characterized by a continuum of anisotropic (zonally elongated) structures qualitatively similar to the ones that have just been examined. But the pattern in Fig. 9 stands out clearly above this background continuum: the negative correlations between its primary and remote "centers of action" are significantly stronger than their counterparts in maps for typical gridpoints.

Just how many teleconnection patterns stand out above the background continuum depends upon one's criterion for "stand out". Barnston and Livezey (1987) assigned triacronymic labels to no fewer than eight patterns, Wallace and Gutzler (1981) labeled five, whereas Kushnir and Wallace (1987) deemed only two patterns worthy of a label. All three studies concur that the PNA pattern and Walker and Bliss's (1932) so called North Atlantic Oscillation (NAO), documented in detail by van Loon and Rogers (1978) clearly qualify as teleconnection patterns. The NAO, shown in Fig. 10, dominates the leading EOF of the sea-level pressure field (Kutzbach 1970, Rogers 1981). There is a hint of it in both the leading EOF's of the 500-mb height field (Fig. 8), but in order to see it clearly it is necessary to rotate (i.e., to form linear combinations of) the EOFs in a manner that tends to simplify their spatial structure (Horel 1981; Barnston and Livezey 1987; Kushnir and Wallace 1987).

In contrast to EOF patterns, which may assume a full range of amplitudes of either polarity, the patterns identified by cluster analysis techniques refer to anomalies of a specified amplitude and sign. Figure 11 shows total fields and anomaly fields for the three dominant clusters of the 10-day lowpass filtered 500-mb height field, as determined by Cheng and Wallace (1993) on the basis of a hierarchical clustering algorithm. Kimoto (1987, 1989) and Kimoto and Ghil (1993) obtained similar patterns using a method of analysis designed to identify local maxima in the probability density function. The "Alaska" cluster resembles the negative polarity of

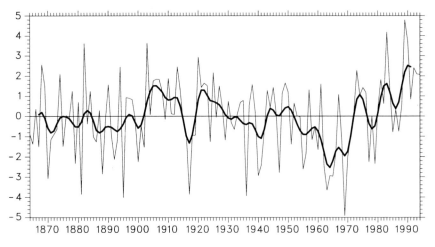

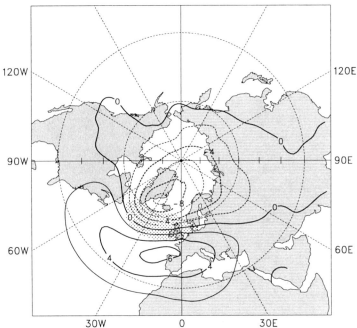

Figure 10: *Upper panel:* Winter (December through March) index of the North Atlantic Oscillation (NAO) based on the normalized difference between normalized sea-level pressure time series at Lisbon, Portugal and Stykkisholmur, Iceland. *Lower panel:* difference in sea-level pressure between "high" and "low index" winters, as defined by +1.0 and -1.0 standard deviations of the same NAO Index, respectively, based on data from 1900 onward. Contour interval 2 mb. From Hurrell (1995).

the leading EOF (Fig. 8) and it contains elements of the PNA pattern in its negative polarity. The "Greenland pattern" resembles the NAO in its negative polarity and the "Rockies" cluster contains elements in common with the PNA pattern (Fig. 9) in its positive polarity. The "Alaska" and "Greenland" clusters are associated with high amplitude ridges or 'blocking episodes' in two regions of the hemisphere that most frequently experience this phenomenon (Dole and Gordon 1983, Knox and Hay 1985), and the "Rockies" cluster is characterized by a strong jet stream, displaced southward of its climatological mean latitude over the central Pacific, with a strongly diffluent flow downstream leading into a longwave ridge that appears to be anchored to the Rockies.

The frequency of occurrence of the clusters is relatively sensitive to the clustering algorithm: in this particular analysis, the "Alaska," "Greenland," and "Rockies" clusters account for roughly 9% , 9% , and 13% of winter days, respectively, leaving 69% of the days unclassified. In Kimoto and Ghil's analysis, the corresponding frequencies of occurrence are 6% , 5% , and 7% and another 10% of the days are assigned to a fourth cluster, not shown here.

The clusters can be identified with distinctive weather anomalies in certain regions of the hemisphere. The "Alaska" cluster is associated with unseasonably mild weather over much of Alaska and it accounts for most of the major cold air outbreaks over the U.S. Pacific Northwest. The "Greenland" cluster is characterized by cold weather over the northeastern U.S. and southeastern Canada and Scandinavia, and the "Rockies" cluster by mild, dry weather over most of the western United States and the interior of western Canada.

The dates of occurrence of the three clusters, shown in Fig. 12, match up well with significant events in the wintertime climate records for these regions. For example, the 1949-50 winter, during which the "Alaska" cluster was prevalent, was unseasonably cold in the U.S. Pacific Northwest. The relatively high frequency of occurrence of the "Greenland" cluster in 1955-70, 1977-80, and 1985-87 matches up well with the dips in the NAO index in Fig. 10, and the increased frequency of occurrence of the "Rockies" cluster after 1911 reflects the ENSO-like interdecadal variablity discussed in section 8.

It is notable that the dominant structures in the wintertime mean 500-mb height field do not resemble pure spherical harmonic functions and their projections upon latitude circles do not resemble pure sinusoidal waves

49

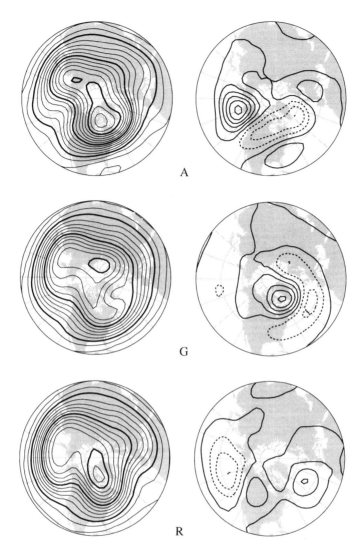

Figure 11: *The three most distinct clusters of the wintertime 500-mb height field based on 10-day lowpass filtered data for 40 winters. Negative contours are dashed. Left: composite 500-mb height fields; contour interval 60m. Right: composite 500-mb height anomaly fields (contour interval 50m) corresponding to the Alaska (A), Greenland (G) and Rockies (R) clusters, as described in the text. After Cheng and Wallace (1993), but updated to 1989.*

like the 50-mb EOFs in Fig. 1 [2]. The obvious geographical dependence of these structures is related, in a rather direct way, to the wintertime

---

[2]A corollary is that time series of the expansion coefficients of the leading spherical or zonal harmonics are not linearly independent.

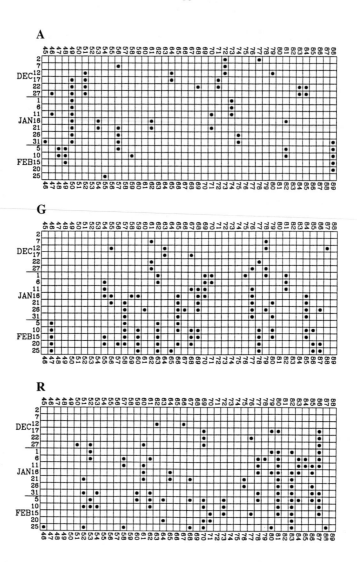

Figure 12: *Times of occurrence of the clusters described in the previous figure, indicated by the dots. Dates at the left represent midpoints of 5-day periods. After Cheng and Wallace (1993), but updated through 1989.*

climatological-mean wind field at the jet-stream level. Both the PNA pattern and the NAO are characterized by dipole structures straddling the 'exit regions' (i.e., the downstream ends) of the climatological mean jet streams. They bear the distinctive signature of the fastest growing mode of instability in a barotropic model, linearized about the climatological mean flow (Simmons et al. 1983) and the patterns most strongly excited by random heat and vorticity sources in a linear multi-level GCM (Branstator 1990). Furthermore, they strongly resemble the dominant modes of low-frequency variability in a variety of GCM's, run with and without time varying bottom boundary conditions. They prevail because they are particularly effective in extracting kinetic energy from the zonally varying climatological mean wintertime flow. The eastward, down-gradient flux of zonal momentum in the jet exit regions is instrumental in the energy conversion process (Simmons et al. 1983, Wallace and Lau 1985, Nakamura et al. 1987). Hence, there can be little doubt that these modes and the related patterns identified by cluster analysis reflect natural, dynamical variability of the atmospheric component of the climate system. They exist independently of any time-varying thermodynamic forcing of the atmosphere such as greenhouse warming or variations in sea surface temperature or land surface properties.

In comparison to the wintertime field, the summertime 500-mb height field is relatively quiescent and featureless. Throughout most of the hemisphere, the root-mean-squared variance of summertime mean 500-mb height is only on the order of half as large as that of wintertime mean 500-mb height (Wallace et al. 1993, Fig. 1), and the leading EOF tends to be of the same polarity throughout nearly the entire hemisphere, in marked contrast to the more wavelike patterns in Figs. 7-10. Yet, despite its unaesthetic appearance when rendered as a contour map (*ibid.*, Fig. 2), this EOF explains almost as much of the variance of the 500-mb height field as its wintertime counterpart. Such bland, monopolar patterns are even more prominent among the EOFs of the hemispheric 1000-500-mb thickness field, which is representative of vertically averaged lower tropospheric temperature; i.e., they account for larger fractions of the total variance than their counterparts in the 500-mb height field (*ibid.*, Table 2). The time series of the expansion coefficients of such EOFs are highly correlated with the time series of the hemispheric-mean of the field itself. The strong seasonality of these structures has implications for the interpretation of hemispheric-mean temperature trends, as will be discussed in the next two sections.

5 Impact of extratropical dynamical variability upon hemispheric-mean temperature

In this section we will consider the premise that the time-mean hemispheric circulation during a particular month, season or decade is capable of influencing the hemispheric-mean temperature observed during that time interval, from which it follows that part of the temporal variability of hemispheric-mean temperature is "dynamically induced": i.e., induced by changing circulation patterns such as those described in the previous section. Two specific examples come to mind:

- the North Atlantic Oscillation (Fig. 10) influences surface air temperatures over a large expanse of the Northern Hemisphere. "High index" periods (as defined by the strength of the westerlies along the node in the dipole pattern in the sea-level pressure field over the North Atlantic) tend to be characterized by above normal temperatures over Europe and most of Russia and the northeastern United States, and below normal temperatures over Greenland and Labrador (van Loon and Rogers 1978, Hurrell 1995). Since the area of positive temperature anomalies is much larger than the area of negative anomalies, one might expect the index of the NAO to be positively correlated with hemispheric-mean surface air temperature.

- the positive polarity of the PNA pattern (as defined by the sign of the 'center of action' in the 500-mb height field over western Canada) is characterized by above normal temperatures over Alaska and western Canada and below normal temperatures over the southeastern United States. Since the positive anomalies tend to be much larger than the negative ones, one might expect the positive polarity of the PNA-pattern to favour above normal hemispheric-mean temperature.

Implicit in both these examples is the presumption that anomalies over the continents are more influential in determining the hemispheric-mean temperature than anomalies over the oceans. The emphasis on land areas is justified by the fact that the temporal variances of surface air temperature over the continental interiors tend to be larger, by a factor of 3 or more, than those over the oceans. This inequality holds for the diurnal cycle, for the climatological-mean annual annual march, and for the year-to-year variability. It is a consequence of the much larger heat capacity of the ocean mixed layer, which renders sea-surface temperature and surface air temperature relatively unresponsive to the thermal variability of the overlying atmosphere on time scales of a few years or less.

If the above reasoning is correct, it should be possible for the hemispheric circulation to induce positive anomalies in hemispheric-mean temperature simply by redistributing heat in a manner so as to make tropospheric temperatures colder than normal over the oceans and warmer than normal over the continents. This hypothesis can be verified by regressing the hemispheric field of 1000-500-mb thickness anomalies, a measure of the temperature anomalies in the lower troposphere, upon the time series of hemispheric-mean surface air temperature anomalies (hereafter denoted T_L), as shown in the middle panel of Fig. 13. The strong regression coefficients are largely restricted to the region poleward of 40°N. The corresponding 500-mb height configuration, shown in the upper panel, is reminiscent of the Rockies cluster in Fig. 11 and the positive polarity of the PNA pattern over that limited sector of the hemisphere. The 1000-mb height (or sea-level pressure) field shown in the lower panel is characterized by negative anomalies over high latitudes, reminiscent of the 'high index' phase of the 'zonal index cycle' defined by Rossby and collaborators during the 1940's in connection with their efforts to develop a method of making long-range weather forecasts. Namias (1980) obtained rather similar looking patterns when he composited Northern Hemisphere 1000-700-mb thickness and 700-mb height anomalies for the nine warmest winters (based on hemispheric-mean thickness) of the period 1951-78.

The patterns in Fig. 13 should not be interpreted as a 'normal mode' or favoured pattern of variability of the hemispheric circulation in its own right, in the same sense as the PNA pattern or the NAO. If it were, one would expect 1000-500-mb thickness anomalies at the two positive 'centres of action' over the Yukon and Siberia to be positively correlated with one another but, in fact, they are not. This pattern has special significance only in relation to hemispheric-mean temperature. The striking correspondence between the thickness pattern and the land-sea distribution poleward of 40°N suggests the notion of a cold ocean/warm land (COWL) index that might be used to characterize the hemispheric circulation pattern observed each month. A positive value of the index for a particular month would be indicative of a thickness pattern characterized by positive anomalies over the continents and negative anomalies over the oceans, and and vice versa. Since the index is supposed to reflect the redistribution of heat by

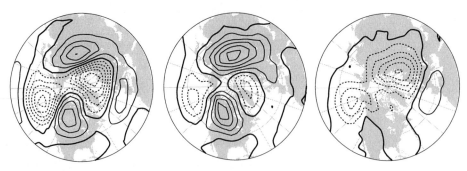

Figure 13: *Regression coefficients between Northern Hemispheric - mean surface air temperature anomalies and (left panel) 500-mb height, (middle panel) 1000-500-mb thickness and (right panel) 1000-mb height anomaly fields based on monthly-mean data for the months of the cold seasons, 1946-93. Contour interval 10 m (equivalent to 0.5 K for thickness) per degree of hemispheric-mean temperature. The reference time series is based on an average of gridded land station data as computed by the Hadley Centre, and the hemispheric fields are based on NMC operational analyses. From Wallace et al. (1995a).*

the circulation, it is based on the 'residual thickness' field T^*: i.e., the difference between the local thickness anomalies at each gridpoint and the thickness anomaly averaged over the polar cap region. A land/ocean 'mask' or weighting function $W(x, y)$ is defined, in which W is a positive constant for land gridpoints and a negative constant for oceanic gridpoints, subject to the constraint that that the average of W over all gridpoints poleward of 40°N be identically equal to zero. The residual thickness field for each month is then projected onto the 'mask' to obtain a COWL pattern index, where the summation is carried out over all gridpoints poleward of 40°N.

The correlation coefficient between the COWL index and the time series of T_L (Fig. 4) based on the 282 cold season months of the years 1946-93, is 0.81. It follows that month-to-month variations in the hemispheric circulation that determine the distribution of lower tropospheric temperature anomalies relative to the underlying land-sea distribution account for $(0.81)^2 = 65\%$ of the month-to-month variability of T_L during the cold season within this particular period of record. By subtracting a 'dynamical adjustment', defined as the product of the corresponding linear regression coefficient (0.51 K per standard deviation of the COWL index) times the monthly value of the normalized COWL index, from the cold season time series of T_L from 1946 onward, the month-to-month variability can be correspondingly reduced, as illustrated in Fig. 14. For further

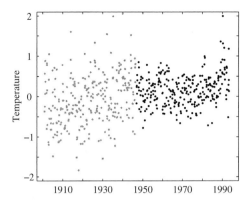

Figure 14: *As in Fig. 3 (right panel), but a dynamical correction linearly proportional to the "COWL index" (a measure of the tendency for cold air masses to lie over the oceans and warm air masses to lie over land) has been applied to the data from 1946 onward (indicated by the blacker dots). From Wallace et al. (1995a).*

details, the reader is referred to Wallace et al. (1995a).

One might inquire whether it is possible to generate a time series analogous to the COWL index, but extending much farther backward in time, making use of monthly, gridded surface air temperature fields based on station data. Since these data are largely restricted to continental regions, it is not feasible to define such an index on the basis of a land/ocean mask. An alternative strategy, employed by Wallace et al. (1995b), is to find, by the method of least squares, the spatially varying pattern whose amplitude (or expansion coefficient) time series is maximally correlated with the time series of T_L, subject to the constraint that the hemispheric-mean value of the pattern be zero. This procedure was applied to monthly data for all calendar months for the period of record 1900-93. Despite the limited space domain the resulting pattern, shown in Fig. 15, bears a striking resemblance to the thickness regression pattern in Fig. 13 [3]. The temporal correlation between the amplitude or expansion coefficient time series of this pattern and the COWL index, based on the overlapping period of record, is 0.76. The expansion coefficient time series is displayed as the middle panel of Fig. 16, and the residual time series (i.e., T_L minus the

---

[3]An analogous result was obtained by Alekseev et al. (1991), who correlated monthly hemispheric-mean surface air temperature with the variance of temperature on latitude circles. They found a strong negative correlation during the months of the cold season: i.e., warm months are characterized by relatively weak contrasts between the cold continents and the warm oceans and vice versa. Such months are characterized by positive temperature anomalies over the continents and negative anomalies over the oceans.

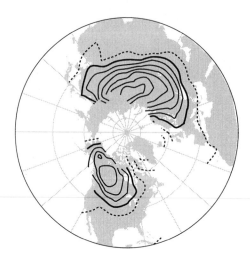

Figure 15: *The spatial pattern in the residual temperature anomaly field that accounts for the maximum possible fraction of the variance of the time series of hemispheric-mean temperature anomalies poleward of 20 °N. Based on monthly-mean, gridded surface air temperature anomalies at land stations, as in Fig. 2. Contour interval 1 deg. K of local residual temperature per deg. K of hemispheric mean temperature. From Wallace et al. (1995b).*

best fit based on the expansion coefficient time series) is shown in the lower panel. The expansion coefficient time series exhibits very little autocorrelation at time lags beyond a month or two, reflecting the relatively short "attention span" of the atmospheric circulation. The residual time series retains virtually all the visually coherent features in the time series of hemispheric-mean temperature shown in the upper panel, including most of the warming since the mid- 1970's, and it exhibits substantially less month-to-month scatter. The remarkable separation between the fitted and residual time series in the frequency domain is in no way contrived: it is inherent in the space-time structure of the temperature field.

Upon close inspection, it is evident that the fitted time series in Fig. 16 is not quite random. Its variance is much larger during the cold season than during the warm season and a disproportionate number of the cold-season months of the past decade or two have been characterized by positive anomalies. The interdecadal variability shows up more clearly in the seasonally partitioned time series in Fig. 17 which have been smoothed, first by averaging them over their respective seasons, and then by applying a 5-year running mean filter. It is evident that the smoothed cold and

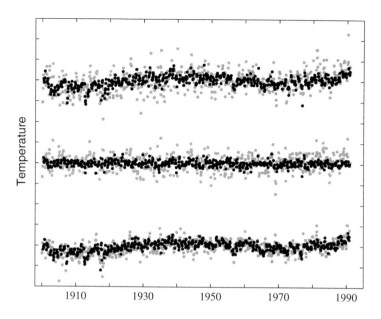

Figure 16: *Top: monthly-mean, hemispheric-mean surface air temperature anomalies poleward of 20 °N. Middle: fitted time series based on the regression coefficients shown in the previous figure. Bottom: Difference between the 'raw' (top) and the fitted (middle) time series. Black dots represent warm season months (May-October) and gray dots represent cold season months (November-April). Based on monthly-mean, gridded surface air temperature anomalies at land stations, as in Fig. 2. From Wallace et al. (1995b).*

warm season time series shown in the upper panel differ by as much as 0.3 K within short segments of the record: the cold season series exhibits substantially more variability on time scales ranging from 3-10 years. A prominent feature of the cold season time series is the upward trend from the late 1970's onward, which is mirrored in the fitted expansion coefficient time series in the middle panel. The residual time series shown in the lower panel exhibits remarkably little seasonality over the 90-year period of record, apart from a slight upward trend in the cold-season series relative to the warm-season series. The resemblance between the cold- and warm-season residual time series is particularly strong after 1950 when the data are most reliable.

These results support the notion that the dynamically induced variability in the time series of hemispheric-mean surface air temperature is largely a high latitude cold-season phenomenon linked to the distribution of warm and cold air masses relative to the continents and oceans, as rep-

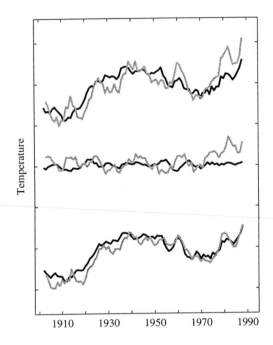

Figure 17: *As in the previous figure, but the time series have been smoothed with a 5-year running mean filter. From Wallace et al. (1995b).*

resented by the amplitude and polarity of the "COWL index". Although the hemispheric circulation exhibits relatively little month-to-month autocorrelation, the prevalence of warm air masses over the continents during the winter months appears to have contributed to the observed rise in hemispheric-mean surface air temperature from the mid 1970s onward. Hurrell (1995, 1996) has argued quite convincingly that the mild winters in Eurasia reflect the prevalence of the "high index" phase of the North Atlantic Oscillation. In section 7 it will be argued that the mild winters in Alaska and western Canada may have been induced by a persistent pattern of SST anomalies over the Pacific.

6   Structure of tropical climate variability

The dominant interannual climate signal in the tropical troposphere and ocean mixed layer is the El Niño / Southern Oscillation (ENSO) phenomenon, which involves coherent fluctuations in equatorial Pacific sea-surface temperature throughout the broad region extending all the way from the South American coast westward to beyond the dateline. As il-

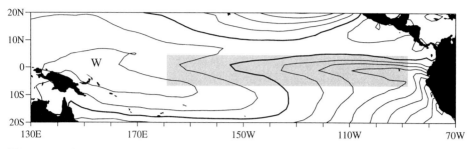

Figure 18: *Annual mean SST over the tropical Pacific, based on the COADS. Contour interval 1 C; the 27 C isotherm is shaded. The cold tongue index is the mean temperature anomaly of the shaded region.*

lustrated in Fig. 18, this region coincides with the so called 'equatorial cold tongue', which is maintained by a narrow band of upwelling along the equator, induced by the prevailing easterly surface winds. An extended record of seasonal-mean SST averaged over this cold tongue region (hereafter referred to as the "cold tongue index") has already been shown in Fig. 8 of chapter 1. It was demonstrated in Fig. 9 of chapter 1 that the fluctuations in this index exhibit a wide range of frequencies, with characteristic periods ranging from 2 to 8 years. Nevertheless, throughout most of the record it is possible to identify a reasonably well defined, albeit irregular "ENSO cycle", whose warm and cold phases can be used as a basis for compositing.

Figures 19-21 show the contrasting atmospheric conditions observed during the warm and cold phases of the ENSO cycle, as defined by the cold tongue index. These composites were constructed by linearly regressing monthly mean time series of each of the variables at each gridpoint upon the normalized version of the time series in Fig. 8 of chapter 1 (repeated in Fig. 22) from 1979 onward to obtain the maps shown in the bottom panels of the figures. The regression coefficients are expressed in dimensional units per standard deviation of the cold tongue index. The anomalies corresponding to a strong positive (negative) departure of the cold tongue index were estimated multiplying these regression coefficients by 1.5 (-1.5) standard deviations of the cold tongue index, which corresponds to an SST anomaly of 1.2 K (-1.2 K), averaged over the shaded region in Fig. 18. These estimated ENSO-related anomalies were added to the climatological mean field to obtain the warm and cold phase composite maps shown in the figures. This procedure may be viewed as a type of

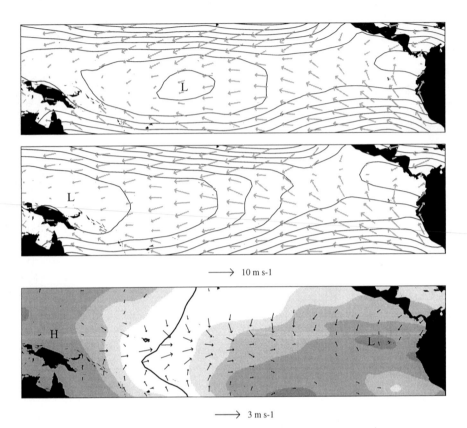

Figure 19: *Sea-level pressure and surface wind over the tropical Pacific during contrasting warm and cold phases of the ENSO cycle, 1979-92. The bottom panel shows the pattern of regression coefficients used in constructing the warm and cold phase composites. Contour interval in the composites 1 mb: the lowest contour corresponds to 1009 mb. The contour in the regression map corresponds to zero, where higher values lie toward the left. The increment for the shading is 0.2 mb per standard deviation of the cold tongue index. All fields in Figs. 19-21 are based on the COADS unless otherwise noted. For further details concerning how the figure was constructed, see the text.*

compositing, in which the observed anomalies for each month are assigned a positive or negative weight proportional to the anomaly in the reference time series.

In comparison to the cold phase composites in Figs. 19-21, the warm phase of the ENSO cycle is characterized by

- a weakening of the easterly trade winds in the equatorial central Pacific,

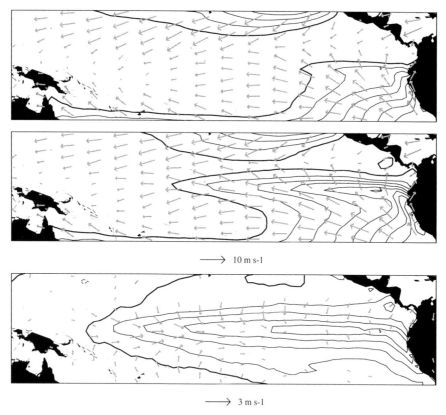

$\longrightarrow$ 10 m s-1

$\longrightarrow$ 3 m s-1

Figure 20: *As in Fig. 19, but for SST and surface wind. Contour interval in the composites 1 K for values up to and including the 27 K contour, which is thickened. (SST in the 'warm pool' where values exceed 27 K is not contoured.) Contour interval in the regression map 0.25 K; the zero contour is thickened; all the other contours are positive.*

- an eastward migration of the belt of low sea-level pressure in the western Pacific,

- a foreshortening and weakening of the equatorial cold tongue in the SST field,

- a dramatic increase in rainfall over the equatorial cold tongue. Desert islands within this region experience torrential rains, month after month, during the warm phases of the ENSO cycle.

The weakening of the equatorial cold tongue is consistent with the pressure falls over the tropical eastern Pacific, which reflect the reduced density of the warmer air in the planetary boundary layer (Lindzen and Nigam

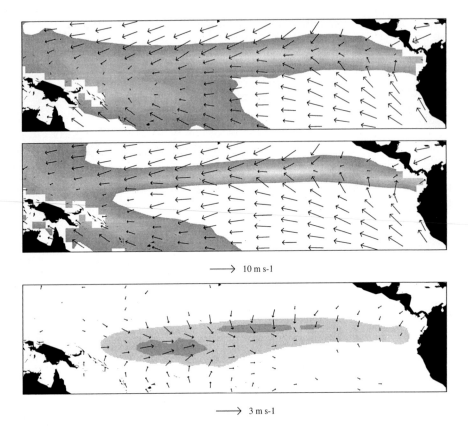

$\longrightarrow$ 10 m s-1

$\longrightarrow$ 3 m s-1

Figure 21: *As in Fig. 19, but for rainfall and surface wind. Rainfall estimates are based on the microwave sounding unit, as described in Spencer (1993). The threshold for the shading is 15 cm of rainfall per month in the composites and +4 and +8 cm per month per standard deviation of the cold tongue index in the regression map.*

1987). This positive feedback loop contributes to the large temporal variability of the climate of the equatorial Pacific. The slackening of the trade winds in the central Pacific during the warm phase of the ENSO cycle is dynamically consistent with the weakening of the east-west pressure gradient in that region. It induces a relaxation of the east-west slope of the equatorial thermocline, leading to a warming of the surface waters on the eastern side of the basin, where the thermocline deepens.

The dramatic increase in rainfall over the cold tongue region during the warm phase of the ENSO cycle can be understood as follows. Convective rainfall tends to be suppressed over regions of the tropics where SST is colder than 27 C because the air within the atmospheric planetary boundary layer is not sufficiently buoyant (i.e., warm and moist) to pene-trate through the overlying layer (i.e., the free troposphere), whose vertical temperature profile tends to be in moist convective equilibrium with the warmest regions of the tropical oceans, where SST is near or slightly above 28 C (Sarachik, 1978). One of the most prominent features of the tropical precipitation climatology is the "equatorial dry zone", a triangular shaped region whose northern part corresponds closely to the cold tongue in the SST field. The weakening of the equatorial cold tongue during the warm phase of the ENSO cycle allows the bands of deep convection that surround the dry zone to encroach upon it: the rain area over the warm waters to the west of the dateline shifts eastward, and the intertropical convergence zone (ITCZ) along $\sim 7 °N$ expands southward, as illustrated in Fig. 21. The encroachment tends to be most pronounced when the warm phase of the ENSO cycle coincides with the season of the year (January through May) when the cold tongue tends to be relatively weak to begin with. At these times intermittent heavy rainfall is observed along the equator across the entire width of the Pacific and even in the coastal deserts of Ecuador and northern Peru. The fluctuations in surface wind and precipitation associated with the ENSO cycle are dynamically consistent: enhanced pre-cipitation is associated with enhanced convergence of the surface wind field. Note that much of the convergence in the surface wind field is associated with the meridional wind component.

The sea-level pressure changes in Fig. 19 are part of a larger pattern that affects the Australia and much of the tropical Indian Ocean as well. The warm (weak tradewind) phase of the ENSO cycle is characterized by above normal pressure to the west, over the tropical Indian Ocean, Indonesia, and northern Australia and below normal pressure to the east, over Pacific Islands such as Tahiti (17 S, 149 W) and Easter Is. (27 S, 109 W) and along the South American coast, and a weakened east-to-west gradient across the central Pacific. The normalized difference between the normalized sea-level pressure anomalies at Tahiti and Darwin (12 S, 131 E) is used as an index of the Southern Oscillation (the SO in ENSO)[4]. This

---

[4]The term Southern Oscillation was coined by Sir Gilbert Walker approximately 75 years ago to describe this recurrent pattern of sea-level pressure fluctuations, which he believed to be related to climate fluctuations over many parts of the globe. However, Walker was not aware of the dramatic contrasts in SST, surface wind, and precipitation over the equatorial Pacific that occur in association with it. The work of Jacob Bjerknes (1966, 1969), half a century later, provided a basis for linking the Southern Oscillation with episodes of abnormally warm SST along the coast of South America, which had hitherto been regarded as a local, primarily oceanographic phenomenon, known as El Niño (Eguiguren 1894, Rasmusson and Carpenter 1982, Deser and Wallace 1987).

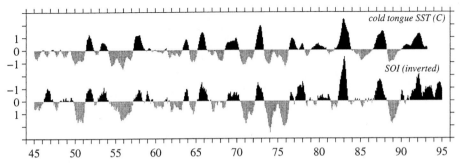

Figure 22: *Inverted time series of the Southern Oscillation Index (SOI), defined as the normalized difference between the normalized sea-level pressure anomalies at Tahiti and Darwin (12 S, 131 E) for the recent period of record. The corresponding segment of the time series of the cold tongue index (Fig. 8 of chapter 1) is repeated here for comparison. Both time series have been smoothed with a 5-month running mean filter.*

so-called "Southern Oscillation Index" (SOI), tends to be negative during the warm phase of the ENSO cycle, which is usually accompanied by El Niño events along the South American coast. An extended time series of the SOI, comparable in length to the SST time series in Fig. 8 of chapter 1 has been compiled by the NOAA Climate Analysis Center. The segment corresponding to the recent period of record is shown in Fig. 22, together with the index of equatorial Pacific cold tongue SST, repeated from Fig 8 of chapter 1. The negative correlation between the SOI and equatorial Pacific SST is clearly evident.

Figure 23 shows the shows the anomalous rainfall pattern depicted in the bottom panel of Fig. 21 superimposed upon contours of vertically averaged (surface to 300-mb) tropospheric temperature anomalies, as revealed by the same MSU instrument. The dipole pattern in tropospheric temperature, with positive anomalies straddling the region of enhanced equatorial precipitation during the warm phase of the ENSO cycle, is a robust feature which has been noted in previous observational studies of El Nino (e.g., Horel and Wallace 1981), and it has been reproduced in numerous GCM simulations of the atmospheric response to tropical SST anomalies. Since geopotential surfaces in the lower troposphere tend to be rather flat in comparison to those in the upper troposphere, the centers of action of the dipole pattern in Fig. 23 are indicative of upper level anticyclones, with anomalous easterly flow over the region of enhanced equatorial rainfall and a strengthening of the climatological mean westerlies in the subtropics of

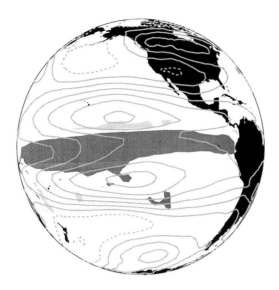

Figure 23: *As in the bottom panel of Fig. 19, but for rainfall (repeated from Fig. 21) and vertically averaged (surface to 300-mb) tropospheric temperature as inferred from channel 2 of the microwave sounding unit. Heavy shading denotes enhanced rainfall, lighter shading reduced rainfall. Contour interval 0.1 K per standard deviation of the cold tongue index: dashed contours denote negative values.*

both hemispheres, along the poleward flanks of the dipole pattern over the eastern Pacific. The anomalous strengthening of the westerlies in these regions during the warm phase of the ENSO cycle is reflected in increased storminess over the southern United States and the corresponding latitude belt of South America (Ropelewski and Halpert 1989).

The MSU tropospheric temperature record, upon which Figure 23 is based, extends back only to the beginning of 1979. Figure 24 shows the corresponding pattern for the Northern Hemisphere wintertime 500-mb height field, based on data extending back to 1946. Unlike the previous figures, it is based upon data for the Northern Hemisphere winter (December through February) when the relationships between the tropical and extratropical circulation tend to be strongest. The typical pattern of ENSO anomalies bears some relation to the PNA pattern discussed in section 4, but the region of positive anomalies over Canada extends farther eastward, and the region of negative anomalies over the North Pacific stretches northwestward toward Kamchatka rather than westward along 45 °N. The positive anomalies over Canada provide an explanation of the finding of Walker and Bliss (1932) that winters over western Canada tend

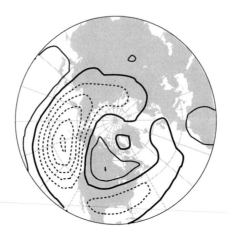

Figure 24: *500-mb height anomalies during the winter months (December through Febru-ary) during a typical warm phase of the ENSO cycle, as inferred from regression coef-ficients upon seasonal mean values of the equatorial Pacific cold tongue index for the period 1946-93. Contour interval 10 m of geopotential height per deg. K of the cold tongue index: dashed contours denote negative values. After Zhang et al. (1996)*

to be warmer than normal during the warm phase of the ENSO cycle[5].

Early results of Walker and Bliss (1932) were suggestive of a warming of the entire troposphere during the negative (weak trade wind) phase of the Southern Oscillation. Consistent with their findings, a distinctive ENSO signature is evident in the recent time series of tropospheric (surface to 300-mb) temperature and surface air temperature averaged over the entire 20°N-20°S latitude belt shown in Fig. 25. The agreement is all the more remarkable, in view of the fact that the surface air temperature time series is based on continental and island stations, only a few of which are located in the equatorial Pacific east of the dateline. The ENSO cycle in these two time series exhibits a phase lag of about a season relative to that in the equatorial Pacific SST index and it is smaller in amplitude by about a factor of 3.

---

[5]Walker predicted that the advent of upper air data would ultimately provide an explanation for the linear correlation between sea-level pressure anomalies in the tropical South Pacific and temperatures over western Canada (Montgomery 1940).

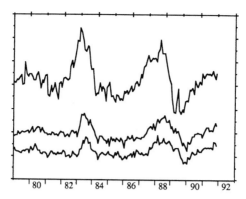

Figure 25: *Time series of (upper) equatorial Pacific sea-surface temperature anomalies averaged over the region 180 °E - 90 °W, 6 °N - 6 °S; (middle) tropical tropospheric (surface to 300-mb) temperature anomalies based on data from channel 2 of the microwave sounding unit; (lower) surface air temperature over all land gridpoints within the tropical belt (20 °N - 20 °S) based on the U.S. Department of Energy dataset (Jones et al., 1985). The temperature scale on the vertical axis is relative: one tick-mark is equivalent to 0.5 K. From Yulaeva and Wallace (1994).*

7 The ENSO signal in tropical-mean and global-mean temperature

Although the atmosphere-ocean coupling that drives the ENSO cycle is concentrated in the equatorial Pacific, the ENSO cycle is clearly manifested in tropical-mean surface and upper air temperatures, as demonstrated in Fig. 26. We will show in this section that it is also evident in tropical-mean SST, exclusive of the equatorial Pacific. Since the tropics occupy nearly half the surface area of the globe, in the absence of a systematic tendency for compensation between tropical and extratropical temperatures, one should expect the ENSO cycle to be reflected in hemispheric- and global-mean temperature time series such as those emphasized in the IPCC reports. The MSU data shown in Fig. 26 indicate that this is, indeed, the case: the ENSO cycle is clearly evident, albeit in a dilute form, in global-mean tropospheric temperature. The absence of a systematic warming or cooling of the extratropics in association with the ENSO cycle is notable. Evidently the regional features in Figs. 23 tend to cancel when averaged over the entire extratropics of either hemisphere.

The signature of ENSO in tropical-mean and global-mean temperature lags the fluctuations in the cold tongue index by about a season. Upon close inspection, this lag is evident in Fig. 25 and it shows up more clearly

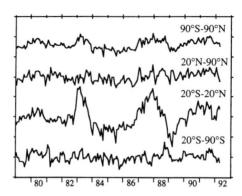

Figure 26: *Tropical tropospheric temperature, averaged over latitude belts as indicated, as estimated from MSU data. The temperature scale on the vertical axis is relative: one tick-mark is equivalent to 0.5 K. From Yulaeva and Wallace (1994)*

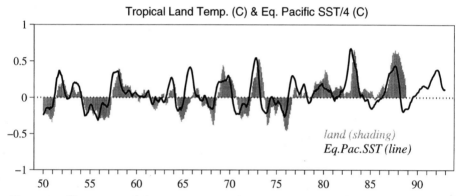

Figure 27: *Time series of the equatorial Pacific cold tongue index (indicated by the line) and tropical (20 °N - 20 °S) surface air temperature as inferred from station data (U.S. Department of Energy: Jones et al., 1985) (indicated by the shading). For ease of comparison, the surface air temperature time series has been divided by 4. The temperature scale on the vertical axis is in deg. C.*

in the more extended time series shown in Fig. 27. A lagged response to the ENSO cycle is also clearly evident in SST in the tropical Indian and Atlantic Oceans (e.g., see Yulaeva and Wallace, Fig. 21). The existence of a systematic phase lag suggests that the changes in the mean temperature of the tropics may be a passive thermodynamic response to the ENSO cycle.

The ENSO cycle may be viewed as perturbing the fluxes at the air-sea interface over the equatorial Pacific cold tongue, which serves as a heat source for a passive system comprised of the the tropical troposphere, plus the mixed layer of the surrounding oceans with which it is thermodynamically coupled. This problem is in some sense equivalent to the one considered by Hasselmann (1976) in his study of the response of a passive ocean mixed layer to stochastic atmospheric forcing. For a discussion of Hasselmann's formalism, the reader is referred to chapter 4 and to Frankignoul (1985). The response is governed by the equation

$$C\frac{\partial T}{\partial t} = F(t) - \alpha T$$

in which T is the temperature anomaly that develops in response to the forcing, t is time, C is the heat capacity per unit area, $F(t)$ is the driving (a time varying heat source / sink), and α is a linear damping rate. In Hasselmann's formalism T refers to the temperature anomaly of the ocean mixed layer: here we will follow the development in Yulaeva and Wallace and think of it as the temperature anomaly of the tropical atmosphere-ocean system. In using a single temperature variable to represent the entire tropics, we are implicitly ignoring the spatial gradients in the temperature field. This simplification is justified, at least to some extent, by the fact that regional temperature time series throughout the tropics exhibit ENSO signatures remarkably similar to the one in the middle panel of Fig. 26 (see Yulaeva and Wallace, Figs. 3 and 4). Strictly speaking, we should think of T as representing the temperature of the tropics, exclusive of the equatorial Pacific where the temperature variability is determined by dynamical processes in the ocean that mediate the upwelling of cold water from below the thermocline. In other words, we are concerned with the dynamically passive part of the tropics, whose temperature variability occurs in response to ENSO: not as an integral part of the ENSO cycle itself.

The anomalous fluxes at the air-sea interface over the equatorial cold tongue are presumed to be what drives the system away from equilibrium. We will assume that the heating of the tropical atmosphere that results from these fluxes, as represented by $F(t)$, is linearly proportional to the cold tongue index itself. Hence, instead of driving the model with a random time series as Hasselmann did, we drive it with a prescribed time series based on historical data. The rate of damping of the tropical temperature perturbation back toward equilibrium is assumed to be linearly proportional to the amplitude of the temperature perturbation itself. The constant of proportionality α, as inferred from the Stefan-Boltzmann law,

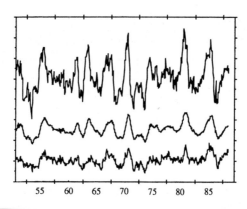

Figure 28: *Time series of (upper) the equatorial Pacific cold tongue index, as in the previous figure; (middle) simulated response, based on a simple thermodynamic model described in the text; and (lower) tropical (20 °N - 20 °S) surface air temperature as inferred from station data (U.S. Department of Energy: Jones et al., 1985) based on the DOE dataset. One small tick-mark on the vertical scale is equivalent to 0.5 K. From Yulaeva and Wallace (1994)*

assuming that the system radiates to space as a black body, is on the order of $4Wm^{-2}K^{-1}$. The only free or 'tunable' parameter in the model is the heat capacity C. For reference, the heat capacity per unit area of the atmosphere, $10^7 Jm^{-2}K^{-1}$, is equivalent to that of a layer of water 2.5 m deep.

In the context that we are using it here, the model can simply be viewed as a filter applied to the time series of the equatorial cold tongue index to smooth it and induce a physically plausible frequency-dependent phase lag. Figure 28 shows the input and output of this filter for a segment of the record, based on an assumed heat capacity of $5 \times 10^7 Jm^{-2}K^{-1}$, which is equivalent to that of the atmosphere plus a layer of ocean 10 m deep. The correlation coefficient (0.80) between the output of the filter and the time series of tropical-mean surface air temperature is quite impressive, in view of the crudeness of the model and the imperfections in the datasets.

8   ENSO-like interdecadal variability

Among the features identified in ENSO-related time series are what some investigators have referred to as "regime shifts". A widely publicized example is the shift toward higher mean values of the cold tongue index and lower values of the SOI in 1976-77 (Quinn and Neal 1984, 1985; Trenberth

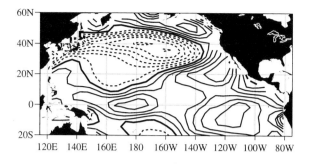

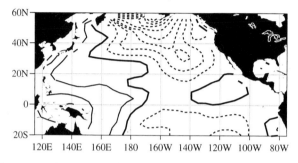

Figure 29: *Changes in (upper panel) SST, contour interval 0.1 K and (lower panel) sea-level pressure, contour interval 0.25 mb from the period 1950-76 to the period 1977-93. based on COADS. Negative contours are dashed. From Zhang et al. (1996).*

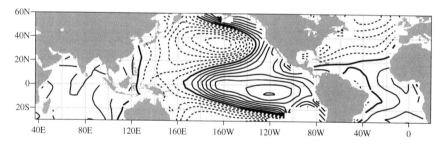

Figure 30: *The global SST field regressed upon the time series of global-mean SST anomalies shown in the next figure, based on the period of record 1950-93. Contour interval 0.25 K per degree K of global-mean SST: the zero contour is thickened and all the other contours are positive. Based on the U.K. Meteorological Office historical sea surface temperature dataset (Folland and Parker 1995), obtained from the Climatic Research Unit, University of East Anglia. After Zhang et al. (1996).*

1990; Trenberth and Hurrell 1994; Miller et al. 1994; Graham 1994, Wang 1995), which is readily apparent in Figs. 22, 27 and 28. The changes in SST and sea-level pressure that took place in association with this feature

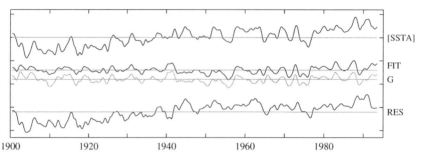

Figure 31: *Time series of (top) global-mean SST anomalies; (middle) expansion coefficient of the regression field shown in the previous figure, scaled as a least squares best fit to the global-mean SST time series (FIT) shown together with the expansion coefficient of the leading EOF of global the SST field (G); and (bottom) residual variations in global-mean SST, obtained by subtracting the fitted curve from the time series of global-mean SST. The interval between tick-marks is 0.5 K. All time series have been smoothed by successive applications of 5-month and 9-month running mean filters. The spacing between the curves is arbitrary. After Zhang et al.(1996).*

are shown in Fig. 29. Within the tropics, both patterns are reminiscent of the differences between contrasting polarities of the ENSO cycle as represented in the bottom panels of Figs. 19 and 20. The SLP falls and cooling over the extratropical North Pacific are also reminiscent of the anomalies associated with the ENSO cycle, but they are more prominent relative to the changes in the tropics. Like the ENSO cycle, this interdecadal-scale change is clearly apparent in the time series of tropical surface air temperature (Fig. 28), and global-mean SST (Fig. 30).

Whether this feature is in any way unique or extraordinary is difficult to assess using conventional measures of statistical significance, for the reasons discussed in section 2.5 of chapter 1. But unique or not, it serves as a clearly discernible marker in many different time series. A similar shift occurred in 1958 but the subsequent "warm regime" persisted for only 3-4 years. Mantua and Graham (1996) have shown evidence that yet another one occurred in 1990. It is only with the benefit of hindsight that it is possible to distinguish between the short-lived "regimes", whose histories are most appropriately viewed in the context of the interannual variability, and the long-lived ones that contribute substantially to the interannual variability in the climate record. One could also question whether the apparent abruptness of features such as these might not be illusory: a product of the superposition of the interannual variations associated with the ENSO cycle upon more gradual interdecadal variations.

The linear regression regression calculation that yielded the temperature pattern Fig. 15 and the fitted expansion coefficient time series in Fig. 16 was repeated, using the time series of global-mean SST in place of T_L and the global SST field in place of the extratropical Northern Hemisphere surface air temperature field. The resulting global SST pattern whose amplitude time series is most highly correlated with the time series of global-mean SST (Fig. 30) exhibits an ENSO-like tropical signature reminiscent of the bottom panel of Fig. 20. Its expansion coefficient time series (Fig. 31) traces out all the major warm and cold episodes of the ENSO cycle, as represented in Fig. 22, as well as the interdecadal-scale features noted above. It also exhibits a shift in the opposite direction (i.e., toward colder equatorial Pacific SST) in 1942. Apart from a few brief positive excursions during the major El Nino events, the time series exhibits negative values throughout the epoch 1942-76.

Also shown in Fig. 31 is the time series of the expansion coefficient of the leading EOF of the global SST anomaly field (hereafter referred to as G; comparable to time series presented by Hsiung and Newell 1983, Nitta and Yamada 1989, and Parker and Folland 1991), defined on the basis of the anomaly deviation field; i.e., the difference between the SST at each gridpoint and the global-mean SST for the same month. G is highly correlated with global-mean SST, with indices of the Southern Oscillation, and with interdecadal climate variability over the extratropical North Pacific (Zhang et al. 1996).

9 Interpretation of the variability of global-mean temperature

From the foregoing, it is evident that regardless of how hemispheric-mean surface air temperature may be responding to increasing concentrations of greenhouse gases and aerosols, it is also varying on the time scale of years to decades in response to the sampling fluctuations associated with the month-to-month variability of the atmospheric circulation patterns that determine the value of the "COWL index" defined in section 5. Superimposed upon this random variability are interdecadal-scale ENSO-like SST variations in the Pacific and Indian Oceans and fluctuations in the amplitude and polarity of North Atlantic Oscillation, which may or may not be deterministic.

The internal variability of the atmospheric circulation, with a characteristic time scale of days to weeks, constitutes an important source of sampling fluctuations in monthly means, seasonal means, annual means, and even in means over averaging intervals up to a few years. It is evident from Fig. 2 that these dynamically induced sampling fluctuations are bound to be even larger (by a factor of 3 or more) in estimates of hemispheric-mean temperature variations based on data from individual stations or localized regions, obscuring even interdecadal scale features in the hemispheric or global 'signal'.

Superimposed upon this background noise are a number of real climate signals that impact hemispheric- and global-mean surface air temperature. The ENSO cycle dominates the variability of tropical-mean surface air temperature and SST and, for lack of a competing interannual extratropical signal, its signature is also clearly evident in time series of hemispheric-mean and global-mean temperature. It shows up more clearly in the time series of global-mean SST than in the time series of global-mean surface air temperature based on station data because the former is less subject to the sampling fluctuations described above. ENSO-like interdecadal variability is evident in SST and cold season circulation patterns over the Pacific sector (Nitta and Yamada 1989, Trenberth 1990, Trenberth and Hurrell 1994, Graham 1994, Zhang et al. 1996). The North Atlantic Oscillation, which impacts wintertime climate over Europe and most of Russia also exhibits what appears to be statistically significant interdecadal variability (Hurrell 1995). Yet another climate signal worthy of mention in this regard is the dramatic rise in SST and surface air temperature in a belt stretching from Labrador across the North Atlantic and the Greenland, Barents and Kara Seas that took place during the decade of the 1920s, followed by a cooling during the 1950s and '60s, as documented documented by Kelly (1982), Ellett and Blindheim (1992), Mysak et al. (1990) and Deser and Blackmon (1993), among others. This very strong regional feature accounts for a substantial fraction of the differences between the Northern and Southern Hemisphere surface air temperature time series.

During the 1980s and early '90s, the North Atlantic Oscillation and the ENSO-like variability over the North Pacific conspired to make wintertime surface air temperatures over the Northern Hemisphere continents (poleward of 40°) warmer, relative to the surrounding oceans, than they were during previous decades. These persistent circulation-related anomalies have contributed substantially to the recent upward trend in hemispheric-mean surface air temperature (Hurrell 1996), and they account for virtually all the accelerated wintertime warming over the high latitude continents reported in the IPCC assessments (Wallace et al. 1995a,b). Hence, it seems unlikely that this 'accelerated warming' can continue much longer. In fact, if the interdecadal component of either the NAO or ENSO undergoes a polarity reversal during the next decade, winter temperatures over the high latitude continents could actually fall, even in the presence of a modest upward trend in global-mean temperature. In view of the much greater sensitivity of cold season temperatures to the vicissitudes of the coupled atmosphere-ocean climate system, warm season data may be a more reliable indicator of current, anthropogenically induced trends in hemispheric-mean temperature.

Some readers may question our premise that interdecadal-scale variations in the atmospheric circulation should be listed among the possible causes of the recent rise in hemispheric-mean surface air temperature. The inverse view (i.e., that the recent rise in hemispheric-mean temperature is responsible for the circulation anomalies that have prevailed during recent years) is certainly not without adherents. For example, Namias (1980) computed correlation patterns analogous to those in our Fig. 13 for the expressed purpose of demonstrating possible regional effects associated with global warming. In defense of our interpretation, we have offered a physically plausible argument as to why the presence of warm anomalies over the continents and cold anomalies over the oceans should favour anomalously warm hemispheric-mean surface air temperature, whereas it is not at all clear how a rise in hemispheric-mean temperature of a mere few tenths of a degree K could force the large, regionally specific atmospheric circulation changes that have been observed during the past two decades. Furthermore, although some records were broken, we are not convinced that the amplitudes of either the Southern Oscillation or the North Atlantic Oscillation during the 1980s and early 90s were large enough to to be considered beyond the range of natural variability. Time will tell which of the two interpretations is more valid.

10 Acknowledgments

Figure 2 was inspired by an workshop presentation by Henry Diaz and Fig. 3 by published results of Bradley et al. (1987). Figures 1, 5, 6, 8, 11 and 12 were kindly provided by Xinhua Chen; Figs. 2-4, 13-17, 24, and 29-30 by Yuan Zhang; Figs. 18-23 and 27 by Todd P. Mitchell; Figs. 7 and 9 by James A. Renwick; and Fig. 10 by James W. Hurrell. Imke Durre proofread the manuscript and offered helpful suggestions. This work was sponsored by the United States National Science Foundation's Climate Dynamics Program under Grant ATM 9215512 and the NOAA Office of Global Programs.

References

ALEKSEEV, G.V., I.A. PODGORNY, AND P.N. SVYASTCHENNIKOV, 1991, Effect of the ocean's heating on global climate fluctuations. *Doklady Akademii nauk SSSR,* **320,** 70-73, 1991 (in Russian)

BARNETT, T. P., 1978, Estimating variability of surface temperature in the Northern Hemisphere., *Mon. Wea. Rev.,* **106,** 1353-1367.

BARNSTON, A.G. AND R.E. LIVEZEY, 1987, Classification, seasonality and persistence of low frequency atmospheric circulation patterns. *Mon. Wea. Rev.,* **115,** 1083-1126.

BATTISTI, D.S., 1988, The dynamics and thermodynamics of a warm event in a coupled atmosphere/ocean model. *J. Atmos. Sci.,* **45,** 2889-2919.

BATTISTI, D.S., AND A.C. HIRST, 1989, Interannual variability in the tropical atmosphere-ocean system: Influence of the basic state, ocean geometry, and nonlinearity. *J. Atmos. Sci.,* **46,** 1687-1712.

BENGTSSON, L. AND J. SHUKLA, 1988, Integration of space and in situ observations to study global climate change. *Bull. Amer. Meteorol. Soc.,* **69,** 1130-1143.

BJERKNES, J., 1966, A possible response of the Hadley circulation to equatorial anomalies of ocean temperatures. *Tellus,* **18,** 820-829.

BJERKNES, J., 1969, Atmospheric teleconnections from the equatorial Pacific. *Mon. Wea. Rev.,* **97,** 162-172.

BLACKMON, M.L., Y.-H. LEE, AND J. M. WALLACE, 1984, Horizontal structure of 500 mb height fluctuations with long, intermediate and short time scales. *J. Atmos. Sci.,* **41,** 961-979.

BOTTOMLEY, M., C.K. FOLLAND, J. HSIUNG, R.E. NEWELL AND D.E. PARKER, 1990, Global Ocean Stuface Temperature Atlas (GOSTA), *Her Majesty's Stationnery Office, London, 24pp + 313 plates.*

BRADLEY, R.S., H.F. DIAZ, P.D. JONES AND P.M. KELLY, 1987, Secular fluctuations of temperature averaged over northern hemisphere land areas and mainland China since the mid-19th century. *In Climate of China and Global Climate, D.Ye, C. Fu, J. Chao and M Yoshino, eds., China Ocean Press, Beijing, 77-87.*

BRANSTATOR, G.W., 1990, Low-frequency patterns induced by stationary waves. *J. Atmos. Sci.,* **47,** 629-648.

CARDONE, V.J., J.G. GREENWOOD, AND M.A. CANE, 1990, On trends in historical marine wind data. *J. Climate,* **3,** 113-127.

CHENG, X. AND J. M. WALLACE, 1993, Cluster analysis of the Northern Hemisphere winter-time 500-hPa height field: spatial patterns. *J. Atmos. Sci.*, **50**, 2674-2695.

DAVIS, R.E., 1976, Predictability of sea-surface temperature anomalies over the North Pacific Ocean. *J. Phys. Oceanogr.*, **8**, 233-246.

DESER, C. AND M.L. BLACKMON, 1993, Surface climate variations over the North Atlantic during winter 1900-89. *J. Climate*, **9**, 1743-1753.

DESER, C. AND J.M. WALLACE, 1987, El Nino events and their relation to the Southern Oscillation. *J. Geophys. Res.*, **92**, 14189-14196.

DOLE, R.M. AND N.D. GORDON, 1983, Persistent anomalies of the extratropical North-ern Hemisphere wintertime circulation: Geographical distribution and regional persistence characteristics. *Mon. Wea. Rev.*, **111**, 1567-1586.

ELLETT, D.J. AND J. BLINDHEIM, 1992, Climate and hydrography variability in the ICES area during the 1980s. *ICES mar. Sci. Symp.*, **195**, 11-31.

EGUIGUREN, D.V., 1894, Las nuivas de Piura, *Bol. Soc. Geogr. Lima*, **4**, 241-258.

FOLLAND, C.K., T.R. KARL AND K.YA. VINNIKOV, 1990, Observed climate variations and change. *In Climate Change: the IPCC Scientific Assessment, J.T. Houghton, G.J. Jenkins and J.J. Ephraums, eds., Cambridge University Press, Cambridge, England, 195-238*.

FOLLAND, C.K., T.R. KARL, N. NICHOLLS, B.S. NYENZI, D.E. PARKER AND K.YA. VIN-NIKOV, 1992, Observed climate variability and change, *In Climate Change 1992: The Supplementary Report to the IPCC Scientific Assessment, J.T. Houghton, B.A. Callander and S.K. Varney, eds., Cambridge University Press, Cambridge, England, 139-170*.

FOLLAND, C.K. AND D.E. PARKER, 1995, Correction of instrumental biases in historical sea surface temperature data. *Quart. J. R. Meteorol. Soc.*, **121**, 319-367.

FRANKIGNOUL, C., 1985, Sea-surface temperature anomalies, planetary waves and air-sea feedbacks in the middle latitudes. *Rev. Geophys.*, **23**, 357-390.

GILL, A.E., 1980, Some simple solutions for the heat induced tropical circulation. *Quart. J. Roy. Meteorol. Soc.*, **106**, 447-463.

GRAHAM, N.E., 1994, Decadal-scale climate variability in the tropical and North Pacific during the 1970's and 1980's: Observations and model results. *Climate Dyn.*, **10**, 135-162.

HASSELMANN, K., 1976, Stochastic climate models, I. Theory. *Tellus*, **28**, 473-485.

HOREL J.D., 1981, A rotated principal component analysis of the interannual variability of the Northern Hemisphere 500-mb height field. *Mon. Wea. Rev.*, **109**, 2080-2092.

HOREL J.D., AND J.M. WALLACE, 1981, Planetary-scale atmospheric phenomena associated with the Southern Oscillation. *Mon. Wea. Rev.*, **109**, 2080-2092.

HSUING, J. AND R.E. NEWELL, 1983, The principal non-seasonal modes of global sea surface temperature. *J. Phys. Oceanogr.*, **13**, 1957-1967.

HURRELL, JAMES W., 1995, Decadal trends in the North Atlantic Oscillation: Regional tem-peratures and precipitation. *Science*, **269**, 676-679.

HURRELL, JAMES W.,, 1996, Influence of variations in extratropical wintertime teleconnections on Northern Hemisphere temperatures. *Geophys. Res. Letters., submitted.*

JONES, P.D., 1988, Hemispheric surface air temperature variations: Recent trends and an update to 1987. *J. Climate,* **1,** 654-660.

JONES, P.D., AND R.S. BRADLEY, 1992, Climatic variations in the longest instrumental records. *In Climate Since A.D. 1500, R.S. Bradley and P.D. Jones, eds. Routledge, London and New York, 246-268.*

JONES, P.D., AND P.M. KELLY, 1983, The spatial and temporal characteristics of Northern Hemisphere surface air temperature variations. *Journal of Climatology,* **3,** 243-252.

JONES, P.D., S.C.B. RAPER, B.D. SANTER, B.S.G. CHERRY, C.M. GOODESS, P.M. KELLY, T.M.L. WIGLEY, R.S. BRADLEY AND H.F. DIAZ, 1985, A Grid Point Surface Air Temperature Data Set for the Northern Hemisphere. *U.S. Dep't. of Energy, Carbon Dioxide Research Division, Technical Report TR022, 251pp.*

KELLY, P.M., P.D. JONES, C.B. SEAR, B.S.G. CHERRY, AND R.K. TAVAKOL, 1982, Variations in surface air temperatures: Part 2. Arctic regions. *Mon. Wea. Rev.,* **110,** 71-83.

KIMOTO, M., 1987, Analysis of recurrent flow patterns in the Northern Hemisphere winter. *M.S, Thesis, Department of Atmospheric Sciences, Univ. of California at Los Angeles, Los Angeles, CA, 104pp.*

KIMOTO, M., 1989, Multiple flow regimes in the Northern Hemisphere winter. *PhD Thesis, Department of Atmospheric Sciences, Univ. of California at Los Angeles, Los Angeles, CA, 210pp.*

KIMOTO, M., AND M. GHIL, 1993, Multiple flow regimes in the Northern Hemisphere winter: Part I: Methodology and hemispheric regimes. *J. Atmos. Sci.,* **50,** 2625-2643.

KNOX, J.L. AND J.E. HAY, 1985, Blocking signatures in the Northern Hemisphere: Frequency distribution and interpretation. *J. Climatol.,* **5,** 1-16.

KUSHNIR, Y. AND J.M. WALLACE, 1989, Low frequency variability in the Northern Hemisphere winter: geographical distribution, structure, and time scale dependence. *J. Atmos. Sci.,* **46,** 3122-3142.

KUTZBACH, J.E., 1970, Large scale features of monthly mean Northern Hemisphere anomaly maps of sea-level pressure. *Mon. Wea. Rev.,* **98,** 708-716.

LAMBERT, S.J., 1990, Discontinuities in the long-term Northern Hemisphere 500-millibar dataset. *J. Clim.,* **3,** 1479-1484.

LATIF, M., E. STERL, E. MAIER-REIMER AND M. M. JUNGE, 1993, Climate variability in a coupled GCM. Part I: The tropical Pacific. *J. Climate,* **6,** 5-21.

LINDZEN, R.S. AND S. NIGAM, 1987, On the role of sea-surface temperature gradients in forcing low level winds and convergence in the tropics, *J. Atmos. Sci.,* **44,** 2418-2436.

MILLER, A. J., D. R. CAYAN, T. P. BARNETT, N. E. GRAHAM AND J. M. OBERHUBER, 1994, Interdecadal variability of the Pacific ocean: Model response to observed heat flux and wind stress anomalies. *Climate Dyn.,* **9,** 287-302.

MANTUA, J. N., AND N. E. GRAHAM, 1996, Recent trends in the climate of the tropical Pacific region. *Submitted to J. Climate.*

MONTGOMERY, R.B., 1940, Report on the work of G.T. Walker, *Mon. Wea. Rev., Suppl. No. 39: Reports on critical studies of methods of long-range weather forecasting., 21pp.*

MYSAK, L.A., D.K. MANAK AND R.F. MARDSEN, 1990, Sea-ice anomalies in the Greenland and Labrador Seas 1900-84 and their relation to an interdecadal Arctic climate cycle. *Climate Dyn.,* **5,** 111-133.

NAKAMURA, H., M. TANAKA AND J.M. WALLACE, 1987, Horizontal structure and energetics of Northern Hemisphere wintertime teleconnection patterns. *J. Atmos. Sci.,* **44,** 3377-3391.

NAMIAS, J., 1980, Some concomitant regional anomalies associated with hemispherically averaged temperature variations. *J. Geophys. Res.,* **85C,** 1585-1590.

NITTA, T., AND S. YAMADA, 1989, Recent warming of tropical sea surface temperature and its relationship to the Northern Hemisphere circulation. *J. Meteorol. Soc. Japan,* **58,** 187-193.

NORTH, G.T., 1975, Theory of energy balance climate models. *J. Atmos. Sci.,* **32,** 2033-2043.

PARKER, D.E. AND C.K. FOLLAND, 1991, Worldwide surface air temperature trends since the mid-19th century. *In Greenhouse-Gas Induced Climatic Change: A Critical Appraisal of Simulations and Observations, M.E. Schlesinger, ed., Elsevier, 173-193.*

PARKER, D.E., P.D. JONES, C.K. FOLLAND AND A. BEVAN, 1995 Interdecadal changes of surface temperature since the late 19th century. *J. Geophys. Res.,* **99D,** 14373-14399.

QUINN, W. H., AND V. T. NEAL, 1984, Recent climate change and the 1982-83 El Niño. *Proc. 8th Annual Climate Diagnostic Workshop, 8, 148-154. Available from Nat'l. Tech. Info. Svc., U.S. Dep't. of Commerce, Sills Bldg., 5285 Port Royal Rd., Springfield VA 22161. Accession #PB84-192418.*

QUINN, W. H., AND V. T. NEAL, 1985, Recent long-term climate change over the eastern tropical and subtropical Pacific and its ramifications. *Proc. 9th Annual Climate Diagnostic Workshop, 9, 101-109. Available from Nat'l. Tech. Info. Svc., U.S. Dep't. of Commerce, Sills Bldg., 5285 Port Royal Rd., Springfield VA 22161. Accession #PB85-183911.*

RASMUSSON, E. M., AND T. H. CARPENTER, 1982, Variations in tropical sea surface temperature and surface wind fields associated with the Southern Oscillation / El Niño. *Mon. Wea. Rev.,* **110,** 354-384.

ROPELEWSKI C AND M. HALPERT, 1989, Precipitation patterns associated with the high index phase of the Southern Oscillation. *J. Climate,* **2,** 268-283.

ROGERS, J.C., 1981, Spatial variability of seasonal sea-level pressure and 500-mb height anomalies. *Mon. Wea Rev.,* **109,** 2093-2106.

SARACHIK, E.S., 1978, Tropical sea surface temperature: an interactive one-dimensional atmosphere-ocean model. *Dyn. of Atmospheres and Oceans,* **2,** 455-469.

SCHNEIDER, E.K., B. HUANG AND J. SHUKLA, 1995, Ocean wave dynamics and El Niño. *J. Climate,* **8,** 2415-2439.

SIMMONS, A.J., K.M. WALLACE AND G.W. BRANSTATOR, 1983, Barotropic wave propagation and instability, and atmospheric teleconnection patterns. *J. Atmos. Sci.*, **40,** 1363-1392.

SPENCER, R.W., 1993, Global oceanic precipitation from the MSU during 1979-91 and comparisons to other climatologies. *J. Climate,* **6,** 1301-1326.

SPENCER, R.W., AND J.R. CHRISTY, 1990, Precise monitoring of global temperature trends from satellites. *Science,* **247,** 1558-1562.

SPENCER, R.W., AND J.R. CHRISTY, 1992, Precision and radiosonde validation of satellite gridpoint temperature anomalies. Part I: MSU channel 2. *J. Climate,* **5,** 847-857.

TRENBERTH, K. E., 1990, Recent observed interdecadal climate changes in the Northern Hemisphere. *Bull. Amer. Meteor. Soc.,* **71,** 988-993.

TRENBERTH, K. E., AND CHRISTY, 1985, Global fluctuations in the distribution of atmospheric mass. *J. Geophys. Res.,* **90D,** 8042-8052.

TRENBERTH, K. E. AND J.W. HURRELL, 1994, Decadal atmospheric-ocean variations in the Pacific. *Climate Dyn.,* **9,** 303-309.

VAN LOON, H. AND J. ROGERS, 1978, The seesaw in winter temperatures between Greenland and northern Europe. Part I: general description. *Mon. Wea. Rev.,* **106,** 296-310.

WALKER, G.T. AND E.W. BLISS, 1932, World Weather V. *Mem. R. Met. Soc.,* **4,** 53-84.

WALLACE, J.M., 1995, Natural and forced variability in the climate record. *In Natural Climate Variability on Decade-to-Century Time Scales, D.G. Martinson, K. Bryan, M. Ghil, M.M. Hall, T.R. Karl, E.S. Sarachik, S. Sorooshian and L.D. Talley eds., National Academy Press, Washington, in press.*

WALLACE, J.M., AND D. S. GUTZLER, 1981, Teleconnections in the geopotential height field during the Northern Hemisphere winter. *Mon. Wea. Rev.,* **109,** 785-812.

WALLACE, J.M., AND N.-C. LAU, 1985, On the role of barotropic energy conversions in the general circulation. *Atmospheric and Oceanic Modeling, a volume in Advances in Geophysics,* **28,** 33-74.

WALLACE, J.M., Y. ZHANG AND K.-H. LAU, 1993, Structure and seasonality of interannual and interdecadal variability of the geopotential height and temperature fields in the Northern Hemisphere troposphere. *J. Climate,* **6,** 2063-2082.

WALLACE, J.M., Y. ZHANG AND L.BAJUK, 1995A, Interpretation of interdecadal trends in Northern Hemisphere surface air temperature. *J. Climate, in press.*

WALLACE, J.M., Y. ZHANG AND J.A. RENWICK, 1995B, Dynamical contribution to hemispheric temperature trends *Science, in press.*

WANG, B., 1995, Interdecadal changes in El Nino onset in the last four decades. *J. Climate,* **6,** 267-285.

WASHINGTON, W.M. AND C.L. PARKINSON 1986, An Introduction to Three-Dimensional Climate Modeling. *University Science Books, Mill Valley, Calif., and Oxford University Press, 422pp.*

WOODRUFF, S.D., R.J. SLUTZ, R.L. JENNE AND P.M. STEURER, 1987, A comprehensive ocean-atmosphere data set. *Bull. Amer. Meteorol. Soc.*, **68**, 1239-1250.

YULAEVA, E, AND J.M. WALLACE, 1994, The signature of ENSO in global temperature and precipitation fields derived from the microwave sounding unit. *J. Climate*, **7**, 1719-1736.

ZHANG, Y., J.M. WALLACE AND D.S. BATTISTI, 1996, ENSO-like decade-to-century scale variability: 1900-93. *J. Climate, submitted.*

PREDICTABILITY OF THE ATMOSPHERE AND OCEANS: FROM DAYS TO DECADES

T.N.PALMER
ECMWF
Reading, UK

Contents

NATO ASI Series, Vol. I 44
Decadal Climate Variability
Dynamics and Predictability
Edited by David L. T. Anderson and Jürgen Willebrand
© Springer-Verlag Berlin Heidelberg 1996

1   Introduction

This paper is concerned with the predictability of the atmosphere and oceans on timescales of days to decades. A variety of phenomena will be discussed, from individual weather events through weather regimes and El Niño, to decadal ocean-atmosphere fluctuations and climate change. However, no matter what timescale or phenomenon is being considered, we shall be studying processes which are believed to be fundamentally chaotic.

A chaotic system can be defined as one whose evolution is sensitive to initial conditions. However, this is not to say that the unpredictability associated with this sensitivity is only of importance for initial-value problems. For example, determining the impact on climate of doubling CO_2 is not primarily an initial value problem. Nevertheless, the fact that the climate is chaotic has fundamental implications for the predictability of this type of question. Just as initial conditions for a weather forecast are not perfectly accurate, and hence can only be specified completely in terms of some probability distribution, so also the formulation of a climate model (associated with the physical parametrisations in particular) is only approximate, and again can only be specifed completely in terms of some stochastic distribution. The instabilities that amplify uncertainties in the initial state, may also amplify uncertainties in model formulation.

We start in section 2 with a discussion on predictability of initial value problems. As suggested above, a fundamental quantity in this discussion is the forecast probability density function (PDF). The evolution of this PDF can be described in the first phase of the forecast by linearised dynamics. The semi-major axes of the PDF are given by the dominant singular vectors of the linear evolution operator (using the so- called Mahalanobis inner product). We relate these singular vectors to more familiar quantities associated with eigenmode growth on the one hand, and to Lyapunov exponent growth on the other. We also discuss the relationship of singular vectors and so-called breeding vectors.

In section 3, we apply the methodology developed in section 2 to study predictability associated with a variety of phenomena on timescales ranging from days to seasons. In particular, the singular vector instability of individual extratropical weather systems, and of the coupled ocean-atmosphere El Niño/Southern Oscillation, is studied. For example, we demonstate the endemic upscale energy cascade associated with extratropical predictability, and indicate that this effect is much less predominant in the tropics. In doing this a new interpretation of the Charney-Shukla paradigm for large-scale tropical predictability is given.

In section 4, we discuss the predictability of forced problems (predictability of the second kind). Explicit examples are given using both the Lorenz 3-component model, and the global circulation models (GCMs). We develop a basic nonlinear paradigm for analysing the response to an external perturbation. The basic notion is that the influence of a weak external forcing is greatest in regions of phase space where the system is particularly unstable; however, the response of the system is greatest in regions of phase space where the system is particlarly stable. The extratropical response of the atmosphere to extratropical sea surface temperature (SST) anomalies is discussed using this paradigm.

In section 5, the predictability of natural fluctuations of the climate system on decadal timescales is discussed on the basis of a number of GCM integrations. One fundamental question concerns the relative contribution of the atmospheric and oceanic dyamics in contributing to observed decadal variability. It is found that, even on decadal timescales, the role of purely internal atmospheric variability is not negligible. This itself implies that the predictability of decadal fluctuations may not be strong. However, in addition, it is suggested that the role of the extratropical oceans is basically to redden the spectrum of atmospheric variability, *eg* by increasing the typical residence time of the atmospheric state vector within a weather regime. As such, decadal atmospheric variability linked directly to decadal SST variability may not itself be strongly predictable.

In section 6, the nonlinear paradigm discussed earlier is applied to the problem of climate change. We assess whether or not the observed warming of the atmosphere can be attributed to the greenhouse effect. It is suggested that the observed warming over the past few decades can be largely interpreted in terms of an increase in the frequency of one of the dominant regimes of the extratropical flow. Singular vector analysis suggests that this regime may be most sensitive to forcing in the tropical west Pacific warm pool region. This therefore may be the most critical area for understanding how enhanced CO_2 may influence global climate. The predictability implications of this analysis are discussed.

In section 7, some remarks are made about the rationalisation of climate and weather prediction models.

In this paper, we shall make some use of low-dimensional chaotic models.

This does not imply that I think that the climate necessarily has a low-dimensional attractor. Rather, for some purposes, I believe the use of relatively simple models can be helpful to illustrate basic processes. On the other hand, as (singular vector) calculations in the body of the text suggest, there are other circumstances where comparison with a turbulent fluid may be more appropriate.

2 Predictability of the first kind

2.1 The forecast probability density function

As mentioned in the introduction, we can distinguish two basic types of prediction. Following Lorenz (1975), predictions of the first kind are initial value problems (*e.g.* medium-range weather forecasts with an atmosphere model, or seasonal forecasts with a coupled ocean-atmosphere model). Predictability of the first kind is therefore concerned with the question of how uncertainties in the initial state evolve during the forecast and limit its skill.

The predictability of a system is strongly dependent on its stability properties. If the system is particularly unstable, then any uncertainty in the initial state that projects significantly onto one of these instabilities will severely limit the skill of an initial-value forecast. Let us try to be more precise. Suppose we represent the uncertainty in the initial condition for a prediction of the first kind in terms of a PDF in some finite m-dimensional phase space. For example, for each direction in phase space, let us assume this PDF to be normally distributed about our best estimate of the initial state. The standard deviation of this normal distribution will, in general, vary with direction. However, we can define a metric on the phase space (local to the initial condition) so that the standard deviations are independent of direction (see section 2.7 for a more explicit description of this). With respect to this metric, the PDF will now be isotropic, and isopleths of the PDF will bound an m-dimensional ball (see Fig 1a).

A quantitative measure of predictability can be defined in relation to the properties of the evolution of this initial PDF. In the early part of the forecast, error growth is governed by linear dynamics. During this period, initially spherical isopleths of the PDF will evolve to bound an m-dimensional ellipsoidal volume (see Fig 1b). The major axis of the ellipsoid corresponds to a phase-space direction which defines the dominant

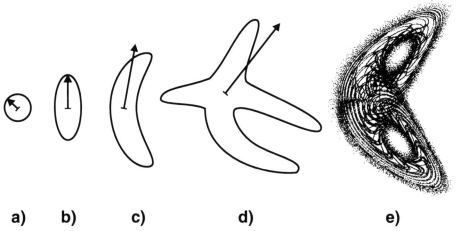

a)   **b)**   **c)**   **d)**   **e)**

Figure 1: *Schematic evolution of the probability density function (PDF) of forecast error. Initially (a) the analysis error distribution is isotropic (with respect to the appropriate Mahalanobis metric). During the linear stage of evolution (b), the error ball evolves into an ellipsoid. A vector pointing along the major axis is shown in (b), and its pre-image at initial time is shown in (a). The weakly nonlinear stage of evolution is shown in (c). During this phase, there is significant agreement between the evolution of the vector in (b) and the principal direction in which the distance of an isopleth of probability from the mode of the distribution is maximal. In the strongly nonlinear stage of evolution (d), the relationship between the PDF and the evolved directions of the major axes of the linear ellipsoid breaks down. Total loss of predictability (e) occurs when the PDF essentially covers the attractor.*

instability of that part of phase space. The ratio of the standard deviation of the PDF along this major axis, compared with the initial standard deviation, is a measure of the amplification rate associated with this dominant instability (and indeed is a measure of the l^∞ norm of the operator which maps initial perturbations to forecast perturbations). In addition to this major axis direction, there may be other orthogonal directions in which the initial PDF has amplified significantly; these clearly define secondary directions of instability.

Before giving a quantitative description of this linear stage, we show (schematically) three further stages in the evolution of the forecast PDF. The growth of the PDF between Fig 1b and Fig 1c could be described as 'weakly nonlinear'. In Fig 1c the PDF has deformed from its ellipsoidal shape in Fig 1b. From the centroid of this PDF, one can define directions D_i for which the distance from the centroid to a chosen isopleth of the PDF is maximised. During the weakly nonlinear period, the evolution of

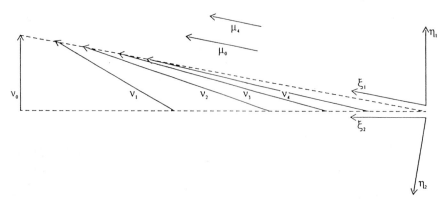

Figure 2: *This diagram illustrates schematically the crucial difference between eigenmode and singular vector growth, and the relationship between singular vectors and adjoint modes. See text for details.*

the major axis directions E_i, from Fig 1b to Fig 1c will be close to the directions D_i.

The growth of the PDF after Fig 1c can be described as 'strongly nonlinear'. In particular, for the PDF in Fig 1d, there may be no correspondence between the dominant directions D_i (defined as above) and the directions corresponding to evolution of the major axis directions in Fig 1b. For example, it might be that one of the dominant directions shown in Fig 1d arose from the evolution of a minor axis direction in Fig 1b. Although the evolution of the PDF is strongly nonlinear, predictability has not necessarily been lost at the stage corresponding to Fig 1d. Rather, predictability is lost when the PDF isopleth has evolved to cover the entire attractor (*cf* Fig 1e).

It should be noted that the timescales associated with the 'linear', 'weakly nonlinear' and 'strongly nonlinear' phases of evolution depend on which isopleth of the PDF one is considering. An isopleth of small probability will bound a larger volume at initial time than an isopleth of large probability. Consequently, the timescales will be shorter for the small-probability isopleth. In practice, for numerical weather prediction, there is evidence that errors of about one standard deviation of the analysis error PDF evolve linearly for 2-3 days, and that the 'weakly nonlinear' phase lasts until about day 7 of the forecast (see section 3.3).

2.2 Singular vectors

Let us try to quantify further the linear stage of evolution of the forecast PDF between Fig 1a and Fig 1b. Let us suppose our basic system is described by the (m-dimensional) nonlinear evolution equation

$$\frac{d\mathbf{X}}{dt} = \mathbf{M}[\mathbf{X}] \qquad (2.1)$$

Consider a small perturbation $\mathbf{x}$ of the state vector $\mathbf{X}$. For sufficiently short time intervals, its evolution can be described by the linearised approximation

$$\frac{d\mathbf{x}}{dt} = \mathbf{M}_l \mathbf{x} \qquad (2.2)$$

of (1). $\mathbf{M}_l \equiv \frac{d\mathbf{M}}{d\mathbf{X}}\big|_{\mathbf{X}(t)}$ is the linear evolution operator evaluated on the nonlinear trajectory $\mathbf{X}(t)$.

Equation (2.2) can be written in the integral form

$$\mathbf{x}(t) = \mathbf{L}(t, t_0)\mathbf{x}(t_0) \qquad (2.3)$$

In practice, we estimate $\mathbf{L}$ in (2.3) by splitting up the trajectory into many short quasi- stationary segments. For each segment we can write

$$\mathbf{L}(t_i, t_j) = e^{(t_i - t_j)\mathbf{M}_l} \qquad (2.4)$$

(Because the full trajectory segment is time-varying, we cannot, in general, write $\mathbf{L}(t, t_0)$ in terms of the operator exponential.)

The operator $\mathbf{L}(t, t_0)$ is referred to as the forward tangent propagator; it maps small perturbations along the (nonlinear) trajectory from an initial time t_0 to some future time t. For the application to weather prediction, if $\mathbf{x}(t_0)$ is the typical error in the initial conditions for a weather forecast, then (2.2) and (2.3) hold for approximately 2-3 days of integration time.

We now define an inner product $(\mathbf{x}; \mathbf{y})$, which in turn defines a metric on the tangent space. Following the discussion above this inner product is not arbitrary, it is defined so that the PDF of the initial conditions is isotropic. As discussed in section 2.7, the atmospheric PDF appears reasonably isotropic using an inner product based on total perturbation energy. Using (2.3), the perturbation norm at time t is given by

$$\|\mathbf{x}(t)\|^2 \equiv (\mathbf{x}(t); \mathbf{x}(t)) = (\mathbf{x}(t_0); \mathbf{L}^*\mathbf{L}\mathbf{x}(t_0)) \qquad (2.5)$$

where $\mathbf{L}^*$ is the adjoint of $\mathbf{L}$ with respect to the energy inner product. Note that if $\mathbf{L}$ is represented in matrix form, then $\mathbf{L}^*$ is just the matrix transpose of $\mathbf{L}$.

Unlike $\mathbf{L}$ itself, the operator $\mathbf{L}^*\mathbf{L}$ (sometimes referred to as the Oseledec operator, *e.g.* Abarbanel *et al.*, 1991) is easily shown to be symmetric. Hence its eigenvectors $\boldsymbol{v}_i(t_0)$ can be chosen to form an orthonormal basis (assumed complete) in the m-dimensional tangent space of linear perturbations, with real eigenvalues $\sigma_i^2 \geq 0$ (*eg* Noble and Daniel, 1977) *i.e.*

$$(\mathbf{L}^*\mathbf{L})\boldsymbol{v}_i(t_0) = \sigma_i^2 \boldsymbol{v}_i(t_0) \tag{2.6}$$

At future time t, these eigenvectors evolve to $\boldsymbol{v}_i(t) = \mathbf{L}\boldsymbol{v}_i(t_0)$ which in turn satisfy the eigenvector equation

$$(\mathbf{L}\mathbf{L}^*)\boldsymbol{v}_i(t) = \sigma_i^2 \boldsymbol{v}_i(t) \tag{2.7}$$

From eqs. (2.5) and (2.6),

$$\|\boldsymbol{v}_i(t)\|^2 = (\boldsymbol{v}_i(t_0); \mathbf{L}^*\mathbf{L}\boldsymbol{v}_i(t_0)) = \sigma_i^2 \tag{2.8}$$

Since, by completeness, any $\mathbf{x}(t)/\|\mathbf{x}(t_0)\|$ can be written as a linear combination of the set $\boldsymbol{v}_i(t)$, it follows that

$$\max_{\mathbf{x}(t_0)\neq 0} \left(\frac{\|\mathbf{x}(t)\|}{\|\mathbf{x}(t_0)\|} \right) = \sigma_1 \tag{2.9}$$

Following the terminology of linear algebra, the σ_i, ranked in terms of magnitude, are called the singular values of the operator $\mathbf{L}$ and the vectors $\boldsymbol{v}_i(t)$ are called the singular vectors of $\mathbf{L}$. Maximum energy growth over the time interval $t - t_0$ is therefore associated with the dominant singular vector: $\boldsymbol{v}_1(t_0)$ at initial time, and $\boldsymbol{v}_1(t)$ at optimisation time. Following the discussion above, the $\boldsymbol{v}_i(t)$ define the directions of the axes of the forecast PDF ellipsoid, with $\boldsymbol{v}_1(t)$ defining the major axis, $\boldsymbol{v}_2(t)$ the second major axis, and so on. The directions at initial time that evolve into these axes are given by $\boldsymbol{v}_1(t_0), \boldsymbol{v}_2(t_0)$ respectively. The amplification of the PDF standard deviations associated with these directions are given by the σ_i.

As far as I am aware, a discussion of singular vector growth in meteorology was first given by Lorenz (1965).

2.3 Correspondence with 'normal' mode instability

Singular vector analysis is in some sense a generalisation of classical normal mode instability analysis. This can be made explicit by linearising about a stationary solution of (2.1), so that normalised eigenvectors $\boldsymbol{\xi}_i$ of $\mathbf{M}_l$ with eigenvalues μ_i give rise to modal solutions $\boldsymbol{\xi}_i e^{\mu_i(t-t_0)}$ of (2.2). The integral operator $\mathbf{L}(t, t_0)$ can be written as $e^{(t-t_0)\mathbf{M}_l}$, with eigenvectors $\boldsymbol{\xi}_i$ and eigenvalues $e^{(t-t_0)\mu_i}$.

For application to atmosphere-ocean dynamics, the linear evolution operators associated with realistic basic state flows are never normal (*ie* $\mathbf{L}^*\mathbf{L} \neq \mathbf{L}\mathbf{L}^*$) because of vertical and horizontal shear (*eg* Farrell and Ioannou, 1996). Now, it is common meteorological parlance to call any modal eigenvectors as 'normal' modes. However, for any meteorological basic state these eigenvectors are not eigenvectors of a normal operator (and hence not normal). In future we refer to such eigenvectors as 'eigenmodes' (hence the quotation marks in the title of this sub-section). However, irrespective of normality, eigenvectors $\boldsymbol{\eta}_i$ and eigenvalues θ_i of the adjoint operator $\mathbf{L}^*$ satisfy the biorthogonality condition

$$(\mu_i - \theta_i^{cc})(\boldsymbol{\eta}_i; \boldsymbol{\xi}_i) = 0 \tag{2.10}$$

where 'cc' denotes complex conjugate. This condition ensures that the eigenvalues of an eigenvector/adjoint eigenvector pair that are not orthogonal, must form a complex conjugate pair. The magnitude of the inner product $(\boldsymbol{\eta}_i; \boldsymbol{\xi}_i)$ for such eigenvector pairs equals the cosine of the angle, α_i, they subtend in phase space.

If an initial disturbance comprises a linear combination of the eigenmodes $\boldsymbol{\xi}_i$ so that

$$\mathbf{x}(t) = \sum_i c_i \boldsymbol{\xi}_i e^{\mu_i(t-t_0)} \tag{2.11}$$

then from the biorthogonality condition (2.10)

$$c_i = (\boldsymbol{\eta}_i; \mathbf{x}(t_0))/(\boldsymbol{\eta}_i; \boldsymbol{\xi}_i) \tag{2.12}$$

From (2.11), the fastest growing eigenmode will ultimately dominate the linear combination. Hence for sufficiently long optimisation times, the dominant singular vector at optimisation time will correspond to the most unstable eigenmode. (Since the singular values are real, whilst the eigenvalues are complex, there is an arbitrary phase factor that has to be defined to make this correspondence precise.)

In order to maximise the contribution of the first eigenmode at optimisation time, c_1 in (2.11) should be as large as possible. If $\mathbf{x}(t_0)$ equals $\boldsymbol{\xi}_1$ then from (2.12), $c_1 = 1$ which could be highly sub-optimal. In fact, if $\mathbf{x}(t_0)$ projects onto $\boldsymbol{\eta}_1$, then c_1 is maximised and is given by the projectibility factor $1/(\cos\alpha_1)$ (Zhang, 1988).

Hence, for indefinitely long optimisation time, the dominant singular vector, at initial time, is determined by the first adjoint eigenmode, whilst the dominant singular vector at optimisation time is determined by the first eigenmode itself. The singular value will depend on both the e-folding time of the dominant eigenmode and its projectibility. For finite optimisation time, the dominant singular vectors will no longer project onto individual eigenmode solutions (and their adjoints), and the amplitude of finite-time instabilities need not be bounded by properties of the dominant eigenmodes alone.

Fig 2 illustrates schematically the crucial difference between eigenmode and singular vector growth, and the relationship between singular vectors and adjoint modes. An idealised 2-D system has two very non-orthogonal decaying eigenmodes $\boldsymbol{\xi}_1$ and $\boldsymbol{\xi}_2$. We take $\boldsymbol{\xi}_1$ to have the larger real eigenvalue component. The adjoint eigenmodes $\boldsymbol{\eta}_1$ and $\boldsymbol{\eta}_2$ are shown with $\boldsymbol{\eta}_1,\boldsymbol{\eta}_2$ orthogonal to $\boldsymbol{\xi}_2,\boldsymbol{\xi}_1$ respectively (according to the biorthogonality condition 2.10). A normalised vector $\boldsymbol{\nu}_0$ is shown parallel to $\boldsymbol{\eta}_1$. Its time evolution can be estimated by mapping the tip and tail of $\boldsymbol{\nu}_0$ along the $\boldsymbol{\xi}_1$ and $\boldsymbol{\xi}_2$ directions (shown as dashed lines) using the modal decay rates. The sequence of vectors $\boldsymbol{\nu}_n$, $n = 1, 2, ...$ giving the time evolution of $\boldsymbol{\nu}_0$ increases in amplitude up to some finite $n = N$ and is aligned almost entirely with $\boldsymbol{\xi}_1$ for large n. The projection of $\boldsymbol{\nu}_n$ onto $\boldsymbol{\xi}_1$ for large n is much larger than that associated with the evolution of a second normalised vector μ_n which is initially aligned along $\boldsymbol{\xi}_1$. The sequence $\boldsymbol{\nu}_n$, $n = 1, 2, ...$ describes singular vector growth over a long time interval. The transient growth of the singular vectors in such systems was first noted by Orr (1907), and a review of this process in plane parallel shear flow (and its relationship to wave overreflection) is discussed in Lindzen (1988).

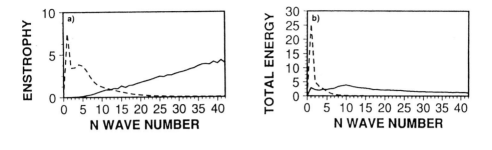

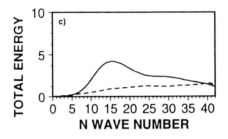

Figure 3: *(a) Average enstrophy (s^{-2}) spectrum of the first 16 enstrophy-SVs at initial (dash, $\times 40\ 10^{16}$) and final (solid, $\times 10^{16}$) time, (b) average energy (per unit mass) spectrum ($m^2\ s^{-2}$) of the first 16 energy-SVs at initial (dash, $\times 40$) and final (solid, $\times 40$) time, and (c) average energy spectrum of the first 16 energy-SVs at initial (dashed, $\times 40$) and final (solid) time, for 6 April 1994. From Molteni et al (1996).*

2.4 Correspondence with Lyapunov exponent growth

If, instead of linearising about a stationary flow, let us consider the other extreme of linearising about a (time-evolving) trajectory portion which is sufficiently long to approximately cover the entire climate attractor.

Specifically, if we apply the forward tangent propagator N times from t_0 so that

$$\mathbf{x}(t_n) = \mathbf{L}(t_n, t_{n-1})\mathbf{x}(t_{n-1}) \tag{2.13}$$

with $t_n - t_{n-1} = \Delta t$ a unit time interval, then according to the Multiplicative Ergodic theorem of Oseledec (1968), the eigenvalues of the Oseledec

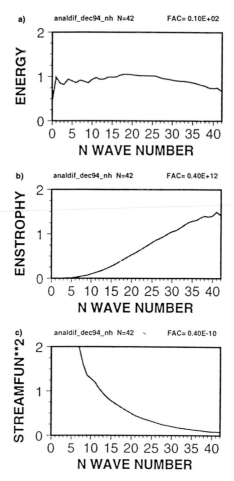

Figure 4: *The mean 2D total wavenumber spectrum of analysis difference fields in terms of a) energy, b) enstrophy, c) streamfunction variance. This mean has been calculated over 42 analysis difference fields, based on experimental operational analysis code and the operational suites at ECMWF. The experimental suites involve both different model formulations and different analysis technique (R. Gelaro, personal communication).*

matrix

$$[\mathbf{L}(t_n, t_0), \mathbf{L}(t_n, t_0)]^{\frac{1}{2n}} \tag{2.14}$$

are independent of initial conditions as $n \to \infty$, and hence are invariants of the dynamical system. The logarithms of these eigenvalues are the Lyapunov exponents l_i.

It is straightforward to see that the singular values σ_i of the propagator

$\mathbf{L}(t_n, t_0)$ are related to the Lyapunov exponents by the equation

$$l_i = \lim_{n \to \infty} [\frac{1}{n} \ln \sigma_i] \tag{2.15}$$

From this, some authors (*eg* Abarbanel *et al*, 1991) define the local Lyapunov exponents $l_i(t_1, t_0)$ of the attractor between t_1 and t_0 as

$$l_i(t_1, t_0) = \frac{1}{t_1 - t_0} \ln \sigma_i(t_1, t_0) \tag{2.16}$$

The corresponding singular vectors can therefore be referred to as local Lyapunov vectors, although it should be noted that this terminology is not universal (*eg* Toth and Kalnay, 1996).

2.5   Projection operators

Singular vectors are, in general, not modal. For the application in section 3.2, their shapes evolve not only in geographical space but also in their spectral distribution of energy. As we shall see for extratropical weather systems, this spectral evolution describes, in a linear context, the upscale energy transfer associated with turbulent processes. In order to study this upscale energy transfer more explicitly we introduce a spectral projection operator $\mathbf{P}_{[n_1,n_2]}$ where $[n_1, n_2]$ denotes the total wavenumber interval $n_1 \leq n \leq n_2$. $\mathbf{P}_{[n_1,n_2]}$ is defined as

$$\begin{aligned} \mathbf{P}_{[n_1,n_2]}\mathbf{x}_n &= \mathbf{x}_n \ \text{ if } \ n\epsilon[n_1, n_2] \\ \mathbf{P}_{[n_1,n_2]}\mathbf{x}_n &= 0 \ \text{ otherwise} \end{aligned} \tag{2.17}$$

Here $\mathbf{x}_n$ is the wavenumber n component of the spherical harmonic expansion of the (atmospheric) state vector. If we wish to find perturbations, initially constrained to be in $[n_3, n_4]$, with maximum energy in $[n_1, n_2]$, these are given by the singular vectors of $\mathbf{P}_{[n_1,n_2]}\mathbf{L}\mathbf{P}_{[n_3,n_4]}$. A similar projection operator can be applied to study singular vectors whose energy is optimised to a specific geographical area (for applications, see Buizza and Palmer, 1995; Hartmann *et al*, 1995).

2.6   Numerical solution

When systems with a large number of degrees of freedom (*eg* $0(10^4)$ or more) are considered, the eigenvalue problem (2.6, 2.7) cannot be solved using direct methods. However, iterative techniques provide an alternative possibility if the adjoint propagator has been coded. The power method, whereby a random initial vector is operated on repeatedly by $\mathbf{L}^*\mathbf{L}$, is an example. A more sophisticated technique such as the Lanczos algorithm (Strang, 1986) is required if more than the largest singular vector is required. More recently, calculations with the Jacobi-Davidson method (Sleijpen and van der Vorst, 1995) has proved both efficient, and allowed estimation of generalised eigenvectors (see section 2.7 below).

2.7 Singular vectors and eigenvectors of the forecast and analysis error covariance operator.

We discussed above the notion that there is a natural inner product defined such that the PDF of the initial state is isotropic with respect to this metric. In this section we shall show this more explicitly, and give evidence that the total energy is a reasonable approximation to this preferred inner product.

Consider an initial state of an operational weather forecast, determined by the operational data assimilation system. We can think of this initial state as a point X in the phase space of the numerical weather prediction model. Now, as mentioned, a complete operational data assimilation system should be able to determine not only the initial state, but also an estimate of the probability that the initial state is in error by a given amount.

To make this idea more precise, let us consider the (linear) vector space T_X tangent to X, and let $d\mu$ denote the probability that the error lies in a small volume at the point $e^i \epsilon T_X$ (ie μ is a measure on T_X). Here we shall use some elementary tensor algebra, with the convention that repeated indices imply summation. Let us assume that the operational analysis is our best unbiased estimate of truth, so that at initial time

$$m^i = \int e^i d\mu = 0 \qquad (2.18)$$

The covariance of analysis error associated with this measure is given by the contravariant second-rank tensor

$$C^{ij} = \int e^i e^j d\mu \qquad (2.19)$$

The linear transformation (2.3) between $T_{X(t_0)}$ and $T_{X(t)}$ can be written (in index form) as

$$\hat{e}^i = L^i{}_j e^j \qquad (2.20)$$

which takes analysis errors e^j to forecast errors $\hat{e}^i$. In terms of this linear mapping, the covariance matrix is transformed as a second rank tensor to the forecast covariance

$$\hat{C}^{kl} = L^k{}_i L^l{}_j C^{ij} \qquad (2.21)$$

Now we are going to define a metric g_{ij} which defines the (scalar) inner product

$$s = g_{ij} x^i y^j \qquad (2.22)$$

between any two vectors x^i and y^j. There are many choices of inner product possible; however, we shall single one out as being special. It is the metric in which the analysis error covariance tensor is isotropic, ie is defined so that

$$C^{ij} g_{jk} = \delta^i{}_k \qquad (2.23)$$

the right hand side being the Kronecker delta. This can be written equivalently as

$$C^{ij} = g^{ij} \qquad (2.24)$$

where

$$g^{ij} g_{jk} = \delta^i{}_k \qquad (2.25)$$

defines the inverse or contravariant metric.

With this choice of metric, the forecast error covariance operator can be written

$$\hat{C}^k{}_l = L^k{}_i L_l{}^i \qquad (2.26)$$

Equation (2.26) can be expressed in matrix form

$$\hat{\mathbf{C}} = \mathbf{L}\mathbf{L}^* \qquad (2.27)$$

where

$$L^{*i}{}_j = L_j{}^i \qquad (2.28)$$

is the adjoint propagator. Hence with the specific choice of metric (2.24), the eigenvectors of $\hat{\mathbf{C}}$ are precisely the evolved singular vectors of $\mathbf{L}$. The vectors at initial time which evolve into these directions at the forecast time are given by the corresponding initial singular vectors of $\mathbf{L}$. This choice of metric is sometimes known as the Mahalanobis metric (*eg* Mardia *et al*, 1979).

How do we define a metric in practice? Three simple choices are based on the enstrophy, energy and streamfunction squared. The spherical harmonic spectrum of typical 48-hour enstrophy and energy-norm singular vectors are shown in Fig 3 at initial (dashed) and at final (solid) time. The enstrophy spectrum (Fig 3a) of the enstrophy SVs are red at initial time, blue at final time. Because of the large enstrophy amplification, values at initial time are multiplied by 40 to plot them on the same scale as final values. (However, Fig 3b, which has no rescaling, shows that there is little energy amplification associated with these enstrophy SVs.) By contrast, the energy spectrum (Fig 3c) of the energy SVs peaks at sub-synoptic scales at initial time and at synoptic scales at optimisation time.

We now want to compare these spectra with spectra of analysis error. As a surrogate for a set of analysis error fields, let us take the differences between analyses made during periods when the ECMWF operational system and some (potentially operational) experimental system were being run in parallel. We have chosen two different periods when such parallel tests were being made. The first corresponded to the testing of an experimental model formulation, the second to the testing of an experimental analysis methodology (3-dimensional variational data assimilation, 3DVAR). Twenty one analysis difference fields were taken from each period. The total horizontal wavenumber spectra of the energy, enstrophy, and streamfunction squared difference fields associated with the northern extratropical component of these difference fields are shown in Fig 4a-c respectively. It is clear that in terms of energy, the difference field is indeed almost white. By contrast, the enstrophy spectrum is blue and the streamfunction squared spectrum is red. (Molteni *et al*, 1996, have demonstrated that these results can be replicated using difference fields from operational analyses from different operational centres.) Comparing 2.4 with 2.3 we can rule out enstrophy as a suitable metric (the initial enstrophy singular vectors have no amplitude on small scales, whilst the analysis errors have all their enstrophy on small scales). A streamfunction metric would also be ruled out (although not shown, an initial streamfunction singular vector is strongly peaked at high wavenumbers). Only the energy metric shows some consistency between singular vector and analysis error structure.

A more accurate estimate of inner product is, in principle, available from the 3DVAR data assimilation system. In 3DVAR a cost function J based on both data and a first guess error, is minimised. The second derivative, or Hessian, of the cost function gives a measure of the analysis error covariance. In terms of the Hessian, the singular vector computation becomes equivalent to a generalised eigenvector problem. In 3DVAR the Hessian is known in operator form (Fisher and Courtier, 1995). Tests are in progress, using the Jacobi-Davidson scheme, to estimate the corresponding singular vectors using this Hessian. Further generalisation, in which the background error covariance in the Hessian is flow dependent, will be possible with the development of Kalman filter schemes (Bouttier, 1993).

2.8 Correspondence with breeding vectors

Toth and Kalnay (1993, 1995) have discussed a technique which they use to generate initial perturbations for the NMC model. The technique is referred to as 'breeding', and is simply described. A random initial perturbation is generated with a specified amplitude, characteristic of a typical uncertainty in the initial state. Two integrations are run over a specified cycle time (eg 12 hours). The first is an integration of the operational weather prediction model from the operational initial state. The second integration is made using the same model, but is initialised by adding the perturbation to the operational analysis. At the end of the integration period, the difference between the two integrations is renormalised using the specified amplitude. The process is repeated for the next cycle time using the renormalised perturbation to generate the perturbed initial state. The process is repeated ad infinitum. Toth and Kalnay (1995) call the breeding vectors, local Lyapunov vectors. Clearly they are not entirely local, depending on the history of evolution of the breeding vector. A more accurate description would be in terms of the local orientation of the global Lyapunov vectors (L. Smith, personal communication).

It is claimed that the breeding method mimics the analysis cycle that is used to generate operational initial conditions. As a result, it is argued that analysis errors will tend to rotate into the direction of the breeding vector. As discussed in section 3.3 below, the singular vectors form the basis for the calculation of initial perturbations for the ECMWF ensemble prediction

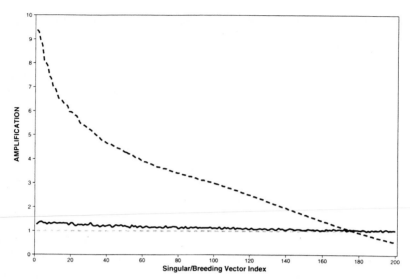

Figure 5: *The growth of orthogonal perturbations produced by the breeding method (solid line) and the singular vector method (dashed line), over a typical 12 hour period, based on the T21L3 quasi-geostrophic model of Marshall and Molteni (1993) (J. Barkmeijer, personal communication).*

system. How do the growth and spectrum of the breeding vectors compare with the singular vectors?

In order to discuss this question, I would like to show some results from a set of integrations of a 3-level quasi-geostrophic model (as formulated by Marshall and Molteni, 1993). In these integrations, the spectrum of breeding vectors has been compared with the spectrum of singular vectors over a 12 hour cycle time (for the breeding method) and a 12 hour optimisation time (for the singular vectors). These calculations have been performed by J. Barkmeijer (personal communication).

Fig 2.5 shows a typical spectrum for the singular vectors and the breeding vectors. The spectrum of the breeding vectors is very flat. The fastest growing perturbation grows by a factor of 1.4, the 200th perturbation grows by a factor of 1.1. In fact there are typically between about 250 and 300 growing directions as determined by the breeding method. By contrast the dominant singular values are much faster growing, and decay more rapidly with singular vector number. The fastest growing singular vector grows by a factor of 9.3, the 200th perturbation is actually decaying. In fact it can be shown that because of the flat spectrum, even small amounts of nonlinearity prevent the breeding vector from converging to a Lyapunov vector.

It can be asked whether the dominant singular vectors and breeding vectors correlate with one another. In general they do not. At initial time the singular vectors are dominated by sub-synoptic scales, whilst the breeding vectors are dominated by synoptic scales (for stationary basic states they would be given by the synoptic-scale eigenmodes). As such their spatial correlation is close to zero. At optimisation time, the singular vectors evolve towards synoptic scales and the correlation with the dominant breeding vectors increases, though only to values of about 0.2 (in the T21QG model).

Using these results, and those from the last section, I believe that it is possible to make the following conclusions about the relationship between the breeding vectors and the PDF of analysis error. Firstly, since the spectrum of breeding vectors is extremely flat, there is no particular reason why the analysis error should project more onto the leading breeding vectors. Of course, if one had enough breeding vectors (and if they were suitably orthogonalised) then any perturbation, including the analysis error, should project into the space spanned by these vectors. However, it appears from the results above that the number of such vectors would have to be a significant fraction of the phase space dimension.

In fact, in my opinion, even if the spectrum of breeding vectors was in fact much steeper, it is questionable whether the analysis error would ever rotate into the direction of a leading breeding vector. The reason for this is to do with the role of observations in the analysis cycle. As discussed *eg* by Hollingsworth (1987) and Daley (1991), operational analyses blend observations with a first-guess field in a scale-dependent manner. On large scales, the observations carry more weight than the first guess field, whilst on small scales the first guess carries more weight than the observations. Therefore, in my opinion, to represent the role of observations, one would require, within each breeding cycle, the breeding vector to undergo a phase-space rotation (and not just a renormalisation as is actually done). This phase-space rotation would continually 'frustrate' the breeding vector's attempt to rotate towards some dominant Lyapunov direction.

The NMC assumption that analysis errors rotate into a selection of preferred directions can be compared with the ECWMF philosophy in which (*cf* section 2g) it is assumed that there are, in fact, no preferred phase-space directions for the analysis error (*i.e.* with respect to a suitable inner product, the PDF of analysis error is relatively isotropic). This assumption has been verified for local European forecasts by Barkmeijer *et al* (1993).

There is one other important difference between the breeding and singular vector methodology. The singular vectors are computed over the period in the immediate future of the initial analysis, whilst the breeding method refers to the period leading up to the initial analysis. This is why singular vector perturbations define, amongst all the *a priori* equally-likely initial directions, those which give rise to the largest possible forecast error.

Nevertheless, perturbations using both breeding vectors and singular vectors have proven useful for medium-range ensemble prediction. It is possible that differences between the two techniques becomes less important as one evolves towards the 'strongly nonlinear' PDF regime (see Fig 1).

3   Examples of singular vectors

3.1   A baroclinic singular vector

Fig 5 shows the dominant singular vector calculated using the ECMWF primitive equation model (Simmons *et al*, 1989; Courtier *et al*, 1991), for a 3-day trajectory portion made from initial conditions on 9 January 1993, at three levels in the atmosphere (200 hPa, 700 hPa and 850 hPa) at initial and optimisation time. The figure illustrates some features which bear qualitative resemblance to an idealised baroclinic eigenmode (Charney, 1947; Eady, 1949): the disturbance clearly amplifies as it propagates through the region of maximum baroclinity (where north-south temperature gradients are strongest), and the disturbance shows evidence of westward tilting phase with height, consistent with a northward flux of heat.

On the other hand, the figure also clearly illustrates the non-modal nature of the disturbance. At initial time the disturbance is localised near the north Atlantic jet entrance region; at optimisation time the disturbance has propagated downstream to Europe. At initial time, maximum disturbance amplitude is located in the lower troposphere, whilst at optimisation time maximum amplitude is located in the upper troposphere at the level of maximum winds (*cf* Farrell, 1989). Finally, the horizontal scale of the initial disturbance is noticeably smaller at initial time than at optimisation time.

As discussed in section 2, singular-vector structure can be related to the

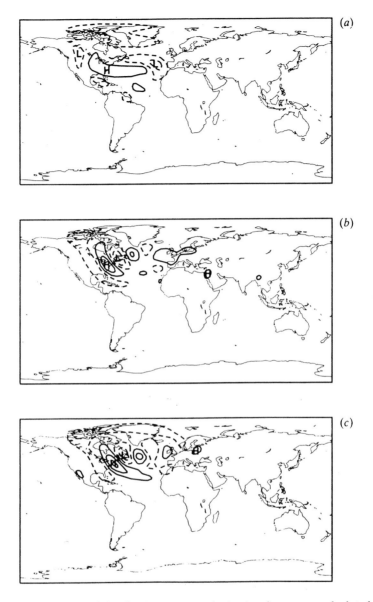

Figure 6: *Streamfunction of the dominant atmospheric singular vector calculated using a primitive equation numerical weather prediction model for a 3-day trajectory portion made from initial conditions of 9 January 1993 at three levels: 200 hPa a) d), 700 hPa b) e), 850 hPa c) f). a)-c) - initial time. From Buizza and Palmer (1995)*

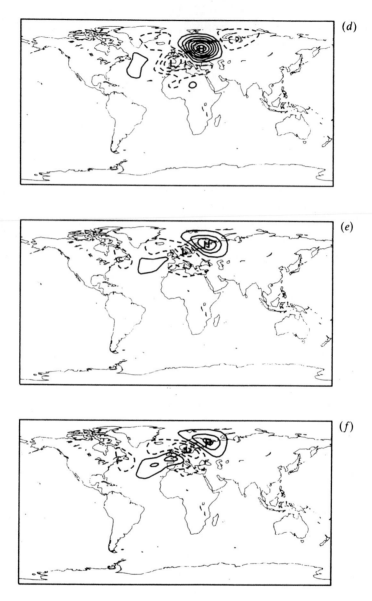

Figure 6: *continued: d)-f) - optimisation time. Contour interval at optimisation time is 20 times larger than at initial time. From Buizza and Palmer (1995)*

Orr (1907) process for transient growth (in asymptotically stable flows). As discussed in Lindzen (1988), the Orr mechanism essentially operates in shear flow from a critical layer. The vertical structure of the singular vectors underpins this relationship. For example, under certain assumptions (steady zonally symmetric basic state flow which is slowly varying in the vertical) the wave-action $E/(\omega - ku_0)$ of a linear disturbance with energy E, zonal wavenumber k and frequency ω will be conserved as it propagates vertically on the background flow u_0. Optimal energy growth will tend to be associated with propagation from a region of small intrinsic frequency (near the baroclinic steering level, to a region of large intrinsic frequency (such as might occur near the jet level).

The horizontal-scale evolution of the singular vector is explored further in Fig 6 which shows the energy distribution of the singular vector at initial and final time, as a function of total wavenumber. Fig 6a shows the spectral distribution of the disturbance shown in Fig 5 peaking near the truncation limit at initial time (dashed line) and at about wavenumber 10 at optimisation time. This upscale energy transfer can occur because the basic state (unlike those in many idealised calculations) is itself an unrestricted solution to the equations of motion, and, in particular contains scales comparable with those in the disturbance field. This allows triad interactions between the disturbance field and the basic state.

Fig 6b,c shows the spectral distribution of two further singular vector calculations made using the same trajectory. In these calculations, the spectral projection operator (2.17) has been applied both at initial and optimisation time. For both calculations, the operator at optimisation time maximises energy between wavenumbers 0 and 10. The initial perturbation is constrained to wavenumbers 0-10 in Fig 6b and to wavenumbers 11-20 in Fig 6c.

The results are quite dramatic. Constraining the perturbation to have the same energy distribution in wavenumber space at initial and final time (which an eigenmode solution, if it exists, must have), severely restricts perturbation growth. On the other hand, constraining the perturbation at initial and final time to have energy in non-overlapping wavenumber intervals hardly restricts energy growth at all (see Hartmann *et al*, 1995, for more details).

These calculations illustrate, in a linear context, the 'butterfly effect' in its original sense (Lorenz, 1963b), *ie* that small-scale initial disturbances can ultimately have an overwhelming influence on large-scale disturbances. This is in addition to the commonly perceived meaning of the butterfly effect that small-amplitude initial disturbances will ultimately have an overwhelming effect on large-amplitude disturbances. (In fact, Lorenz refers to

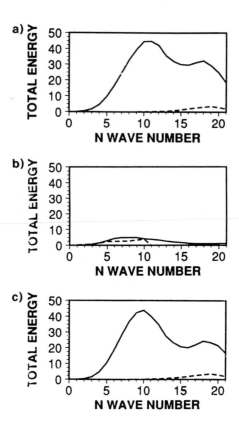

Figure 7: *Energy distribution of 3-day singular vector from 9 January 1993 as a function of (total) wavenumber. Dashed - at initial time (x20). Solid - at optimisation time. a) For singular vector shown in Fig 5 b) For singular vector with energy optimised for wavenumbers 0-10, and constrained to wavenumbers 0-10 at initial time. c) For singular vector with energy optimised for wavenumbers 0-10, and constrained to wavenumbers 11-20 at final time. From Palmer et al (1994).*

the influence on the weather of a flap of a sea-gull's rather than a butterfly's wings. In view of the location of the initial singular vectors over the west Atlantic ocean, rather than, say, over Amazonia, this original metaphor is in this case fortuitously appropriate!). The upscale cascade associated with the butterfly effect has been modelled in a turbulent rotating fluid context by Lilly, 1983 and Metais *et al*, 1994 (see also section 3.2).

The accumulation of initial energy towards the truncation limit suggests that a significantly more accurate estimation of singular vector growth should be obtainable using a higher resolution model. In fact studies using a T42 resolution tangent model confirm this (Hartmann *et al*, 1995; Buizza *et al*, 1996). With this resolution, a significant fraction of perturbation energy is located at subsynoptic scales at initial time, cascading to synoptic scales at optimisation time (see also section 3.2). One practical consequence of this result is that the predictability of synoptic scale weather may be determined more by uncertainties in the initial state on scales much smaller than the disturbance itself, and less by uncertainties on the scale of the disturbance. In particular, the notion that the predictability of synoptic scale weather is determined by the e-folding time of a characteristic eigenmode is incorrect (confirming earlier results of Farrell, 1990).

As mentioned, although the structures of these baroclinic singular vectors are not modal, their geographical locations are more prevalent in regions of strong baroclinity. Fig 7a for example, shows the 'Eady index'

$$\sigma_E = 0.31 \frac{f}{N} \frac{du}{dz} \tag{3.1}$$

based on the a winter mean static stability and wind shear (from ECMWF data; for details see Buizza and Palmer, 1995). (Of course the functional relationship between the wind shear, Coriolis parameter and static stability in 3.1 are not particular to the Eady model). Fig 7b shows the location of the dominant singular vectors at initial time (based on their vorticity maxima) from daily calculations over a whole winter. It can be seen that the singular vectors tend to be positioned in regions of strong Eady index, over the east Asian/west Pacific region, the northeast American/west Atlantic region, and the northern subtropical African region. The tropics and southern hemisphere extratropics also appear in these northern winter statistics, though to a lesser extent.

3.2 Pseudo-inverse analysis

In terms of the singular vectors, we can decompose the forward tangent propagator **L** as,

$$\tilde{\mathbf{L}} = \mathbf{U}\boldsymbol{\Sigma}\mathbf{V}^* \tag{3.2}$$

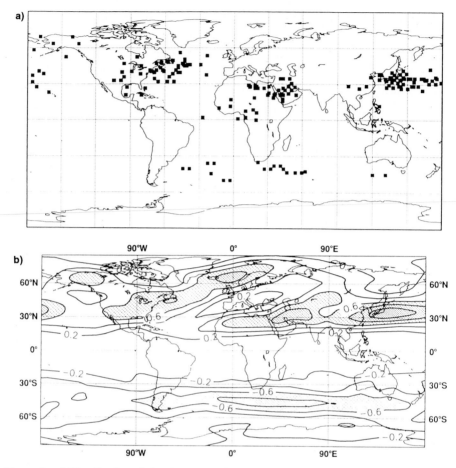

Figure 8: *a) Distribution of dominant singular vectors for a winter season. The position of a singular vector is denoted by a black dot at the vorticity maximum at initial time. b) Eady index (3.1) based on seasonal mean flow. From Buizza and Palmer (1995).*

where $\mathbf{V}$ contains the initial singular vectors, $\mathbf{U}$ the final singular vectors, and $\mathbf{\Sigma}$ is a diagonal matrix with elements σ_i. Within the subspace spanned by the dominant singular vectors, we can define the inverse

$$\tilde{\mathbf{L}}^{-1} = \mathbf{V}\mathbf{\Sigma}^{-1}\mathbf{U}^*$$ (3.3)

In terms of the full phase space, $\tilde{\mathbf{L}}^{-1}$ is related to the Moore-Penrose pseudo-inverse (for more details, see Buizza *et al*, 1996).

By operating on a given forecast error field with the pseudo-inverse operator, an estimate of the unstable component of initial error is obtained.

Some examples of this pseudo-inverse field are shown in Buizza *et al* (1996) based on a 30 singular vector truncation, and a 36-hour forecast error. By adding this estimate to the initial analysis, a much superior forecast can be obtained, not only at 36 hours, but throughout the entire medium-range. It should be noted that the pseudo-inverse component of initial error may be small compared with the total error. Indeed, associated with the non-invertibility of the full operator **L**, the forecast error will be insensitive to estimates of initial error in strongly decaying directions.

This technique is closely related (but not identical) to the sensitivity analysis technique (see section 3.5) in which the forecast error is integrated (backwards) using the adjoint model itself (see Rabier *et al*, 1996; Buizza *et al*, 1996).

3.3 Comparison of tropical and extratropical singular vector growth: a paradigm for tropical predictability.

It is clear from numerous studies that the large-scale tropical circulations are much more strongly coupled to the underlying SST than are large-scale extratropical circulations. As such, the prospects for extended-range prediction in the tropics are greater than for the extratropics. Charney and Shukla (1981) were one of the first to rationalise this disparity in behaviour between the tropics and extratropics. These authors argued that, from a dynamical point of view, the principal relevant difference between the large-scale tropical and extratropical flows lay in their instability properties. The absence of generic large-scale exponentially growing eigenmodes in the tropics makes more reproducible the effects of slowly-varying lower boundary conditions associated, for example, with SST anomalies.

Whilst these considerations are clearly important and relevant, they cannot be considered complete. For example, the tropical flow is convectively unstable, and these in turn contribute to the growth of synoptic-scale disturbances, *eg* associated with tropical cyclones. However, it is known that inevitable errors in the initial conditions for atmospheric seasonal integrations associated with tropical synoptic and subsynoptic flow, do not infect the planetary scale circulations in the same ubiquitous way that is found in the extratropics.

It is manifestly impossible to describe this difference in upscale energy cascade using modal instability theory. On the other hand, as we have seen in the last section, singular vector analysis is able to describe, within

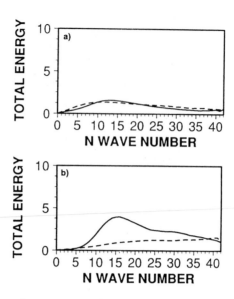

Figure 9: *Total wavenumber spectrum of mean energy the dominant 18 48-hour singular vectors a) optimised over the tropics, b) optimised over the northern hemisphere extratropics. The initial spectrum (x40) is given by the dased lines, the final spectrum is given by the solid lines. The tangent propagator for these calculations had a T42 resolution.*

a completely linear framework, the inverse cascade process associated with extratropical baroclinic growth.

Let us compare the spectrum of singular vectors calculated for extratropical and tropical growth respectively. Fig 8 shows the spectrum of the first 18 singular vectors for a particular (but typical) 36-hour period based on the T42 tangent model. Fig 8a shows growth optimised for the tropics; Fig 8b shows growth optimised for the extratropics. The solid line shows the energy of the singular vectors at optimisation time, the dashed lines show the energy at initial time (multiplied by a factor of 40 to make them more visible on the diagram).

The singular values for the extratropics are larger than those for the tropics. In addition, the spectrum of tropical singular vectors is considerably more modal than the extratropical spectrum. The spectrum peaks in the large synoptic scale, and therefore indicates that error growth at these

111

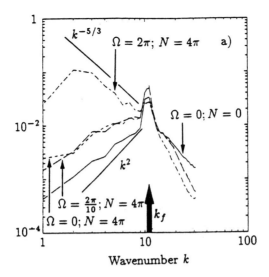

Figure 10: *Wavenumber spectrum of kinetic energy of a laboratory turbulent fluid forced at an intermediate wavenumber, for various rotation and stratification values. From Metais et al (1994).*

scales will not be dominated by error growth from smaller scales.

These results appear consistent with studies of inverse energy cascades in fully-developed turbulence models, in which the Coriolis parameter is varied. Fig 9 (from Metais *et al*, 1994) shows the spectrum of kinetic energy in a stratified turbulent fluid forced at an intermediate wavenumber (shown by the heavy arrow), for various rotation rates and stratifications. Results show that in a strongly rotating regime, the energy spectrum is an increasing function with increasing scale. However, in the nonrotating case, the energy decreases with decreasing wavenumber. The influence of rotation appears to be associated with the dimensionality of the associated turbulence fields, and this dimensionality determines the upscale cascade.

Hence, singular vector analysis is apparently able to distinguish between the different cascade processes in low and high rotation rate regimes, and this, as much as the intrinsic modal instability properties of the flow, may be important in distinguishing between the different predictability regimes in the tropics and extratropics. This may be important in understanding the predictability of ENSO. Sarachik (1990), for example, discussing an intermediate coupled ocean-atmosphere model of the tropical Pacific (see section 3.6 for more details) notes that '..whenever the [model's] mean state is unstable, the resulting cycle is perfectly regular: instability (in this model) is therefore related to perfect predictability'. Sarachik then notes that '..This paradigm of ENSO predictability is radically different from our classical concepts of mid-latitude predictability'. However, Sarachik does not explain why there should be two paradigms: one for the tropics, the other for the extratropics. A possible explanation could be given in terms of the modality of the associated singular vectors. In contrast with tropical singular vectors, the extratropical singular vectors are profoundly non-modal and perturbation growth is therefore susceptible to the 'butterfly effect' (*cf* above). As discussed above, this non-modality and associated upscale cascade can cause extratropical weather forecasts to fail before the timescale set by any characteristic e-folding rate.

3.4   Singular vectors for ensemble forecasting

Singular vectors are used as the basis of initial perturbations for medium-range ensemble forecasting. The basic rationale for this is that we cannot sample explicitly, analysis uncertainties in all the phase space directions, the dimension of a numerical weather prediction model (at least $O(10^6)$) is too large. Hence we sample explicitly directions in which analysis error is likely to occur and can lead to significant departures from the unperturbed forecast. In the linear regime these are given by the dominant singular vector directions. Hartmann *et al* (1995) have shown that up to about day 7, perturbations using singular vectors have significantly larger spread than unstable synoptic-scale perturbations. This would suggest that the 'weakly nonlinear' timescale (*cf* section 2.1) lasts until about day 7.

Other directions (for example associated with decaying perturbations) can in principle be taken implicitly into account in an ensemble forecast by giving the unperturbed forecast a higher than average *a priori* weight compared with other members of the ensemble.

Fig 10 shows two examples of ensemble forecasts made from initial conditions one week apart. The thin lines show the spread of the members of the ensemble forecast relative to the unperturbed control forecast (using a correlation measure of spread). The thick line shows the skill of the control forecast (using a correlation measure of skill). The examples illustrate the desirable occurrence of low spread indicating high skill, and high spread indicating relatively poor skill.

Obviously one cannot make any definitive conclusions based on just two results. The interested reader is directed to a more complete description and validation of the ECMWF ensemble prediction system in Molteni *et al* (1996).

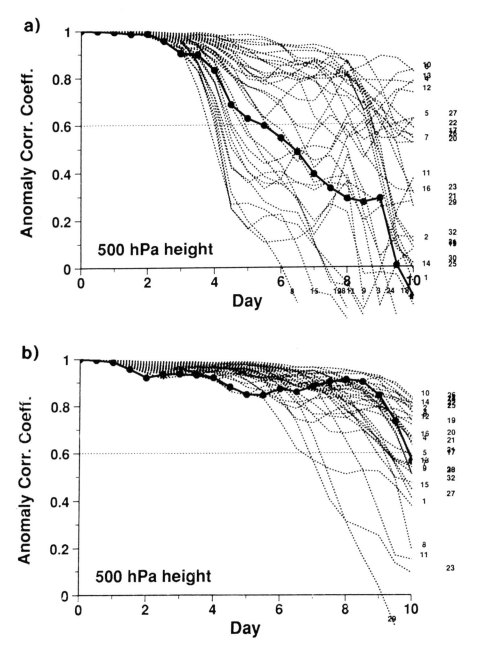

Figure 11: *Ten-day ensemble forecast dispersion as measured by anomaly correlation between 500 hPa geopotential height of control and perturbed forecasts, over Europe (light lines). Skill of 10-day control forecast (heavy line), a) forecasts from 30 October 1993, b) forecasts from 13 November 1993. From Palmer et al (1994).*

3.5 Weather regimes, singular vectors and sensitivity patterns

The concept of weather regimes is a long standing one (*cf* Grosswetter-lagen; Hess and Brezowsky, 1977), and is based on the notion that the large-scale flow may evolve around various recurrent configurations. This notion was made more precise in modelling studies (Reinhold and Pierre-humbert, 1982) who related the onset, maintenance and decay of regimes to interactions of the large-scale flow with synoptic-scale variability. The existence of such weather regimes in the real atmosphere has been inferred through observational studies (*eg* Hansen and Sutera, 1986, 1995; Mo and Ghil, 1988, Molteni *et al* , 1990, Cheng and Wallace, 1993, Kimoto and Ghil, 1993), though the existence of unambiguous multimodality is still a matter of debate (Wallace *et al*, 1991). In many observational studies, (*eg* Yang and Reinhold, 1991; Dole and Gordon, 1983 and Toth 1992), it is suggested that baroclinic instability sets the timescale for the transition process betweeen regimes. This timescale is much shorter than a typical residence timescale (on the order of weeks). This two-timescale behaviour is consistent with the regime structure in the 3-component Lorenz model (see Fig 19 below).

Fig 11 shows two of the large-anomaly cluster centroids found by Mo and Ghil (1988) (11a and 11b) and by Molteni *et al* (1990) (11c and 11d). Despite different clustering algorithms, Figs 11a and c correspond to one another quite well (as do Fig 11b and d). The regime centroids have signifi-cant projection onto opposite phases of the Pacific North American (PNA) pattern (Wallace and Gutzler, 1981), though also have structure over the Atlantic and EurAsia (*cf* section 6 on climate change). By convention, the PNA index of the fields in Fig 11a and c is positive, the PNA index of the fields in Fig 11b and d is negative. For future reference, Molteni *et al* (1990) refer to Fig 11c as cluster 2, and Fig 11d as cluster 5.

Regimes can also be found in atmospheric model integrations. Fig 12a shows the PDF from a 100 consecutive winter sample of a 1200 perpetual-winter integration of a 3-level T21 quasi- geostrophic model (Marshall and Molteni, 1993; Corti, 1994; Palmer *et al*, 1994). The PDF is estimated in a phase-space plane spanned by two of the dominant empirical orthogonal functions of the model (shown in Fig 12b). During this chosen 'century', the PDF is bimodal along an axis that corresponds to fluctuations in the North Atlantic Oscillation.

As shown in Palmer (1988), low-frequency intraseasonal variability is

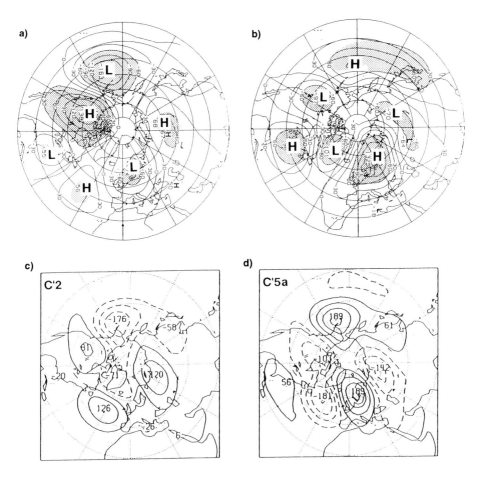

Figure 12: *Anomalies corresponding to the centroids of clusters 1 and 2 (a and b) found by Mo and Ghil (1988), and of clusters 2 and 5 (c and d) found by Molteni et al (1990). In a) and b), shading indicates statistical significance at the 95% confidence level.*

stronger over the PNA area during periods when the atmosphere resides in negative PNA states, than positive. Consistent with this, medium-range forecasts are more skilful during positive PNA periods (Molteni and Tibaldi, 1990). On longer timescales, it has been found from both modelling and observational studies (Von Storch, 1988; Chen and Van Den Dool, 1996) that there is less low frequency intraseasonal variability over the PNA region during warm-phase ENSO winters than cold-phase ENSO

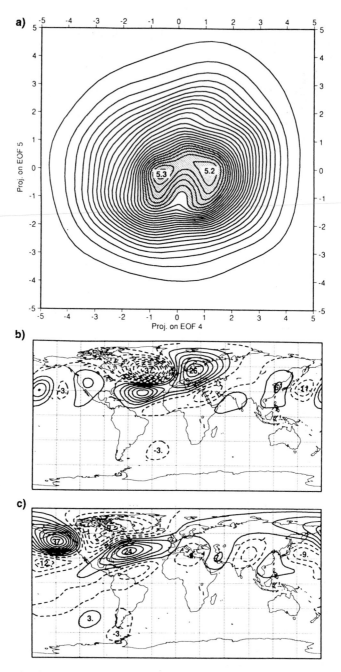

Figure 13: a) 2-D cut of the PDF from 100 consecutive winters of a multi-decadal integration in the T21L3 quasi-geostrophic model of Marshall and Molteni (1993). b), c) The model empirical orthogonal functions, used to define the axes in a). From Palmer et al (1994).

117

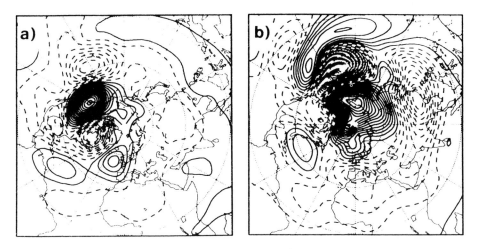

Figure 14: *500hPa stream function of dominant barotropic day-8 singular vector at day 8 for (a) cluster 2 and (b) cluster 5 of Molteni et al (1990). From Molteni and Palmer (1993).*

winters.

Evidence of variations in the instability characteristics of positive and negative PNA states was discussed in Palmer (1988), who found that the growth of linear perturbations in a barotropic model depended strongly on the sign of the PNA index of the basic-state flow. However, it was noted that the difference in these growth rates could not be explained in terms of differences in the eigenmode growth associated with the basic states. Molteni and Palmer (1993) compared the growth rates of the fastest-growing eigenmodes and the optimal singular vectors for the cluster 2 and cluster 5 flow in the barotropic model. For the most unstable eigenmode (which is stationary in both cases) there is little difference in growth rates for the two basic states. On the other hand, the dominant singular values are quite different. In particular, for 8-day optimisation, the dominant singular value for the cluster 2 flow is almost a factor of 2 smaller than that of the cluster 5 flow. The streamfunction of the final dominant singular vector for an 8-day optimisation is shown in Fig 13 for the cluster 2 and cluster 5 basic states.

There are two important approximations that have been made in these calculations. The first is the use of a barotropic model. The second relates to the fact that we have chosen stationary basic states which are not themselves solutions of the equations of motion. With regard to the second approximation, the error involved in using a stationary basic state should at least be smallest for those states corresponding to observed cluster centroids, since, by construction, these are, in some ensemble sense, closest to stationary.

The approximation involved in using the barotropic model is actually not as bad as might be imagined at first sight. According to results in Molteni and Palmer (1993), 8-day singular vectors of a realistic time-varying baroclinic basic state are (at final time) more accurately represented by a time-averaged barotropic basic state than by a time-averaged baroclinic basic state. The reason for this is that the instability of a time-averaged baroclinic flow will tend to overly dominated by the baroclinic instabilities which relate directly to the meridional thermal contrasts. Moreover, since the Rossby-wave structures of such time-averaged flows are under-represented, the upscale cascade process discussed in section 3.1 is weak. As a result, singular vectors from time-averaged baroclinic flows tend, at final time, to be dominated by smaller scales than would occur with a time-varying flow. By contrast singular vectors from time-averaged barotropic basic states (which also evolve through upscale energy evolution) are dominated at final time by larger, more realistic scales.

It is of interest to ask, however, how one might go about calculating regime instabilities without using such approximations. Let us represent the regime centroid by the a given normalised large- scale pattern $E(x, y, z)$. One possibility is to compute the pseudo-inverse of $E(x, y, z)$ using the dominant singular vectors, for a variety of finite-time trajectories. A less computationally demanding, but largely equivalent calculation can be achieved with a single integration of the adjoint model. Suppose we want to find a perturbation, with unit norm at initial time, and maximum projection onto the given pattern E at final time. In symbols, we want a perturbation e at $t = t_0$ which at $t = t_0 + \Delta t = t_1$ maximises $< Le, E > / < e, e >$. This perturbation will be given at $t = t_0$ by $e = L^* E$. We can refer to e as the sensitivity pattern (cf Marchuk, 1974; Cacuci, 1981; Rabier et al, 1995) for E, and the growth $\|Le\|/\|e\|$ as the instability index of E. Such work is currently in progress using a three-level quasi-geostrophic model whose climatology, as shown above, has realistic regime structures (Susanna Corti, personal communication). These calculations could be of particular interest in studying the sensitivity of observed climate change patterns (see section 6).

3.6 Singular vectors from a coupled tropical ocean-atmosphere model

ENSO appears to be predictable up to a year or so in advance using relatively simple coupled models of the atmosphere and ocean (Zebiak and Cane, 1987). According to Münnich *et al* (1991), long-term variability of ENSO is intrinsically chaotic (independent of the chaotic nature of weather itself). The skill of ENSO forecasts made with coupled ocean-atmosphere models is seasonally dependent (Cane *et al*, 1986; Webster, 1995). Typically, seasonal forecasts beginning in spring tend to be less skilful than forecasts beginning, for example, in autumn. This is sometimes referred to as the 'spring barrier' effect.

Blumenthal (1991) has analysed the behaviour of the eigenfunctions of a linear Markov model approximation to a nonlinear coupled ocean-atmosphere model. He finds that in summer, the eigenfunction best describing ENSO has larger eigenvalues than at other times of year. In spring this ENSO eigenfunction is least orthogonal to other modes, *ie* is associated with large projectibility (see section 2). Both types of error growth (modal and non-modal) are implicit in singular vector analysis which has been performed on this Markov model by Xue *et al* (1994).

We show some results of a singular vector analysis applied to the coupled ocean- atmosphere model of Battisti (1988). Preliminary results were described by Palmer *et al* (1994), more extensive results are given in Chen *et al* (1996). A similar study has been performed independently by Moore and Kleeman (1996). The ocean component of the model used here is a single vertical mode tropical Pacific basin anomaly model, governed by linear shallow water wave dynamics. The nonlinear thermodynamics are only active in a surface mixed layer. The atmospheric component is a thermally -forced steady linear model with single vertical mode (Gill, 1980). Air-sea interactions are nonlinear: given by surface wind stress, heat flux, and sea surface temperature, SST. The number of independent degrees of freedom in the coupled model is reduced to 420 by considering only the equatorial oceanic Kelvin mode, and first 3 symmetric Rossby modes. With such a reduction, singular vectors can be computed using conventional matrix algorithms. For these calculations, the inner product is based on the spatial variance of the SST anomaly.

The first results have been made using a climatological basic state flow. Perturbation growth is strongly dependent on the annual cycle in the Pa-

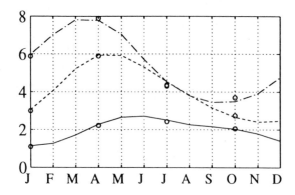

Figure 15: *Dominant singular values for 3-month (solid), 6-month (dashed) and 9-month (dashdot) from the Battisti (1988) intermediate coupled ocean atmosphere model. The basic state trajectory is a climatological annual cycle. From Chen et al (1996).*

cific, and on the duration of the integration. Fig 14 shows the dominant singular values for 3-, 6-, and 9-month optimisation times. (The optimisation time is greater *eg* than in section 3.1 because of the longer predictability time associated with ENSO.) The maximum singular vector for the 6-month integration ranges from 6 for the April start, to less than 3 for the October start. The results point to a period of greater sensitivity during the boreal spring and summer.

In contrast with the sensitivity of the singular values, the pattern of both initial and final state singular vectors are relatively insensitive to the month in which the initial perturbation is applied, and to the duration over which the error is allowed to grow. For example, Fig 15 shows the SST of the singular vectors (at initial and optimisation time) for 6-month optimisation based on starting conditions for January, April, July and October. The initial pattern consists of an east-west dipole spanning the entire tropical Pacific basin, superimposed on a north-south dipole in the eastern tropical Pacific. The pattern at optimisation time resembles the ENSO mode.

Of course, in keeping with the other results discussed in this section, it is likely that these singular vectors will depend on the state of ENSO itself, as well as on the annual cycle. In order to understand this further, singular vectors have been calculated with respect to a 'free trajectory' of the Battisti model which includes ENSO variability but (for this integration

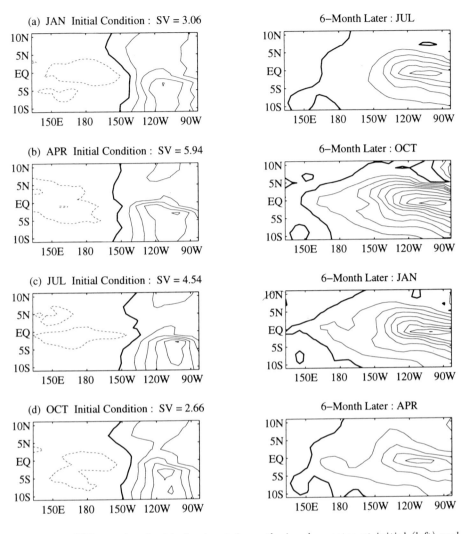

Figure 16: *SST associated with dominant 6-month singular vector at initial (left) and final (right) time, for a) January, b) April, c) July, d) October starts. The basic state trajectory is a climatological annual cycle. From Chen et al (1996).*

in particular) has no annual cycle. In fact, when the Battisti model is run freely, the extrema of the model (NINO3) SST anomalies range from over 2C in the warm phase to less than -1C in the cold phase. The solid line in Fig 16 shows a timeseries of Nino 3 SST anomaly from year 9-

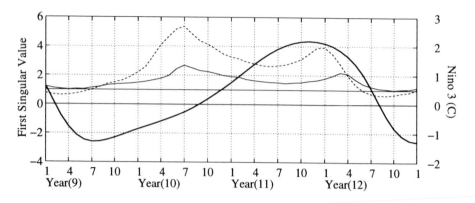

Figure 17: *Dominant singular values for 3-month (light solid) and 6-month (dashed) from the Battisti intermediate coupled ocean atmosphere model. The basic state trajectory is a free integration of the model without climatological annual cycle. The NINO3 index from the basic state trajectory is shown as the heavy solid line. From Chen et al (1996).*

year 12 of a model climate integration. Also shown are the dominant 3- and 6- month singular values associated with singular vector calculations made using this freely evolving basic state. It can be seen that the growth rates do indeed depend on the state of ENSO in the basic state trajectory. For example, largest growth occurs during the transition phase between the cold and warm event. A secondary maximum is associated with the trajectory portion starting near the peak of the warm event.

The dependence of the dominant singular value on ENSO is of relevance when we consider the decadal variability of ENSO predictability. In particular, as discussed by Balmadesa *et al* (1995), the 'spring barrier' effect is more prominent in the 1970s than in the 1980s. In order to study whether there is any decadal variability in the dominant ENSO singular values, Y-Q. Chen (personal communication, 1995) has estimated the mean and standard deviation of the dominant singular values over the periods 1960-1975 and 1975-1990 respectively. Results indicate that there is a less significant annual cycle in the dominant singular values in the period 1975-1990 than in the period 1960-1975. This is consistent with the fact that ENSO itself was more active in the latter period.

4 Predictability of the second kind

4.1 Uncertainty in forcing

In section 2, initial value problems were referred to as predictions of the first kind. In a prediction of the second kind, we estimate how (the attractor of) a given dynamical system responds to a change in some prescribed parameter or variable. The response of climate to doubling CO_2, or of the stratosphere to an increase in CFCs, or of an atmospheric GCM to a prescribed change in SST, are all predictions of the second kind. Uncertainties in such predictions may arise from the accuracy in the prescribed change itself, or from uncertainties in model formulation. (In practice, of course, many forecasts do not fall exclusively into either of these two categories).

Even though predictions of the second kind are, by construction, not sensitive to initial conditions, the underlying instabilities of the flow play an important role in determining the associated predictability. To see this, let us apply the singular vector analysis discussed in section 2 to a forced problem. Consider then the generalisation of (2.2) to

$$\frac{d\mathbf{x}}{dt} = \mathbf{M}_l \mathbf{x} + \mathbf{f}(t) \tag{4.1}$$

As before, we let us integrate this equation over the finite time interval $[t_1, t_0]$. Using the tangent propagator $\mathbf{L}(t, t_0)$ (cf equation 2.3), then the solution to (4.1) can be written as

$$\mathbf{x}(t_1) = \mathbf{L}(t_1, t_0)\mathbf{x}(t_0) + \int_{t_0}^{t_1} \mathbf{L}(t_1, s)\mathbf{f}(s)ds \tag{4.2}$$

From (4.2) it can be seen that the effect of an initial error $\mathbf{x}(t_0)$ can be replicated by the action of the impulsive forcing $\mathbf{f}(t) = \mathbf{x}(t_0)\delta(t - t_0)$. Hence, the maximum response $\|\mathbf{x}(t_1)\|$ from such a normalised impulsive forcing occurs when $\mathbf{f}(t) = \boldsymbol{\nu}_1(t_0)\delta(t - t_0)$, where $\boldsymbol{\nu}_1(t_0)$ is the dominant singular vector at initial time associated with the interval $[t_1, t_0]$. More generally, if $\mathbf{f}(t)$ is a normalised impulsive forcing $\mathbf{f}(t) = \mathbf{x}(s)\delta(t - s)$, $t_1 < s < t_0$ then the maximum response $\|\mathbf{x}(t_1)\|$ can be induced by choosing $\mathbf{x}(s)$ to be the dominant initial singular vector for the interval $[t_1, s]$. Putting this together we can see that if $\mathbf{f}(t)$ is any spatially normalised forcing, then the maximum response at t_1 will be obtained by setting $\mathbf{f}(t) = \boldsymbol{\nu}_{[t_1,t]}(t)$, the dominant initial singular vector over the interval $[t_1, t]$.

Now let us interpret the forcing $\mathbf{f}(t)$ as an uncertainty in model formulation which, for the sake of argument, we shall assume arises principally from the physical parametrisations. We have argued that a complete specification of the initial state would include not only the best estimate of the initial conditions, but also a probability distribution of the error associated with that best estimate. Similarly, given the inherent uncertainties in parametrising sub-gridscale processes, a complete specification of the diabatic tendency in a grid box should include not only our best estimate of the diabatic tendency (ie the parametrised tendency), but also a probability distribution of the error associated with that estimate.

When considering the probability distribution of parametrised diabatic tendency, it is certainly not permissable to ignore the first moment. This first moment can be thought of as defining what is generally referred to as 'systematic error'. However, in addition to its systematic component, the parametrised diabatic tendency will certainly have a stochastic component of error. Consider, for example, the parametrisation of convective heating in terms, say, of resolved moisture fluxes or temperature profiles. If convectively-driven mesoscale circulations occur on scales which are not substantially smaller than the resolution of the model, then the usual assumptions of a quasi-equilibrium of convective heating elements within a grid box will fail. This failure will generate a second moment of the PDF.

The basic message behind (4.2) is that the system response to forcing errors depends very much on the convolution of this forcing with the dynamical instabilities of the flow itself. This can be seen clearly if we put $\mathbf{f}(s) = \mathbf{f}_0$, $\mathbf{x}(t_0) = 0$ so that

$$\mathbf{x}(t_1) = \mathcal{L}(t_1, t_0)\mathbf{f}_0 \tag{4.3}$$

where

$$\mathcal{L}(t_1, t_0) = \int_{t_0}^{t_1} \mathbf{L}(t_1, s)ds \tag{4.4}$$

4.2 A simple chaotic model paradigm for predictability of the second kind

As a simple example of the impact of a fixed forcing on a heterogeneous attractor, consider the Lorenz (1963a) model

$$\begin{aligned} \dot{X} &= -\sigma X + \sigma Y \\ \dot{Y} &= -XZ + rX - Y \end{aligned} \tag{4.5}$$

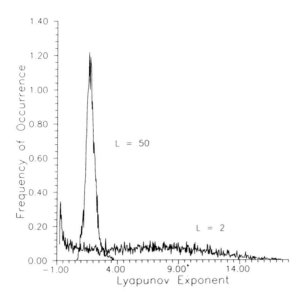

Figure 18: *The distribution of the largest singular value,σ, (shown as its exponent* ln σ*)
for the Lorenz model (4.5) for two trajectory lengths L (L=50, L=2). Each distribution
is normalised to unity. From Abarbanel et al (1991).*

$$\dot{Z} \;=\; XY - bZ$$

Singular values for the Lorenz model have been computed by a number of
authors (*eg* Mukougawa *et al*, 1991; Abarbanel *et al*, 1991; Trevisan, 1993).
Fig 17 shows the distribution of exponents of dominant singular values for
two choices of the trajectory length. For relatively long trajectory portions,
the distribution of singular values (expressed as an equivalent exponent)
is relatively narrow and is clearly asymptoting to the appropriate largest
Lyapunov exponent (here about 1.5). For short trajectory portions the
distribution of maximum exponents is broad, varying from negative values
(*ie* decaying singular vectors) to values over one order of magnitude greater
than the fastest growing Lyapunov exponent.

Let us now examine some time traces of one of the state variables of the
modified Lorenz model

$$
\begin{aligned}
\dot{X} &= -\sigma X + \sigma Y + f_0 \\
\dot{Y} &= -XZ + rX - Y + f_0 \\
\dot{Z} &= XY - bZ
\end{aligned}
\qquad (4.6)
$$

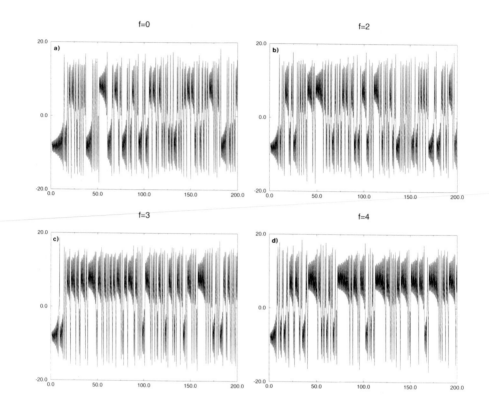

Figure 19: *Timeseries of the X component of the modified Lorenz equation (4.6) for a) $f_0=0$, b) $f_0=2$, c) $f_0=3$, d) $f_0=4$.*

as the time invariant forcing f_0 increases from zero (see Fig 18). Notice that the X values do not simply translate to larger values as f_0 increases, rather the probability that the state vector resides in the regime with positive X increases, and the probability that the state vector resides in the regime with negative X decreases. The X-values of the regimes themselves are largely unchanged.

Fig 19a shows the state vector PDF of the Lorenz model (4.5) computed from a long integration. It is effectively symmetric with maxima corresponding to the centroids of the butterfly-wing regimes. Fig 19b shows the PDF of (4.6) with non-zero f_0; the PDF is now biased to one of the regimes; however, the phase space position of the regime centroids remains essentially unchanged.

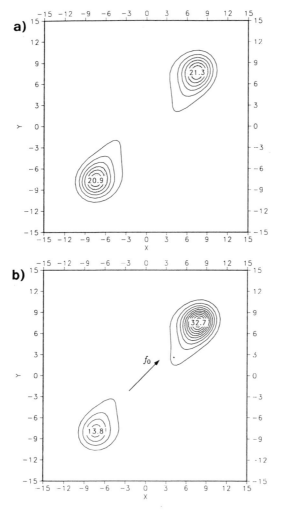

Figure 20: *PDF of the Lorenz model in the X-Y plane, low-pass filtered to remove oscillations around a regime centroid (a) from the unforced model (b) with a constant f_0.*

This behaviour can be understood in terms of the heterogeneity of finite-time singular values on the attractor. In particular, at the regime centroids (PDF maxima for the state vector time averaged over the fast oscillation time scale) the singular values are rather small, corresponding to the fact that the attractor is rather stable in this part of phase space. On the other hand, in other parts of the attractor, particularly near the origin, the singular values are particularly large, corresponding to a very unstable saddle point instability. In the Lorenz model (4.5), the outset (*eg* Thompson and Stewart, 1991) of the saddle is symmetric with respect to the two Lorenz regimes. In the modified model (3.2) it becomes biased towards one of the regimes.

A statement of these results in general terms leads to the following nonlinear paradigm (Palmer, 1993). The *influence* of a weak forcing f_0 on a nonlinear system (such as the climate) is greatest in regions of phase-space where the dominant singular value is large. On the other hand, the *response* of the system to f_0 is greatest in regions where the local PDF is a maximum, and singular values are small. To first order, the response to f_0 will be a change in the value of the PDF at the maxima.

This analysis is broadly consistent with recent studies of systematic errors in weather prediction and climate models. For example, as shown by Molteni and Tibaldi (1990), medium-range systematic errors tend to have large-scale equivalent barotropic structures which correspond reasonably well with the regime structures discussed in section 2. In order to understand how such errors arise, *ie* what systematic forcing errors these structures are most sensitive to, we need to estimate either a pseudo-inverse or a sensitivity gradient of that error pattern over an ensemble of trajectory portions, that, together, cover the climate attractor. Such studies are underway (see section 3.5 and section 6).

It should be noted that this paradigm is very hard to prove from first principles; even the notion of existence and uniqueness of a PDF is hard to establish. Nevertheless below we shall use the paradigm as a possible means of interpreting climate variability.

4.3 Application to atmospheric forcing by tropical SST anomalies

Up to now I have imagined f_0 to represent a model uncertainty. However, we can apply the paradigm more generally. Consider a prediction of the second kind, mentioned earlier; the response of an atmospheric GCM to imposed SSTs. Fig 20 a,b shows the 1000hPa temperature and 500hPa height difference in the wintertime climate of (a recent version of) the ECMWF model using firstly SSTs from the El Niño winter 1986/87, and secondly using SSTs for the winter 1988/89.

Applying the nonlinear paradigm, the forcing associated with the imposed SST changes will have increased certain atmospheric regimes, and

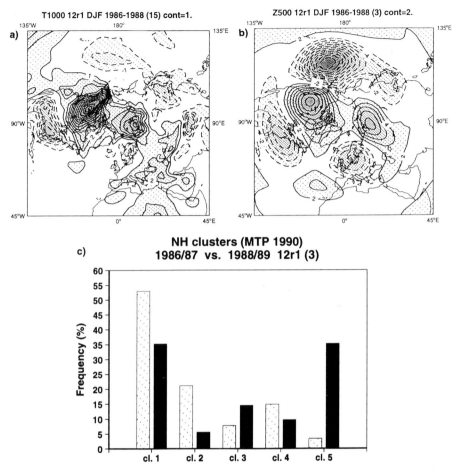

Figure 21: a), b) Difference between DJF 1000hPa temperature and 500hPa height (respectively) from two (3-member) 120-day ensembles of the T63L19 ECWMF atmosphere model. The first ensemble was run with observed SSTs for the winter 1986/87, the second ensemble was run with observed SSTs for the winter 1988/89. c) Cluster frequency of pentad fields with the DJF period for 1986/87 ensemble (left hand bars) and 1988/89 ensemble (right hand bars). Clusters are based on Molteni et al (1990) analysis.

decreased others. Fig 20c shows the impact of the SST changes on the climatic frequency of the Molteni et al (1990) regimes. These frequency statistics have been based on 5-day mean 500hPa height fields. It can be seen, comparing 1988/89 with 1986/87, that the frequency of both clusters 1 and 2 have decreased, and the frequency of cluster 5 has increased. This

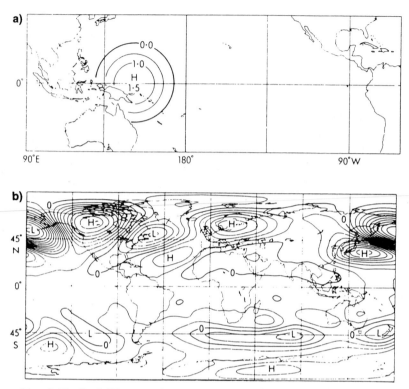

Figure 22: a) Idealised West Pacific SST anomaly. b) 200hPa height response of the
UKMO GCM to the imposed anomaly in a). Contour interval 2 dam. From Palmer and
Mansfield (1986).

change in frequency corresponds to what actually occurred.

Note from Fig 20a, the 1000hPa high latitude zonal mean temperature
difference is positive. The reason for this is not that the SST forcing has
a high latitude zonal mean positive component, it is that the temperature
difference reflects the change in frequency of the cluster 2 and 5 regimes.
We will return to this basic idea in section 6.

Fig 21b shows the time-mean 200hPa response of the UKMO GCM to a
relatively small warm SST anomaly in the western Pacific, as shown in Fig
21a. Again, there is a PNA like response in the northern hemisphere. It
can easily be seen that this response would also have significant projection
onto the Molteni et al regimes.

However, why should the imposed west Pacific SST differences have
changed the frequencies of the Molteni et al regimes so effectively? Now
Fig 22a shows the dominant stationary eigenmode of the time-averaged

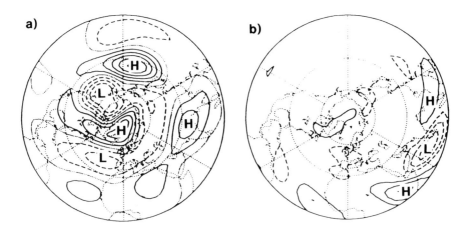

Figure 23: *a) Streamfunction of dominant stationary barotropic eigenmode based on climatological winter mean basic state. b) Streamfunction of adjoint of the eigenmode shown in a). From Zhang (1988).*

climatological flow in a barotropic model (from Zhang, 1988). Fig 22b shows the adjoint of this eigenmode. As discussed in section 2, these can be thought of as final and initial singular vectors (respectively) for an indefinitely long trajectory, in the approximation where we treat the trajectory as stationary (see remarks above on the relationship between this approximation and the barotropic model approximation). The eigenmode structure (very roughly) approximates the difference field between cluster 5 and cluster 2 (having a negative PNA index). The adjoint eigenvector has no correspondence to the eigenmode itself, but rather has most of its structure in the tropics and subtropics, over Indonesia, south-east Asia and the Indian Ocean.

From this perspective, the PDFs of the dominant extratropical regimes may be most sensitive to external forcing from the tropical western Pacific area.

A second application of the nonlinear paradigm concerns the predictability of the monsoons. As discussed in Palmer (1994), the break and active phase of the Asian summer monsoon can be thought of as representing regional flow regimes in which the ITCZ has a predominantly oceanic or continental position. Intraseasonal monsoon variability can be thought of as comprising essentially chaotic fluctuations between these regimes. During warm-phase ENSO years, the PDF of the oceanic regime is enhanced leading to a greater probability of a poor overall monsoon. During cold-phase ENSO years the PDF of the continental regime is favoured, leading to greater probability of a good monsoon. In this picture, the predictability of seasonal-mean monsoon fluctuations to imposed external SST anomalies is only partial, a result demonstrated in GCM ensemble integrations (Sperber and Palmer, 1996).

One simple consequence of the picture put forward is that if the circulation patterns are dominated by a few basic regimes, and if the system basically responds to weak forcing through changes in regime frequency, then the EOFs of the system will match the patterns of these regimes no matter how long a timescale they are computed over. Hence, the fact that the PNA and NAO patterns emerge as dominant EOFs on weekly, seasonal, and multi-decadal timescales is entirely consistent with the nonlinear paradigm put forward here.

We will return to the paradigm put forward here in section 6 when we discuss climate change.

5   Predictability of interdecadal fluctuations

5.1   Internal atmospheric variability

A basic theme underlying this paper is the chaotic nature of climate. Much of this derives from the atmosphere. The ubiquitous growth of atmospheric perturbations, as revealed by singular vector analysis above, together with the underlying nonlinear structure of the atmosphere, suggested, for example, by potential vorticity diagnosis, (eg the wave-breaking and wave, mean-flow interaction processes illustrated in Hoskins et al, 1985), is itself supporting evidence of chaotic variability.

Since chaotic processes are inherently aperiodic, a spectral analysis of a chaotic time series will reveal power over a range of timescales, possibly strongly removed from the principal timescale of the dominant instability process (Lyapunov exponent timescale). For the atmosphere, it is possible that chaotic variability associated with the 'fast' baroclinic timescale, together with the 'medium' timescale processes associated with regime dynamics, may generate a significant component of 'long' timescale, interannual and interdecadal fluctuations. One could define the word 'significant' through an f-test, comparing the fraction of low-frequency variance explained by internal atmospheric chaotic dynamics, with the fraction of variance explained by 'external' forcing (*eg* from the ocean). From a practical point of view, if chaotic variations are significant, then the predictability of the atmosphere will be limited on these timescales.

Simple model estimates (*eg* James and James, 1989) suggest that internal chaotic processes could be significant even on decadal timescales. In order to be able to have a more quantitative estimate, one needs to resort to more comprehensive models. To start to address these questions I will show results from some decadal timescale integrations made using the UKMO unified model (D. Rowell, personal communication). These integrations were made as part of a coordinated study using a number of GCMs worldwide. Results are based on an ensemble of 6 integrations in which the model was run for 45 years with observed prescribed SSTs from 1948-1993 using the UKMO GISST (Folland and Rowell, 1995) data set. The ensemble members differ only in terms of their initial conditions.

Figs 23 and 24 show the percentage of variance of simulated surface pressure that can be attributed to the time-varying SSTs for seasonal and decadal averages. This diagnostic was estimated by taking the ratio of the temporal variance of the ensemble-mean fields, to the total variance of all surface pressure fields within the ensemble. The temporal variance of the ensemble mean is assumed to be attributable to the underlying SST variability, hence in regions where the ratio is large, we can assume that the influence of time-varying SST dominates over internal atmospheric variability. Figs 23 and 24 show results for DJF and JJA respectively. For each figure the top panel shows results for interannual fluctuations (*ie* seasonal means), the bottom panel shows results for interdecadal fluctuations (*ie* based on 10-year running means of a given season).

Results for seasonal timescale fluctuations are consistent with many previous studies. In particular, over much of the tropics, the total ensemble variance is dominated by the effects of SST variability, whilst in the (more chaotic) extratropics the percentage is generally smaller. In the extratropics, the maximum percentage occurs over the north-east Pacific and is probably associated with the PNA response to El Niño SST anomalies. In summer, the percentage of variance in the north Pacific is smaller, consistent with a reduction in tropical-extratropical teleconnectivity associated with weak potential vorticity gradients.

Interestingly, the percentage of variance of the decadal fluctuations explained by SST variability is, in many areas, smaller than that associated

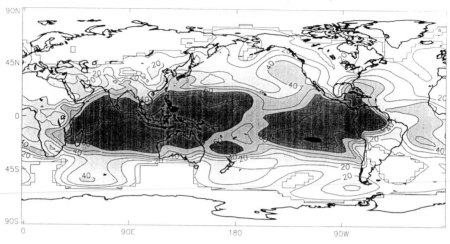

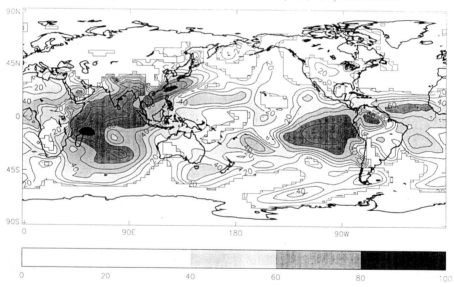

Figure 24: Percentage of variance for DJF due to SSTs based on analysis of surface pressure from a 6-member ensemble of 45-year integrations of the UKMO GCM made with observed specified SSTs. Top panels : seasonal mean fields. Bottom panels: running 10-year average of seasonal mean fields. (D. Rowell, personal communication).

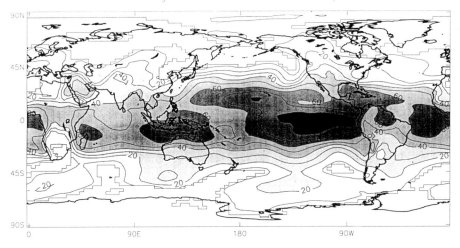

Percentage variance due to SSTs. JJA mean pmsl.

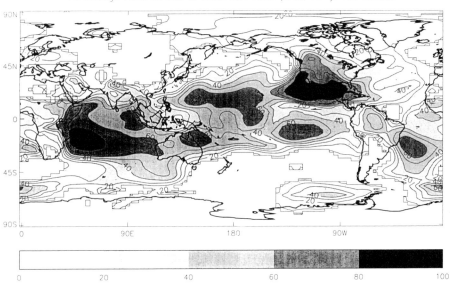

Percentage variance due to SSTs. JJA mean pmsl, 10-yr run means.

Figure 25: *As fig 23 but for JJA.*

with seasonal fluctuations. In winter, for example, there are localised regions over the Indian Ocean, eastern Pacific, tropical Atlantic and so on, where the percentage of variance exceeds 60%, but in other areas, such as the north Atlantic, it appears negligible. This latter result might cause surprise to some people, given the debate about the role of the thermohaline circulation on atmospheric decadal variability. In summer, the percentage of SST-explained variance of decadal fluctuations is somewhat larger in the extratropics. Of course, one has to bear in mind that these results have been obtained by integrating with SSTs over a specific 45 year period.

5.2   The role of the oceans on decadal predictability

Although the results in Figs 23 and 24 suggest that internal atmospheric dynamics can explain a considerable amount of the observed decadal variability (especially in winter), there is modelling evidence that ocean dynamics might enhance the amplitude of decadal fluctuations. For example, Manabe and Stouffer (1996), have studied the geographical distribution of the standard deviation of 25-year mean surface air temperature from 1000 year integrations of three different models: a coupled GCM, an atmospheric GCM coupled to a ocean mixed layer model, and an atmospheric GCM with fixed SSTs. Results show that over continents, the standard deviations in all three runs are broadly comparable. Over much of the oceanic regions, the coupled and mixed layer models produce comparable standard deviations which in turn are larger than the fixed SST run. Over specific regions such as the Denmark Strait and in some regions over the circumpolar ocean of the Southern Hemisphere, the standard deviation of the coupled model is larger than the mixed layer model.

There are in fact reasons to suspect that the influence of the oceans on the atmosphere may be somewhat larger than suggested in Manabe and Stouffer (1996) analyses. In particular, if the feedback from the transient eddies onto the mean flow is an essential component in accounting for the impact of mid-latitude SSTs on the atmospheric flow (see below), then this impact might be under-represented in a model in which the transient eddy covariances were weak. It is known that low-resolution GCMs, and the R15 GFDL GCM in particular (Held and Phillipps, 1993) does suffer from excessively weak eddy momentum fluxes. Secondly, if eddy-mean-flow interaction is important, then if observed SST anomalies are added to a model in which the storm track position has systematic error, the potential impact of midlatitude SST anomalies may be underestimated. In fact significant interdecadal midlatitude air-sea interaction has been found in a coupled GCM integration by Latif and Barnett (1994). In this case, the atmosphere model was integrated at T42 resolution, and, moreover,

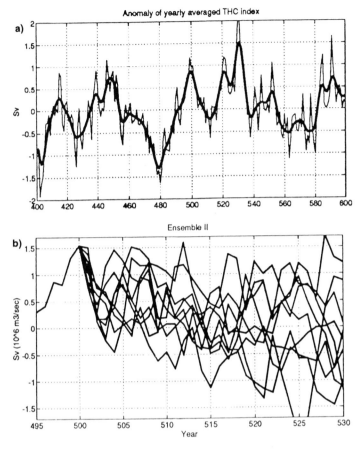

Figure 26: a) 200 years of linearly detrended anomalous yearly averaged thermohaline circulation index from the central portion of a coupled GCM experiment (thin solid line). The time series after a 10-year low-pass filtering has been applied is also shown (thick solid line). b) a 9-member ensemble of yearly averaged anomalous thermohaline circulation index from the coupled model. The ensemble starts from year 500 of the climatology. (From Griffies and Bryan, 1996.)

the SST anomalies were 'in balance' with the dynamics of the overlying atmosphere model.

On the other hand, just because the ocean may contribute to decadal atmospheric variability, this does not mean that the oceanically-forced component is predictable.

For example, according to diagnosis of experiments by Palmer and Sun (1985), the atmospheric circulation can respond to prescribed midlatitude SST anomalies near the western ocean boundaries through processes which intimately involve baroclinic eddies. In broad terms, by warming SST, for

138

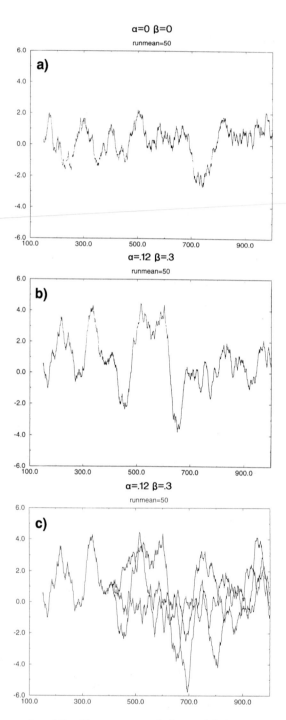

Figure 27: *Time series of the X-component of the modified Lorenz model (5.1) with a)*
$\alpha = 0, \beta = 0$ *b)* $\alpha = 0.12, \beta = 0.3$, *c)* *an ensemble of integrations of the Lorenz model with*
$\alpha = 0.12, \beta = 0.3$ *in which very small perturbations are introduced at t=400.*

example near Newfoundland, the baroclinic gradient near the climatological region of cyclogenesis is weakened, which reduces storm-track activity. Through wave-mean flow interaction processes (diagnosed using the E-vector), the mean flow over and immediately downstream of this region is weakened throughout most of the depth of the troposphere. By weakening the mean flow, the heat loss from ocean to atmosphere is reduced and the SST anomaly can be maintained. However, as simple mixed-layer ocean models suggest (*eg* Daly, 1978) if the surface wind forcing is weakened over the North Atlantic, then SSTs will warm (*cf* Bjerknes, 1964). Hence there is the possibility of a weak positive feedback between ocean and atmosphere over the north Atlantic.

The net effect of this feedback is to redden the spectrum of atmospheric variability by making atmospheric weather regimes (*eg* associated with the North Atlantic Oscillation in the case of Atlantic SSTs, the PNA pattern in the case of Pacific SSTs) somewhat more stable. However, there is no evidence from these experiments to suggest that the regime transitions are themselves any more predictable through coupling to an ocean. Indeed, if such transitions are ultimately associated with fast baroclinic processes, it is unlikely that coupling to the oceans would alter the predictability timescale of these transitions.

An initial study of coupled-model ensemble experimentation appears to support this view about decadal predictability. Fig 25a (from Griffies and Bryan, 1996) shows 200 year of linearly detrended anomalous yearly-averaged thermohaline circulation (THC) index from a run of the GFDL coupled model. Also shown is the time series after a 10-year low pass filter has been applied (thick solid line). Fig 25b (also from Griffies and Bryan, 1996) shows the THC index for a 9-member ensemble made by adding atmospheric perturbations at year 500 of the main run. Although the THC index clearly has substantial variability on multi-decadal timescales, its predictability appears to be determined by the overlying chaotic atmospheric variability, which is much shorter than the dominant THC timescale.

5.3 A simple chaotic 'coupled' model paradigm for decadal fluctuations and predictability

To explore these ideas further, I want to put forward another simple extension of the Lorenz model. As before, the basic Lorenz model (4.5) describes the atmosphere, now it is coupled to a simple ocean model, which here is simply a storage device. The 'coupled model' equations are:

$$\dot{X} = -\sigma X + \sigma Y + \alpha \int_{-\infty}^{t} X(t')e^{-\beta(t-t')}dt'$$
$$\dot{Y} = -XZ + rX - Y + \alpha \int_{-\infty}^{t} Y(t')e^{-\beta(t-t')}dt' \qquad (5.1)$$
$$\dot{Z} = XY - bZ$$

Here α and β crudely parametrise the midlatitude ocean SST and mixed-layer depth respectively. We presume that α is small, so that the storage terms do not destroy the chaotic nature of the dynamical system. If the state vector has resided in one regime of the Lorenz attractor over the period β^{-1} preceding the current time, then the storage effect will generally increase the probability that the system stays in that regime. If the state vector has made a regime transition during the time β^{-1} preceding the current time, then the storage term will have little effect on the subsequent evolution.

Interested readers can experiment for themselves by noting that (5.1) can be readily transformed into a 5th order set of ordinary differential equations by introducing two additional independent variables to represent the ocean storage terms. The (linear) dynamical equations that govern the storage variables are reminiscent of the schematic model put forward by Hasselmann (1976) for obtaining a red noise output from a stochastic white noise atmospheric input. However, unlike the Hasselmann model, the ocean is coupled to a deterministic 'atmosphere' and hence partially stabilises this atmosphere.

One interesting aspect of (5.1) is that even though β^{-1} may be just a few Lorenz time units, the storage terms can induce substantial variability on a timescale of hundreds of Lorenz time units. For example, Fig 5.4a shows a timeseries of the X component of (4.5) with a running mean of 50 Lorenz time units applied to X. It should be noted that the regime centroids are at about $X = \pm 9$. Fig 5.4b shows the same timeseries but from (5.1) (with $\alpha = 0.12, \beta = 0.3$). This reddening is found in the surface air temperature spectra in Manabe and Stouffer's (1995) study.

The intrinsic predictability of the 'coupled' model (5.1) is governed by the internal dynamics of (4.5). This can be seen in Fig 5.4c where two extra integrations are made by adding extremely small perturbations to X at t=400. The predictability timescale O(10 time units) is much shorter than the low-frequency timescale O(100 time units) apparent in Fig 5.4b.

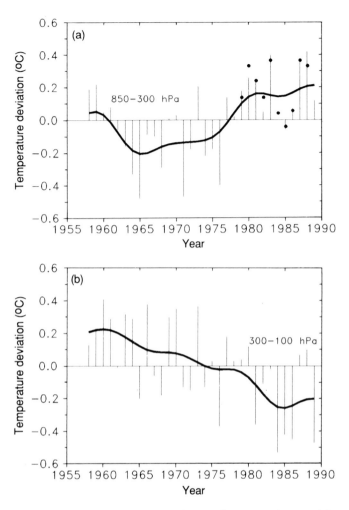

Figure 28: *Observed temperature anomalies in the troposphere and lower stratosphere from 1958- 1989, based on Angell (1988). a) annual global values for 850-300hPa. Dots are values from Spencer and Christy (1990). b) 300-100hPa. From IPCC (1992).*

This appears to be consistent with the predictability of the THC index in Griffies and Bryan's (1996) study, shown in Fig 25.

It is worth noting that a model such as (5.1) could never be considered a representation of ENSO. For example, the tropical atmosphere is not, by itself, chaotic, and internal atmospheric dynamics alone cannot account for the Southern Oscillation (Dix and Hunt, 1995). Moreover, it is well known that ocean dynamics are essential for the ENSO mechanism, and therefore cannot be represented by a simple storage effect.

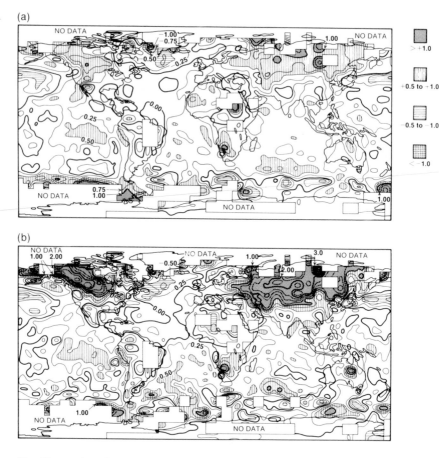

Figure 29: *Observed surface temperature anomaly for the decade 1981-1990 relative to the 1951-80 average (a) annual average (b) winter average. From IPCC (1992).*

6 Climate change

6.1 The vertical and horizontal structure of observed climate change

In this section I would like to make some remarks about the predictability of climate change, bearing in mind the nonlinear paradigm put forward in section 4. To do this I want to focus on a very specific aspect in the validation of GCM climate change integrations. At first sight it might appear that this aspect is rather too specific for an overview discussion on

PNA

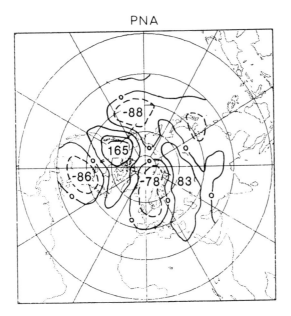

Figure 30: *1000-500hPa thickness anomalies associated with the PNA 500hPa height teleconnection pattern; for winter months only. From Gutzler et al (1988).*

predictability of climate change, but in my opinion it is symptomatic of how climate may respond to weak imposed perturbations.

The original IPCC (1990) report notes an apparent discrepancy between observations of decadal variability and GCM simulations of the impact of doubled CO_2. Fig 26 (reproduced from the IPCC report) shows observed temperature anomalies in the troposphere and lower stratosphere based on Angell (1988). IPCC notes that 'in the upper troposphere (300-100hPa), Fig 26 shows that there has been a rather steady decline in temperature since the late 1950s and early 1960s, in general disagreement with model simulations that show warming at these levels when the concentration of greenhouse gases is increased'.

To reiterate, the basic notion of the nonlinear paradigm in section 4 is that the system response to a small external forcing will, to first order, alter the residence frequency associated with naturally-occurring atmospheric regimes. Changes to the geographical structure of the regimes would be a second-order effect. This paradigm arises because the regions of phase-space in the vicinity of the regime centroids are relatively stable. The dominant instabilities are, in part, likely to be associated with saddle points of the system. An imposed time-invariant forcing would have the effect of biasing the outset of these saddles towards one or more of the regimes.

Fig 27 shows the geographical distribution of change in surface temperature between the 1950s and 1980s, as reported in IPCC (1992). The increase in surface temperature is largest in winter, where it is concentrated over North America and northern Eurasia, and is partially offset by cooling over the North Atlantic and the North Pacific.

The tendency for hemispheric-mean lower tropospheric temperatures to be warm during periods in which the PNA index is positive has already been noted by Gutzler et al (1988). For example, Fig 28 shows the 1000-500hPa thickness anomalies associated with the PNA 500hPa height teleconnection patterns. At the surface, the temperature anomalies associated with these thickness patterns will tend to be accentuated over land relative to the ocean, because of the smaller heat capacity of the active land surface. These positive land surface temperature anomalies will therefore dominate over the negative SST anomalies if a zonal mean temperature anomaly is calculated.

More particularly, the observed change in (northern winter) mean temperature is consistent with an increase in the frequency of the regimes shown on the left hand side of Fig 13, as suggested by the nonlinear paradigm outlined in section 4. One further consequence of this is that, just as the horizontal structure of decadal change should reflect the horizontal structure of the regime centroids, so also the vertical structure of decadal change should reflect the vertical structure of the regimes. Now, to some degree of approximation, the regimes themselves have an equivalent barotropic structure, with the equivalent barotropic level at about 300hPa. From the hydrostatic relationship, if there is a tendency for warming below the equivalent barotropic level, then there should be a tendency for cooling above this level. This is what Angell observes, as shown in Fig 26. (It can be noted that this figure is constructed from radiosonde data which has a land bias. However, according to our discussion above, most of the observed warming has occurred over land regions. Hence it is likely that, according to our nonlinear paradigm, the temperature in the upper troposphere will be cooler in a radiosonde-based estimate of hemispheric mean temperature, than in a satellite-based estimate with its more uniform coverage.)

If this interpretation of the upper tropospheric cooling is correct, then one can ask the following question.

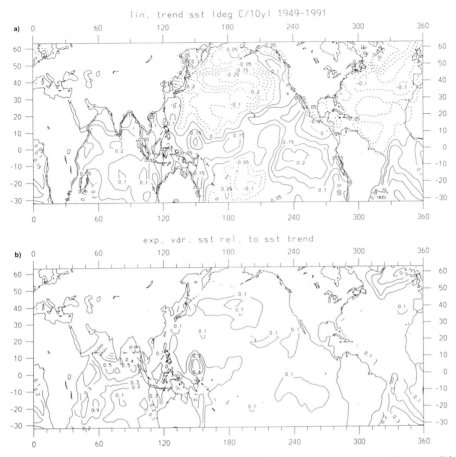

Figure 31: *Spatial distribution of linear trend coefficients of observed SST (degrees C/ decade) for the period 1949-1991. b) Variances explained in SST by the linear trends. From Latif, et al (1996).*

6.2 Is the observed surface warming due to the greenhouse effect?

If the analysis in the preceeding section is correct, then the answer to this question is, in large measure, no. The greenhouse effect is a purely radiative mechanism, whilst much of the observed warming is, according to our nonlinear paradigm, the result of changes in large-scale dynamics in which certain regimes have become anomalously populated. In this picture, the warming is a property of the anomalous surface temperature associated with the regimes in Fig 11a,c.

However, this does not imply that the observed surface warming is not associated with increases in CO_2. To see this, consider first a more parochial argument. If it is anomalously warm over London, then the chances are that the winds have a more southerly than normal component. If it is persistently warmer over London then the chances are the winds are persistently more southerly. This may or may not be due to enhanced CO_2. However, if one wants to know whether the fact that London is warmer is due to enhanced CO_2 then one must ask whether enhanced CO_2 will cause the wind to be more southerly.

Hence, similar to the 'London' argument, if one wants to know whether the fact that the hemispheric mean temperature anomaly has been persistently positive is due to enhanced CO_2, then one should ask whether enhanced CO_2 will increase the probability of the regimes in Fig 11a,c being observed. The question then arises; how could such an increase in the PDF of these regimes be most effectively achieved? Certainly, the adjoint analysis that we have discussed in this paper suggests that the most effective way to force an increase in the frequency of a regime is not achieved by a forcing which projects directly on the regime's geographical structure.

Precisely this question arose in section 4.2 in trying to understand what processes model systematic error was most sensitive to. One way to study this is through the sensitivity analysis described in section 3.5. We can imagine taking an atmospheric model trajectory over, let's say, a decade. For every trajectory segment say of order 10 days or so, we can compute the sensitivity of the positive PNA regime by integrating the regime backwards with the adjoint model. Each trajectory segment will have its own sensitivity pattern. By compositing over many different trajectory segments we get a mean sensitivity for the regime of interest.

Such a calculation is currently in progress (Susanna Corti, personal communication). However, the adjoint eigenmode results shown in Fig 20 already gives us an approximate idea of the difference between the sensitivity structure and the regime structure. In particular, Fig 20 suggests that extratropical quasi-stationary Rossby wave patterns are likely to be sensitive to forcing in the Indonesian area. This result is consistent with the SST anomaly experiment shown in Fig 21, where a localised SST anomaly of about 1K could induce a substantial extratropical response, not altogether different from the positive PNA regime patterns shown in Fig 11.

Hence, even though we might only be interested in anthropogenic changes in hemispheric mean temperature, the large-scale (nonlinear) dynamics of the atmosphere appear to lead us to focus attention onto the possible impact of enhanced CO_2 on key physical processes in specific localised sensitive regions of the globe; the very antithesis of the global greenhouse effect.

It can be noted that in fact SSTs have increased over the west Pacific and Indian Ocean over the last few decades. Fig 29a (from Latif *et al*, 1996) shows the linear trend in SST (deg C/yr) over the period 1949-1991, Fig 29b shows the percentage of variance that this linear trend explains. It can be seen that the trend over the West Pacific and the Indian Ocean explains a relatively large part of the observed variance in SST in those areas.

Now what has this discussion to do with the predictability of climate change? According to these ideas, if we are to be able to simulate accurately the global atmospheric response to enhanced CO_2, it may be necessary to simulate very accurately how CO_2 influences the atmosphere in very specific regions of the globe (specifically over the warm pool area), and perhaps less accurately elsewhere. At present, uncertainties in basic radiative flux parametrisations in the warm pool area certainly exceed the basic $4W/m^2$ associated with radiative effect of doubled CO_2.

Despite these remarks, the predictability of climate change may be greater in the northern summer. In summer, the role of regime dynamics may be weakest, and the radiative greenhouse mechanism may be most significant. Moreover, according to the results of Wallace *et al* (1996) the amorphous component of hemispheric temperature change has its largest amplitude in summer.

Let us conclude this section by asking why the IPCC GCM simulations have not replicated the Angell observations? It is possible that the large-scale dynamics of many GCMs is still inadequate, and that the simulated warming in the 300-100hPa region is a manifestation of this inadequacy. For example, stationary-wave amplitudes in many GCMs are still poor, although recent diagnoses have suggested that weather regime structure does exist in the latest generation of atmospheric climate models (Haines and Hannachi, 1995). My guess is that as models improve, particularly in their representation of low-frequency atmospheric variability, the vertical structure of the response of GCMs to observed CO_2 will correspond more and more closely to the observed vertical structure of warming. (Of course I am aware that there are other complicating factors in discussing these issues, related to the observed depletion of ozone and enhancement in sulphate aerosols.)

7 Unification and rationalisation of climate and weather prediction models

In this paper the notion of predictability has been reviewed, incompletely, on timescales of days, seasons and decades. In doing so we have used both simple and comprehensive models. Simple models are useful in order to formulate hypotheses and to study basic mechanisms, but quantitative assessment of predictability can only be achieved with comprehensive model studies. On timescales of days, the comprehensive models were based on those used for numerical weather prediction; on timescales of decades, global climate models were clearly most appropriate.

One of the successes of our subject in recent years has been the blurring of divisions between these two classes of model. Models that were primarily developed as weather prediction models are now used for climate prediction. Indeed some institutes develop unified models which are used for both climate and weather forecasting in equal measure.

At ECMWF, I have long taken the view that if a weather prediction model is to be successful in predicting blocking activity, for example, it should be able to simulate blocking activity satisfactorily in terms of its own climate. Similarly, it would seem reasonable to expect that a coupled model used for forecasting El Niño should be able to simulate El Niño variability in climate mode.

However, the converse is also true. As IPCC (1990) state, '..confidence in a model used for climate simulation will be increased if the same model is successful when used in a forecasting mode'. This conclusion is consistent with the analysis of the climate predictability problem in section 7, where it was suggested that in order that climate change be predictable, errors in flux representation must be locally less than 4 W/m^2 in regions such as the warm pool. Tolerable errors elsewhere might exceed 4W/m^2 significantly. This moves the persective of climate change away from the global point of view, towards the local point of view. In my opinion, the best way to reduce these local flux errors is through comparison with detailed observations over the relevant area, *eg* from the TOGA COARE experiment (Webster and Lucas, 1992). However, such data are only made over limited periods of time, and therefore only cover specific synoptic situations. A comparison of COARE fluxes with some mean climatology from a long integration of a climate model is clearly inadequate. In comparing model with data, the model should be run in forecast mode. Short-range forecast errors in fluxes can then be compared directly with the observations, and the impact of revisions to the model physics (*eg* to cloud parametrisation schemes) can be readily validated. By validating the model for predictions of the first kind, we should have improved the model's reliability in making predictions of the second kind.

In addition to these concerns, we need to be able to pool modelling resources in order to test hypotheses in a fashion in which problems of model dependence can be minimised. The Atmospheric Model Intercomparison Project (Gates, 1993) is a good example of collaborative research that highlights the successes and limitations of the current generation of models.

However, it has been more difficult, on the basis of these intercomparison studies, to provide corrective prescriptions that will reduce the overall level of model bias. In my opinion, a fundamental question that the climate community needs to address, is whether we can achieve the required level of model accuracy for reliable climate prediction through the development of a diverse range of GCMs, maintained on an individual institute basis, or whether resources should be pooled into the detailed investigation and development of a smaller subset of proven models. I have my views (Palmer and Webster, 1995), I would like to hear yours!

Acknowledgements

I would like to thank Jan Barkmeijer, David Battisti, Roberto Buizza, Ying-Quei Chen, Ron Gelaro, David Rowell and Stephen Griffies for providing me with material which has yet to be published. I am also grateful to Andy Moore and David Stephenson for very helpful comments on an earlier version of the manuscript.

References

ANGELL, J.K., 1988, Variations and trends in tropospheric and stratospheric global temperatures. 1958-87. *J.Clim.*, **1**, 1296-1313.

ABARBANEL, H.D.I., R.BROWN, AND M.B.KENNEL, 1991, 'Variation of Lyapunov exponents on a strange attractor', *Journal of Nonlinear Science*, **1**, 175-199.

BALMADEDA, M.A., M.K.DAVEY, AND D.L.T.ANDERSON, 1995, Decadal and seasonal dependence of ENSO prediction skill. *J.Climate.; In press.*

BARKMEIJER, J., P.HOUTEKAMER AND X.WANG, 1993, Validation of a skill prediction method. *Tellus,* **45A,** 424-434.

BATTISTI, D.S., 1988, Dynamics and thermodynamics of a warming event in a coupled tropical atmosphere-ocean model. *J Atmos Sci,* **45,** 2889-2919.

BJERKNES, J., 1964, Atlantic air-sea interaction. *Advances in Geophysics., Academic Press,* **10,** 1-82.

BLUMENTHAL, B., 1991, Predictability of a coupled ocean-atmosphere model. *J Climate,* **4,** 766-784.

BOUTTIER, F., 1993, The dynamics of error covariances in a barotropic model. *Tellus,* **45A,** 408-423.

BUIZZA, R. AND T.N.PALMER, 1995, The singular vector structure of the atmospheric general circulation. *J Atmos Sci.* **52,** 1434-1456.

BUIZZA, R., R. GELARO, F.MOLTENI AND T.N.PALMER, 1996, Predictability studies with high resolution singular vectors. *Q.J.R.Meteorol.Soc., submitted.*

CACUCI, D.G., 1981, Sensitivity theory for nonlinear systems: I: Nonlinear functional analysis approach. *J.Math.Phys.,* **22,** 2794-2802.

CANE, M.A., S.ZEBIAK, AND S.DOLAN, 1986, Experimental forecasts of El Niño. *Nature,* **321,** 827-832.

CHARNEY, J.G., 1947, The dynamics of long waves in a baroclinic westerly current. *J Meteor,* **4,** 135-163.

CHARNEY, J.G., AND J.SHUKLA, 1981, Predictability of monsoons. Monsoon Dynamics, *J.Lighthill and R. Pearce, Eds., Cambridge University Press, 735pp.*

CHEN, W.Y. AND H.M.VAN DEN DOOL, 1996, Asymmetric impact of ENSO on atmospheric internal variability over the north Pacific. *J.Clim. Submitted.*

CHEN, Y-Q., D.S.BATTISTI, T.N.PALMER, J.BARSUGLI AND E.S.SARACHIK, 1996, A study of the predictability of tropical Pacific SST in a coupled atmosphere/ocean model using singular vector analysis: the role of the annual cycle and the ENSO cycle. *Mon. Wea. Rev. Submitted.*

CHENG, X., AND J.M.WALLACE, 1993, Cluster analysis of the northern hemisphere wintertime 500-hPa height field: spatial patterns. *J. Atmos. Sci.,* **50,** 2674-2696.

CORTI, S., 1994, Modellistica dei regimi di circolazione atmosferica alle medie latitudini durante la stagione invernale. *PhD thesis. Available from University of Bologna, Physics Department. pp233.*

COURTIER, P, C.FREYDIER, J-F.GELEYN, F.RABIER, AND M.ROCHAS, 1991, The Arpege project at Mto-France. Proceedings of ECMWF Seminar on 'Numerical methods in atmospheric models', Shinfield Park, Reading RG2 9AX, UK, 9-13 September 1991, 2, *193-231.*

DALY, A.W., 1978, The response of North Atlantic sea surface temperature to atmospheric forcing processes. *Q.J.R.Meteorol.Soc.,* **104,** 363-382.

DALEY, R., 1991, Atmospheric data analysis. *Cambridge University Press. 457pp.*

151

DIX, M.R., AND B.G.HUNT, 1995, Chaotic influences and the problem of deterministic seasonal predictions. *Int. J. Climatol.*, **15**, 729-752.

DOLE, R.M. AND N.D.GORDON, 1983, Persistent anomalies of the Northern Hemisphere wintertime circulation: Geographical distribution and regional persistence characteristics. *Mon.Wea.Rev.*, **106**, 746-751.

EADY, E.T., 1949, Long waves and cyclone waves. *Tellus*, **1**, 33-52.

FARRELL, B.F., 1989, Optimal excitation of baroclinic waves. *J Atmos Sci*, **46**, 1193-1206.

FARRELL, B.F., 1990, Small error dynamics and the predictability of atmospheric flows. *J.Atmos.Sci.*, **47**, 2409-2416.

FARRELL, B.F. AND P.J.IOANNOU, 1996, Generalised stability theory. Part I: Autonomous operators. *J.Atmos.Sci., submitted.*

FISHER, M. AND P.COURTIER, 1995, Estimating the covariance matrices of analysis and forecast error in variational data assimilation. *ECMWF Technical Memorandum, 220.*

FOLLAND, C.K. AND D.P.ROWELL, 1995, Workshop on simulations of the climate of the twentieth century using GISST. Climate Research Technical Note 56. Hadley Centre. Meteorological Office. Bracknell UK. pp 111

GATES, W.L., 1993, AMIP: The Atmospheric Model Intercomparison Project. *Bull. Amer. Meteor. Soc.*, **73**, 1962-1970.

GILL, A.E., 1980, Some simple solutions for heat-induced tropical circulation. *Q J R Met Soc*, **106**, 447-462.

GRIFFIES, S.M. AND K.BRYAN, 1996, North Atlantic thermohaline circulation predictability in a coupled ocean-atmosphere model. *J.Climate. In press.*

GUTZLER, D.S., R.D.ROSEN, AND D.A.SALSTEIN, 1988, Patterns of interannual variability in the northern hemisphere wintertime 850mb temperature field. *J.Clim.*, **1**, 949-964.

HAINES, K. AND A.HANNACHI, 1995, Weather regimes in the Pacific from a GCM. *J.Atmos.Sci.*, **52**, 2444-2462.

HANSEN, A.R. AND A.SUTERA, 1986, On the probability density distribution of the planetary-scale atmospheric wave amplitude. *J.Atmos.Sci.*, **43**, 3250-3265.

HANSEN, A.R. AND A.SUTERA, 1995, Large amplitude flow anomalies in northern hemisphere midlatitudes. *J. Atmos. Sci*, **52**, 2133-2151

HARTMANN, D.L., R.BUIZZA, AND T.N.PALMER, 1995, Singular vectors: the effect of spatial scale on linear growth of disturbances, *J. Atmos. Sci.*, **52**, 3885 3894.

HASSELMANN, K., 1976, Stochastic climate models. Part I: Theory. *Tellus*, **28**, 473-485.

HELD, I.M. AND P.J.PHILLIPPS, 1993, Sensitivity of the eddy momentum flux to meridional resolution in atmospheric GCMs. *J. Climate*, **6**, 499-507.

HESS, P. AND H.BREZOWSKY, 1977. Katalog der Grosswetterlagen, *Ber. Dtsch. Wetterdienst, Offenbach*, **113**, Bd 15, 39pp.

HOLLINGSWORTH, A., 1987, Objective analysis for numerical weather prediction. In: Short and Medium Range Numerical Weather Prediction. Collected papers presented at WMO/IUGG NWP symposium , Tokyo, 4-8 August, 1986, ed by T.Matsuno, *Special Volume of the J.Meteor.Soc.Japan.* **11,** 59.

HOSKINS, B.J., M.E.McINTYRE AND A.W.ROBERTSON, 1985, On the use and significance of isentropic potential vorticity maps. *Q.J.R.Meteorol. Soc.,* **111,** 877-946.

IPCC, (INTERGOVERNMENTAL PANEL ON CLIMATE CHANGE) 1990, Climate Change: the IPCC Scientific Assessment, J.T.Houghton, G.J.Jenkins and J.J.Ephraums (eds). *Cambridge University Press, Cambridge, UK, 198pp.*

IPCC, 1992, Climate change, the supplementary report to the IPCC scientific assessment, J.T. Houghton, B.A.Callander and S.K.Varney (eds). *Cambridge University Press, Cambridge, UK, 365pp.*

JAMES, I.N. AND P.M.JAMES, 1989, Ultra-low-frequency variability in a simple atmospheric model. *Nature,* **342,** 53-55.

KIMOTO, M. AND M.GHIL, 1993, Multiple flow regimes in the northern hemisphere winter. Part I: Methodology and hemispheric regimes. *J. Atmos. Sci.,* **50,** 2625-2643.

LATIF, M. AND T.P.BARNETT, 1994, Causes of decadal climate variability over the North Pacific and North America. *Science,* **266,** 634-637.

LATIF, M., R.KLEEMAN AND C.ECKERT, 1996, Greenhouse warming, decadal variability, or El Niño? An attempt to understand the anomalous 1990s. *J.Clim. Submitted.*

LILLY, D.K., 1983, Stratified turbulence and the mesoscale variability of the atmosphere. *J.Atmos.Sci.,* **40,** 749-761.

LINDZEN, R.S., 1988, Instability of plane parallel shear flow (towards a mechanistic picture of how it works). *PAGEOPH,* **126,** 103-121.

LORENZ, E.N., 1963A, Deterministic nonperiodic flow. *J Atmos Sci,* **20,** 130-141.

LORENZ, E.N., 1963B, The predictability of hydrodynamic flow. *Trans New York Acad Sci, Ser 2,* **25,** 409-432.

LORENZ, E.N., 1965, A study of the predictability of a 28-variable atmospheric model. *Tellus,* **17,** 321-333.

LORENZ, E.N., 1975, Climate predictability: The physical basis of climate modelling. *WMO, GARP Pub.Ser.,* **16,** 132-136.

MANABE, S. AND R.J.STOUFFER, 1996, Low-frequency variability of surface air-temperature in a 1000 Year integration of a coupled ocean-atmosphere model. *J.Clim. To appear.*

MANSFIELD, D.A., 1993, The storm of 10 January 1993. *Met Mag,* **122,** 140-146.

MARCHUK, G.I., 1974, Osnovnye i soprazhennye uravneniya dinamiki atmosfery i okeana. *Meteor. Gidrol.,* **2,** 9-37.

MARDIA, K,V., J.T.KENT, AND J.M.BIBBY, 1979, Multivariate Analysis. *Academic Press, London. 518pp.*

MARSHALL, J. AND F.MOLTENI, 1993, Toward a dynamical understanding of planetary-scale flow regimes. *J Atmos Sci*, **50**, 1792-1818.

METAIS, O., J.J.RILEY, AND M.LESIEUR, 1994, Numerical simulations of stably-stratified rotating turbulence. From 'Stably stratified flow and dispersion over topography.' Eds I.P. Castro and N.J.Rockliff. *Clarendon Press. Oxford.*

MO, K.C. AND M.GHIL, 1988, Cluster analysis of multiple planetary flow regimes. *J. Geophys. Res., 93D*, **10**, 927-10 952.

MOLTENI, F., S.TIBALDI AND T.N.PALMER, 1990, Regimes in the wintertime circulation over northern extratropics I: Observational evidence. *Q.J.R.Meteor.Soc.*, **116**, 31-67.

MOLTENI, F. AND S.TIBALDI, 1990, Regimes in the wintertime circulation over northern extratropics. II: Consequences for dynamical predictability. *Q.J.R.Meteor.Soc.*, **116**, 1263-1288.

MOLTENI, F. AND T.N.PALMER, 1993, Predictability and finite-time instability of the northern winter circulation. *Q J R Met Soc*, **119**, 269-298.

MOLTENI, F., R.BUIZZA, T.N.PALMER, AND T.PETROLIAGIS, 1996, The ECMWF ensemble prediction system: methodology and validation. *Q.J.R. Meteorol. Soc.*, **122**, 73-120.

MOORE, A.M. AND R.KLEEMAN, 1996, The dynamics of error growth and predictability in a coupled model of ENSO. *Q.J.R.Meteorol.Soc., submitted.*

MUKOUGAWA, H., M.KIMOTO, AND S.YODEN, 1991, A relationship between local error growth and quasi-stationary states in the Lorenz system. *J Atmos Sci*, **48**, 1231-1237.

MÜNNICH, M., M.A.CANE, AND S.E.ZEBIAK, 1991, A study of self-excited oscillations in a tropical ocean-atmosphere system. Part II: Nonlinear cases. *J Atmos Sci*, **48**, 1238-1248.

NOBLE, B. AND J.W.DANIEL, 1977, Applied Linear Algebra. *Prentice-Hall Inc. 477 pp.*

ORR, W.M.F., 1907, The stability or instability of the steady motions of a perfect liquid and of a viscous liquid,. *Proc. Roy. Irish Acad.*, **A27**, 9-138.

OSELEDEC, V.I., 1968, A multiplicative ergodic theorem. Lyapunov characteristic numbers for dynamical systems. *Trudy Mosk Mat Obsc*, **19**, 197.

PALMER, T.N., 1993, Extended-range atmospheric prediction and the Lorenz model. *Bull Am Met Soc*, **74**, 49-65.

PALMER, T.N., 1988, Medium and extended-range predictability and stability of the Pacific/North American mode. *Q.J.R.Meteor.Soc.*, **114**, 691-713.

PALMER, T.N., 1993, A nonlinear dynamical perspective on climate change. *Weather*, **48**, 313-348.

PALMER, T.N., 1994, Chaos and predictability in forecasting the monsoons. *Proc.Indian Natn. Sci. Acad.*, **60A**, 57-66.

PALMER, T.N., AND D.A.MANSFIELD, 1986, A study of wintertime circulation anomalies during past El Niño events using a high resolution general circulation model. II: Variability of the seasonal mean response. *Q.J.R.Meteor.Soc.*, **112**, 639-660.

PALMER, T.N. AND Z.SUN, 1985. A modelling and observational study of the relationship between sea surface temperatures in the northwest Atlantic and the atmospheric general circulation. *Q.J.R.Meteor.Soc.*, **111**, 691-713.

PALMER, T.N., R.BUIZZA, F.MOLTENI, Y-Q.CHEN AND S.CORTI, 1994, Singular vectors and the predictability of weather and climate. *Phil.Trans.R.Soc.Lond.A*, **348**, 459-475.

PALMER, T.N. AND P.J.WEBSTER, 1995, Towards a unified approach to climate and weather prediction. Global change. Proceedings of the first Demetra meeting held at Chianciano Terme, Italy. Published by the European Commission. *EUR 15158 EN. 429pp.*

RABIER, F., E.KLINKER, P.COURTIER, AND A.HOLLINGSWORTH, 1996, Sensitivity of two-day forecast errors over the northern hemisphere to initial conditions. *Q.J.R. Meteor.Soc.*, **122**, 121-150.

RATCLIFFE, R.A.S. AND R.MURRAY, 1985, New lag associations between North Atlantic sea temperatures in the north-west Atlantic and the atmospheric general circulation. *Q.J.R.Meteorol.Soc.*, **96**, 226-246.

SARACHIK, E.S., 1990, Predictability of ENSO. In 'Climate-Ocean Interaction' ed M.E.Schlesinger, *Kluwer Academic Publishers. Dordrecht, the Netherlands. pp385.*

SIMMONS, A.J., D.M.BURRIDGE, M.JARRAUD, C.GIRARD, AND W.WERGEN, 1989, The ECMWF Medium-Range Prediction Models Development of the Numerical Formulations and the Impact of Increased Resolution. *Meteorol Atmos Phys*, **40**, 28-60.

SLEIJPEN AND VAN DER VORST, 1995, A Jacobi-Davidson iteration method for linear eigenvalue problems. *Universiteit Utrecht, Department of Mathematics, preprint 856.*

SPENCER, R.W., AND J.R.CHRISTY, 1990, Precise monitoring of global temperature trends from satellites. *Science*, **247**, 1558-1562.

SPERBER, K.R. AND T.N.PALMER, 1996, Interannnual tropical rainfall variability in general circulation model simulations associated with the atmospheric model intercomparison project. *J.Clim. submitted.*

STRANG, G., 1986, Introduction to applied mathematics. *Wellesley-Cambridge press, 758 pp.*

THOMPSON, J.M.T. AND H.B.STEWART, 1991, Nonlinear dynamics and chaos. *John Wiley. Chichester. pp376*

TOTH, Z., 1992, Quasi-stationary and transient periods in the Northern Hemisphere winter. *Mon.Wea. Rev.*, **119**, 1602-1611.

TOTH, Z. AND E.KALNAY, 1993, Ensemble forecasting at NMC: The generation of perturbations. *Bull. Amer. Meteor. Soc.*, **74**, 2317-2330.

TOTH, Z. AND E.KALNAY, 1996, Ensemble forecasting at NMC and the breeding method. *Monthly Weather Review. Submitted.*

TREVISAN, A., 1993, Impact of transient error growth on global average predictability measures. *J Atmos Sci*, **50**, 1016-1028.

VON STORCH, H., 1988, A statistical comparison with observations of control and El Niño simulations using the NCAR CCM. *Beitrge sur Physik der Atmosphre*, **60**, 464-477.

WALLACE, J.M. AND D.S.GUTZLER, 1981, Teleconnections in the geopotential height field during the northern hemisphere winter. *Mon Wea.Rev.*, **109**, 784-812.

WALLACE, J.M., X.CHENG AND D.SUN, 1991, Does low-frequency atmospheric variability exhibit regime-like behaviour? *Tellus*, **43AB**, 16-26.

WALLACE, J.M., Y.ZHANG AND J.A.RENWICK, 1996, Dynamically-induced variability in hemispheric mean surface air temperature. *Science. To appear.*

WEBSTER, P.J., 1995, The annual cycle and the predictability of the tropical coupled ocean-atmosphere system, *Meteor. and Atmos. Phys.*, **56**, 33-35.

WEBSTER, P.J. AND R.LUCAS, 1992, TOGA-COARE: The Coupled Ocean-Atmosphere Response Experiment. *Bull. Amer. Meteor. Soc.*, **73**, 1377-1416.

XUE, Y., M.A.CANE, S.E.ZEBIAK AND M.B.BLUMENTHAL, 1994, On the prediction of ENSO. A study with a low order Markov. *Tellus*, **46**, 512-528.

YANG, S. AND B.REINHOLD, 1991, How does low-frequency variance vary? *Mon.Wea.Rev.*, 119-127.

ZEBIAK, S.E. AND M.A.CANE, 1987, A model El Niño - Southern Oscillation. *Mon Wea Rev*, **115**, 2262-2278.

ZHANG, Z., 1988, The linear study of zonally asymmetric barotropic flows. *PhD Thesis. University of Reading, UK.*

MECHANISMS FOR DECADAL-TO-CENTENNIAL CLIMATE VARIABILITY

E.S. SARACHIK, M. WINTON, AND F.L. YIN.
University of Washington
Seattle, USA

Contents

NATO ASI Series, Vol. I 44
Decadal Climate Variability
Dynamics and Predictability
Edited by David L. T. Anderson and Jürgen Willebrand
© Springer-Verlag Berlin Heidelberg 1996

1   Introduction

We really don't understand what causes decadal-to-centennial climate variability although we have some rudimentary ideas about the possible mechanisms. Which mechanism operates implies the existence and range of
possible predictability. These lectures, therefore, have two purposes: 1) to
look at the evidence for the various proposed mechanisms and see if any can
be chosen as most likely (or if any can be rejected and thereby eliminated
from further consideration) and 2) to see which of the mechanisms imply
predictability of the decadal variability and which are intrinsically unpredictable. The variety of mechanisms is diverse, and, to save the reader
from turning to the end to see how the story turns out, none can be totally
eliminated at this stage. Indeed, more than one mechanism may apply
at any given time at various places on the globe. It is impossible to be
exhaustive in lectures of this type, so it is hoped that these lectures will
serve as a guide to the ideas and literature of a field that is rapidly catching
on as the realization takes hold that natural variability of the climate is
omnipresent and important, and may be predictable.

We will begin by looking at the nature of decadal-to-centennial variability in the observational record to see what it is we wish to explain
and predict. We will move on to the paleo-record and then to the possible mechanisms that may explain decadal-to-centennial variability in these
records.

1.1   The Instrumental Record

Fig. 1a shows the historical record of averaged air temperature over Northern Hemisphere land for the last 100 or so years. While the trend is clearly
upward (unless that trend is part of a still longer cycle which clearly cannot be determined from this record alone) we will be interested, in these
lectures, more in the wiggles of the curve than its trend. Clearly visible are
variations of multi-decadal time scale. If the warming since the mid 1970's
are part of a natural fluctuation, then the amplitude of such a fluctuation
is of order 0.5^o. If the warming in the last twenty years or so is taken as
part of the trend, then there are modulations on the trend of order 0.1^oC.
The ocean's sea surface temperature (SST), Fig. 1b, is equally variable
on decadal time scales but the large warming of the 1970's and beyond
is clearly less, at least in the Northern Hemisphere SST. While there are

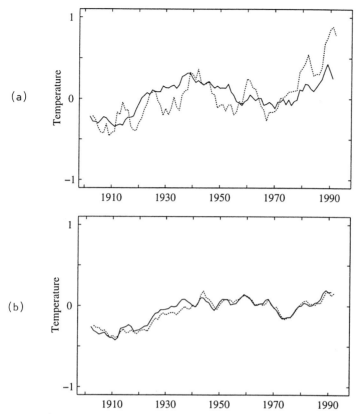

Figure 1: a) *Air temperature variations averaged over Northern Hemisphere land and* b) *Sea surface temperaure averaged over the Northern Hemisphere ocean. Warm season is solid, cold season dashed. (From Wallace, Zhang, and Bajuk, 1995).*

greater difficulties in measuring SST, the smaller warming of the northern hemisphere ocean relative to the land since the 70's is robust and generally agreed to. Further, the large difference between cold and warm season northern hemisphere temperatures so evident for the land is not apparent for the ocean temperatures.

A confirmation of these trends over the last decade or so for wintertime conditions is given in Fig. 2 where we see that the northern land masses have warmed but the north Pacific and north Atlantic have cooled. We see, therefore, why the SST shown in Fig. 1 is not rising as fast as the land temperatures: the land is heating over the last two decades while the northern Atlantic and Pacific are cooling. We will return to this in section 3 below.

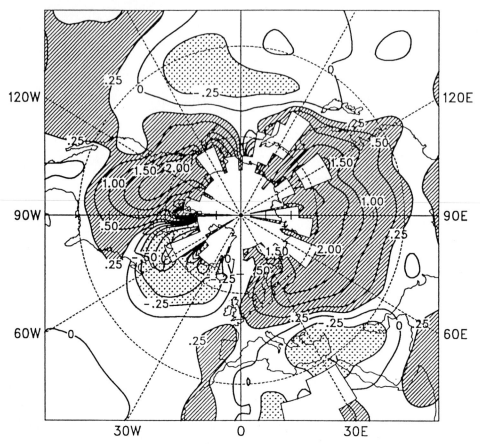

Figure 2: *Winter surface temperature changes of the thirteen year period 1981- 1993 relative to the 1951-1980 mean. (From Hurrell, 1995).*

Regional, rather than global, indications of decadal-to-centennial change are numerous–a large variety of indications are given in the report by the Intergovernmental Oceanographic Commission (1992) and in the compilation of papers in Martinson et al., 1995. In particular, the work of Deser and Blackmon (1993) and Kushnir (1994) show that interdecadal variability of North Atlantic SST can be considerably larger than 1^oC.

1.2   The Paleo-Record

Over the past twenty five years, remarkable records of the climate changes that occur over the glacial/interglacial cycles have been amassed (e.g. Fig. 3 from the Greenland ice cores). The Greenland ice sheet is ideally situated to record variations in the formation of North Atlantic Deep Water (NADW) which occurs in the adjacent Greenland, Norwegian, and Labrador seas. Climate changes on time scales of decades to hundreds of thousands of years are present in this record.

The Greenland ice cores show that the cold climate of the last ice age was punctuated with occasional warm ("interstadial") periods lasting from several hundred to several thousand years (Dansgaard et al., 1993). The transitions between the warm and cold periods were abrupt – the most recent warmings following the last glacial and Younger Dryas cold periods took place in less than fifty years (Alley et al., 1993). These Dansgaard-Oeschger (D-O) events have been well correlated with foraminiferal indicators of sea surface temperatures buried in the North Atlantic ocean sediments (Bond et al., 1993) implying that these oscillations occurred over the entire North Atlantic and were not simply confined to the region of Greenland. The global nature of the events is revealed by their methane signal (Chappelez et al., 1993), also present in the Greenland ice cores, and the presence of muted features corresponding to many of the events in the Vostok Antarctic ice core (Dansgaard et al., 1993). Recently, the foram record has been extended back to the Eemian period (the interglacial previous to our own; McManus et al, 1993; Keigwin et al., 1994). While confirming the D-O events during glacial times, it appears that the last warm period (the Eemian), which perhaps most closely corresponds to our own interglacial period, exhibited stable North Atlantic sea surface temperatures.

1.3   Mechanisms

It will be the basic purpose of this set of lectures to evaluate the current theories for the mechanisms for decadal-to-centennial variability. Any hope of eventually predicting aspects of the climate years in advance will depend on understanding the precise mechanism the real climate system uses to generate this variability.

We may categorize these mechanisms as follows:

1. Decadal-to-centennial variability of the forcing functions external to

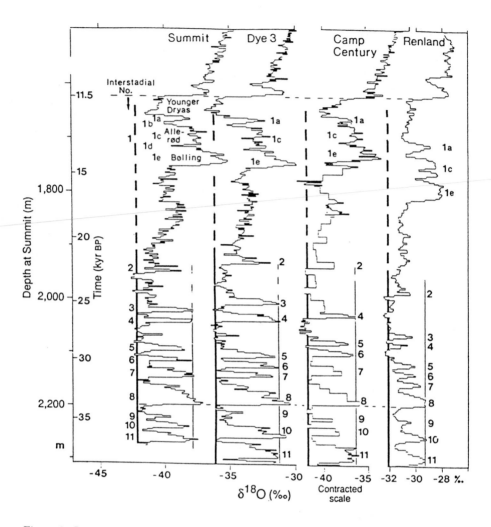

Figure 3: *Ice core records from four sites in Greenland (from Johnsen et al., 1992).*

the climate system: variability of the sun's output, variability of volcanic injections of aerosols to the atmosphere, variability of emissions of anthropogenic aerosols, land use changes, etc.

2. Stochastic forcing of the ocean (and land) by the white noise synoptic scale variability of the atmosphere, i.e. by weather (the so-called Hasselmann mechanism).

3. ENSO variability, namely the decadal-to-centennial variability induced by nonlinear processes modulating essentially seasonal-to-interannual processes and, by extension, variability of normal patterns affected by ENSO (i.e. the PNA pattern) and perhaps normal oscillatory patterns not known to be affected by ENSO (e.g. the North Atlantic Oscillation).

4. Internal ocean variability, where the variability arises from internal processes in the ocean without reinforcement from a changing atmosphere. This includes variability from decadal to millennial time scales.

5. Coupled atmosphere-ocean modes: variability that would not exist either in the atmosphere or in the ocean but arises solely from their joint interaction.

6. Abrupt transitions of climatic regimes, for example the return from the Younger Dryas cooling eleven thousand years ago which seemed to have happened on exceedingly rapid (decadal) time scales.

These are the same mechanisms cited by CLIVAR (1995) and it is one of the purposes of these lectures to look more critically at each of the mechanisms to decide which are the most plausible and what are the implications of each mechanism for the predictability of climate. A review of some of these decadal mechanisms, in a spirit similar to that taken in this article, is given in Rind and Overpeck (1993).

2 Forcing Function Variability

2.1 Solar Output

The sun radiates as a black body with surface temperature of about 5770 K. The variability of the sun's output cannot accurately be measured by earth-based measurements since clouds, aerosols, ozone, and other radiatively active components of the atmosphere in the visible and ultraviolet spectrum interfere with the incoming solar beam on its way to the earth's surface. Satellite measurements of solar output, however, have been going on for the last 15 or so years. It is observed (Lean, 1991) that total solar output does vary on very short time scales (of no particular interest to us here since the atmosphere cannot respond, especially to ultraviolet radiation, on these short time scales) and varies also with the 11 year sunspot number (Williams and Hudson, 1988). The magnitude of this decadal variation is on the order of 1 or 2 W/m^2 compared to the solar constant of 1367 W/m^2, a variation of order 0.1% .

It should be noted that the direction of output variability with sunspot number is opposite to what intuition tells us: when the sun's disc is clear (no sunspots) the output is reduced, not increased. This occurs because the so-called bright features (networks, plages, and faculae) covary with sunspots and dominate the output variability. Due to interactions in the sun's upper atmosphere, of order 20% of this variability with sunspot cycle lies in the ultraviolet and therefore gets absorbed by ozone relatively high in the earth's atmosphere. The basic question therefore is whether or not the troposphere is sensitive to the remaining one or so W/m^2: this is a general question about the sensitivity of the climate system to small changes of radiation. Because the radiation reaching the intercepted disc at the outer edge of the earth's orbit must be spread over the entire area of the earth, this number must be divided by 4 and because the earth's albedo in the visible is .3 (mostly due to clouds) the number must further be multiplied by 0.7. Thus a solar output variability of 1 W/m^2 translates to a variability of 0.2 W/m^2 at the top of the global atmosphere.

On the basis of CO_2 doubling experiments with numerical GCMs, it has been found that 4 W/m^2 radiative input (in this case long wave) at the top of the atmosphere will lead to a warming of global surface temperature from 1.5 to 4.5^{o}C (e.g. IPCC 90), giving a sensitivity of 0.4 to 1.1^{o}C$W^{-1}m^2$ to radiation at the top of the atmosphere. If we accept this model estimate of climate sensitivity, it is clear that the solar output variability over decadal time scales could be responsible for a few tenths of a degree response in globally averaged surface temperature.

The existence of long periods when sunspots were absent from the face of the sun (for example, the so called "Maunder Minimum," in the late 1600's, see Eddy, 1976) are well established. To estimate the reduction of solar output implied by sunspot minima, and analysis of the observed variability of other main sequence stars similar to the sun has been made (Lean, Skumanich, and White, 1992) and implies a reduction of solar output as much as 5W/m^2 (with a more probable value of half that) which at the top of the atmosphere would be less than 1W/m^2 . These estimates are uncertain because the measured variability of the sun seems to be generally less that expected an anlysis of other sun-like stars (Lockwood et al., 1992). This is of the right order of magnitude to (by itself) account for the Little Ice Age which was coincident with the Maunder Minimum (1645 to 1715) and involved a global cooling of order 1 degree (see Webb, 1989). (This number is much in doubt and is inferred from ^{18}O ratios taken at isolated ice cores around the globe–see e.g. the ice core papers in Bradley and Jones, 1992).

We may conclude, on the basis of admittedly fragmentary and indirect evidence, that the variability in the sun's output is likely to be the cause of decadal to centennial variability in the earth's global surface temperature at a level of a few tenths of a degree. The ability to say more about the effect of the future effects of the solar output on climate depends on measuring solar radiation over long periods of time to better characterize its characteristics and variability, on understanding the sun better so as to better predict its output, and on better understanding the sensitivity of the earth's climate system to small changes in solar radiation. A complete and readable review of the effect of solar variations on the earth is given in National Research Council (1994).

2.2 Volcanoes

It has been suggested that aerosols, both volcanic and anthropogenic, may have a significant effect on the earth's radiation balance (e.g. Charlson and Wigley, 1994).

Volcanic aerosols (mostly sulfate) have global impact only when significant amounts are injected into the stratosphere, where, due to lack of effective removal mechanisms, they remain for several years . Smaller volcanic eruptions inject their aerosols into the troposphere where they are rapidly rained out. As it turns out, the frequency of volcanic eruptions large enough to affect the global climate is low. Pinatubo, for example, was a once-in-a-century eruption.

The effects of the Pinatubo volcanic eruption were predicted in 1992 (Hansen et al., 1992, 1993, 1995) and later verified. The net global surface A temperature response was a spike of cooling, of amplitude 0.5^oC, which was gone in about two years. Thus the emission of the volcanic aerosol, while having a significant effect on global climate, was a transitory phenomenon unlikely to be the cause of major variability. It is not out of the realm of possibility, however, that future clusters of large volcanic eruptions would have a major effect on decadal to centennial variability. Lindzen (1994) has argued that the sequence of large volcanic eruptions, the 1883 eruption of Krakatoa and the 1912 eruption of Katmai, produced more lasting effects than would a single volcanic eruption. Since the aerosol output of these volcanic eruptions was not measured, the argument must be taken as indicative rather than definitive. The possibility of clusters of volcanic eruptions is, according to our current capabilities, unpredictable.

2.3   Anthropogenic Aerosols

In contrast to volcanic aerosols which are injected into the stratosphere sporadically but reside there for a long time, anthropogenic aerosols are constantly injected into the troposphere and last only a few days. They are a major component of the surface radiation balance, and can affect temperature in regions close to the sources of emission (Charlson et al., 1991). These sulfate aerosols have a lifetime of only 5 days or so and therefore do not reach laterally far from their origin or vertically into the stratosphere. The net effect of these aerosols is to considerably decrease the solar radiation reaching the ground (by as much as 5 W/m^2 locally near large sources of emissions) and leads to trends of net surface cooling in these regions, even when the warming effect of greenhouse gas emission tends to warm the rest of the earth (Karl et al., 1995).

Anthropogenic aerosols are now estimated to have a global cooling impact of order 0.5 W/m^2 (at the top of the atmosphere) compared to the warming impact of about 2.3 W/m^2 due to the increases of radiatively active greenhouse gases (e.g. IPCC, 1995). Decadal variations in emissions, as in the oil stoppages of the early 1970's, could have significant impact on local decadal variations and cannot be neglected as a decadal climate variability mechanism. Recent calculations of anthropogenic climate change have included direct aerosol effects, i.e. the direct effect on the radiative budget, but cannot yet estimate the indirect effect, i.e. the effects of aerosols on clouds through their actions as cloud condensation nuclei. A complete state- of-the-art review is given in Chapter 3 of the IPCC 95 report. We must conclude that decadal variations of emissions of anthropogenic aerosols are a possible mechanism for local decadal climate change, but one that is predictable only if the social and economic factors that determine their emissions are themselves predicted.

2.4 Other External Forcings

In the 1600's, the entire part of the US east of the Rockies was covered by virgin forest. By the early 1900's, essentially all of this forest was removed to provide agricultural land and urban living space. Over these very long time scales, the ground albedo of the eastern US was severely changed. Similarly, population increases in other parts of the world have impacted the nature of the land surface and changed the nature of the surface vegetation. Quantifying changes in land cover over time on a global scale is fraught with difficulties (e.g. Williams, 1994) and there is no general agreement even on the current area covered by forest.

The time scale for land surface changes can be decadal in local areas but is centennial and longer on global scales. These changes are clearly human induced (although climate variations are affected by means of species replacements and through the influence of forest fires). While these human induced changes can alter natural variability, their study is at such an early stage that they will not be discussed further.

3 Stochastic Forcing

The atmosphere is constantly in motion and, while we tend not to think of atmospheric motions as decadal in nature, it is now well understood that rapid atmospheric motions can induce decadal and longer motions in the more slowly varying systems that couple to the atmosphere. This section will describe some, by now, well known results.

3.1 Some Simple Statistical Preliminaries

3.1.1 Coin Tossing

Lets start with the simplest of all possible situations: coin tossing. Let us assume the coin is unbiased so that the probability of heads or tails on the n'th toss is always 50%. We might assume that the sum of the number of heads minus tails would stay close to zero but this is not true (Feller, 1968). Fig. 4 shows the result of a coin tossing experiment of 5000 tosses.

Let the outcome of a toss be $P_n = \pm 1$ and let us define a sum of the number of heads and tails as:

$$X(N) = \sum_{n=1}^{N} P_n \quad .$$

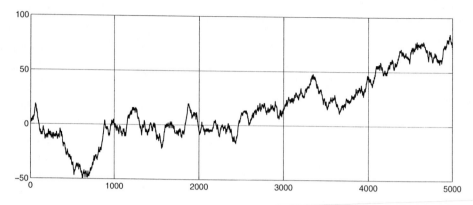

Figure 4: *Coin tossing experiment. The results of tossing a coin 5000 times. (Courtesy Jim Renwick).*

The average $< X(N) >$ is clearly zero for large N but $< X^2(N) >$ goes as N for large N and the number of zero crossings go as $N^{1/2}$. We see that we have generated, simply by summing, some very low frequencies from a process that has basically a white spectrum (the spectrum is the Fourier transform of the correlation function and since the correlation function is a delta function for a unbiased coin, the spectrum is white). We may understand these results in terms of Taylor's theory of dispersion based on the random walk (Taylor, 1921):

Let $v = \frac{dy}{dt}$ and define the correlation function R by:

$$< v(t)v(t+\tau) >=< v^2 > R(\tau) .$$

In general, $R(\tau)$ will fall to something like half value in a time T_1 (or will have an integral which will be of order T_1). We can then manipulate:

$$< v^2 > \int_0^t R(\tau)d\tau = \int_0^t < v(t)v(t+\tau) > d\tau$$

$$=< v(t) \int_0^t v(t+\tau) > d\tau =< v(t)y(t) >= \frac{1}{2}\frac{d}{dt} < y^2(t) >$$

so that $< y^2(t) >= 2 < v^2 > \int_0^t dt' \int_0^{t'} R(\tau)d\tau .$

For t very small, $R(\tau) \approx 1$ and $< y^2(t) > \approx < v^2 > t^2$, while for t large, $\int_0^\infty R(\tau)d\tau = T_1$ so that $< y^2(t) > = 2T_1 < v^2 > t$ and the rms distance from zero, $y_{rms} \propto \sqrt{t}$.

The analogy to coin tossing puts v as the rapid variable (analogous to P_n) and y as the integral of v (analogous to X) : just as in coin tossing, after a very long time the sum y (analogous to X) is unlikely to return to zero. While the spectrum of the coin tossing is white, the spectrum of the sum is clearly red. As the record gets longer and longer, the likelihood of $X(N)$ returning to zero for large N becomes vanishingly small.

3.1.2 The variance of sample means

Consider a stationary time series with variance σ^2 and with correlation time T_1. Consider means over a time $T \gg T_1$. The variance of the means over time T is given by:
$$\sigma_T^2 = \frac{\sigma^2}{N} \text{ where } N = \frac{T}{T_1}$$ is the number of correlation times in the interval over which the mean is taken which is equivalent to the number of independent samples. We see that $\sigma_T \propto \frac{1}{\sqrt{T}}$ so that it takes a long time to reduce the variance of the long term means. As an example, if we consider the annual mean temperature at a point in the presence of diurnal and other variations, if the correlation time is of order of a week, then the annual mean temperature will have a variance only a seventh of the variance of the original temperature time series. We will use this later when the variance of decadal sample means is discussed in coupled general circulation models.

3.2 The Hasselmann Mechanism

Hasselmann (1976) proposed a stochastic climate model that can most simply be understood in terms of the coin tossing example given above (also see Wunsch, 1992). Consider a model thermal equation of the form:

$$\frac{dT}{dt} = Q$$

where the fluxes Q are taken to be some sort of a random forcing on the rate of change of temperature (this is like the forcing of a mixed layer by random fluxes of heat at the surface). The spectra of the forcing and the response is given by:

$$S_T(\omega) = \frac{S_Q(\omega)}{\omega^2}$$

and we see that the response spectrum is reddened by the factor of $1/\omega^2$ and that the forcing at low frequencies must have some energy at the low frequency ω if the response is to have energy at this frequency.

If the temperature response is damped with coefficient α:

$$\frac{dT}{dt} = Q - \alpha T \,,$$

and the spectrum becomes:

$$S_T(\omega) = \frac{S_Q(\omega)}{\omega^2 + \alpha^2}$$

and we see that at very low frequency $\omega \ll \alpha^{-1}$, the spectrum becomes flat.

Again the analogy to coin tossing is clear: Q is the rapid variable and T the integral of Q over time. T would develop increasingly long periods were it not for the feedback. In the presence of a perturbation of mixed layer temperatures (and in the absence of ocean processes that change the sea surface temperature) the ocean mixed layer acts to delay the sea surface temperature from returning to the value the atmosphere would like to maintain. Therefore the ocean mixed layer acts as a damper, with the damping time scale determined by fluxes at the surface and the mixed layer heat capacity: usually on the order of 2 or 3 months. When forced by high frequency (time scale less than a week) fluxes from the atmosphere, which can be considered as "white noise", the resulting power spectrum is "red" down to this damping frequency a^{-1} and white again at frequencies lower than this. When spatial coherence of these high frequency fluctuations are taken into account, decadal variability in sea surface temperature can be generated through oceanic processes, which then induce decadal variability in the atmosphere. This mechanism can be taken as a null hypothesis for decadal-to-centennial variability, and can be assumed to apply unless proven otherwise, as in Wunsch (1992).

Ocean models can be used to examine the dominant variability when forced with "white" noise fluxes. The first such calculation to look at long ' term ocean variability was conducted by Mikolajewicz and Maier-Reimer (1990) using the Hamburg geostrophic ocean model–"loop oscillations" of several hundred year period will be discussed in Sec. 5. First, the global world ocean model was spun up, driven by Hellerman and Rosenstein wind stress, and was restored to COADS air temperature and Levitus salinity (the model included sea ice). Then the fresh water flux was diagnosed, and a "white" noise component with standard deviation of 16 $mm/month$ was added.

Weisse et al. (1994) analyzed the same run using Principal Oscillation Patterns (POPs). They found that the dominant POP mode showed decadal to interdecadal frequencies (10 to 40 year time scales) and is centered near the Labrador Sea. Salt anomalies accumulate first in the Labrador Sea, and then are advected into the northern North Atlantic Ocean. The decadal time scale seems to be set by the flushing time of the upper layers of the Labrador Sea. Once the salt anomalies leave the enclosed Labrador Sea, their properties and lifetimes are governed by advection and convection in the open Atlantic. It is interesting that the Labrador Sea seems to be chosen both by oceanic processes and, as we will see in the next subsection, by atmospheric processes.

3.3 Atmospheric Variability

Because the atmosphere has a great deal of high frequency variability, there is no question that it can force a slowly responding ocean (or land or ice) and generate lower frequency variability. In analogy to the coin-tossing paradigm, it is clear that sudden events in the atmosphere (squalls, weather, fronts, etc.) produces energy at all lower frequencies and will therefore generate a white spectrum on periods longer than a few days. It has been indicated that atmospheric motions can, through non-linear interactions, transfer energy directly to slower time scale (James and James, 1989) tending to redden the spectrum at the lower frequencies. As we have seen, the Hasselmann mechanism also gives energy at low frequencies due to the action of short impulsive weather events on the oceanic mixed layer but, in this case, the spectrum would be white. In either case, the question naturally arises as to whether the atmosphere can organize the variability spatially and, in particular, whether the spatial distribution of these low frequency variations, induced by topography and orography, can explain ocean decadal variability. The Hasselmann theory does not explain spatial patterns directly: something else must be going on.

Explicit calculations (G. Nitsche, personal communication) using the

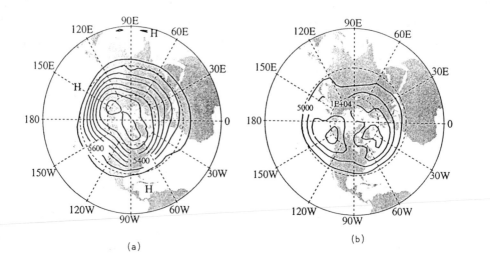

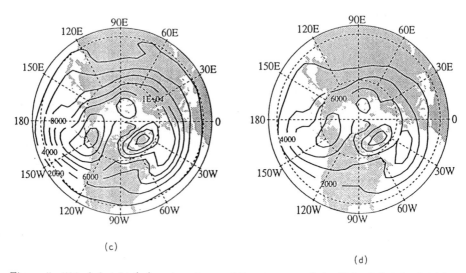

Figure 5: *500mb height (m) and variance of the averages of the 500 mb height field for a 100,000 day run for perpetual January R15 GFDL AGCM (a) Mean 500mb field (b) Variance over the entire run (c) Variance of 6 day averages (d) Variance of 16 day averages (Courtesy Gregor Nitsche).*

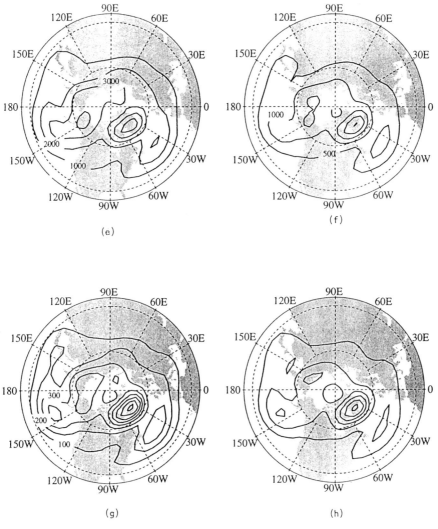

Figure 5: *continued:* (e) *Variance of 50 day averages* (f) *Variance of 150 day averages* (g) *Variance of 500 day averages* (h) *Variance of 1500 day averages.* (*Courtesy Gregor Nitsche*).

GFDL course resolution (R15) atmospheric general circulation model in a 100,000 day run using SST fixed at perpetual January throughout the run show a number of interesting features. The variance of time averages begins to go inversely with the length of the averages (see section 3) for intervals greater than 150 days, indicating independent intervals, and therefore a purely white spectrum, beyond this time scale. The spatial organization is clear by examining Fig. 5. The variance of the long term averages becomes spatially organized and, for the longest intervals (beyond 500 days), clearly has organized itself around Greenland and the Labrador Sea looking very much like the North Atlantic Oscillation with hints of the Pacific North American pattern clearly visible.

3.4 Variability in Coupled Models

There have been, at this writing, only two long coupled simulations of long term natural variability known to us, that of Schneider and Kinter (1994) and that of Manabe and Stouffer (1995).

Schneider and Kinter compare a 400 year atmosphere-only run with seasonal variation in solar forcing to a 400 year coupled run (without flux correction). While the coupling increases the variability, it doesn't increase it that much and these authors conclude that the basic mechanisms for long term variability is the Hasselmann mechanism, with high frequency forcing by weather disturbances in a way that the "...integrating effects of components with long memories, i.e., soil moisture, snow cover and heat storage by the ocean appear to be important in determining the spectral characteristics of the variability."

The Manabe and Stouffer calculation has a flux adjustment in both heat and momentum in order to maintain the thermohaline circulation (see Manabe and Stouffer, 1988) and is the same calculation that shows Atlantic variability identified as internal ocean variability (Delworth, Manabe, and Stouffer, 1993, see section 5). The global decadal averages are roughly consistent with random samples chosen at random: the 25 year samples had a variance about $5^{1/2}$ times that of the five year samples. They also showed that over much of the globe, a mixed layer model of the ocean gives much of the excess variability over that attained by fixing SST at its climatological march. The overall conclusion we draw is that much of the modeled variability is consistent with the Hasselmann mechanism for climate and that the additional role of the ocean is confined to local regions. It should be pointed out that the ENSO phenomenon is not well represented by these coarse resolution atmospheres and therefore that its variability and contribution to climate must be investigated by special purpose models.

4 Decadal ENSO Variability

As a result of the increased attention devoted to ENSO (El Niño/Southern Oscillation) in the last decade or so, we know that ENSO is a Pacific-wide phenomenon involving major eastward expansion of warm SSTs (of the normally warm western Pacific) into the central and eastern Pacific. This expansion occurs on interannual time scales and is accompanied by westerly wind anomalies in the equatorial Pacific (to the west of the warm SST anomalies) and anomalous rainfall, mostly over the expanded warm water. We also know that the warm and cold phases of ENSO have global teleconnections, especially during winter in each hemisphere and affect Northwest North America. Northeast Brazil, Southern Africa, etc. (e.g. Ropelewski and Halpert, 1987)

Although the mechanism for ENSO is believed to be interannual (e.g. the so-called "retarded oscillator" mechanism, Battisti and Hirst, 1989) there are known to be decadal variations in ENSO (Fig. 6). These decadal variations have been associated with decadal changes in the circulation over the northern Pacific (e.g. Trenberth and Hurrell, 1994) and it has been noted that the shift in northern Pacific circulation in 1976 (an intensification of the PNA pattern) has been associated with an increase of the intensity and regularity of ENSO in the tropical Pacific. Hurrell (1995) has also noted that the winter warmth over Europe and Asia over the last decade or so coincides with an abnormally strong North Atlantic Oscillation which guides the mid- latitude jet into a more southerly position over Europe and continuing on to the Asia landmass.

Wallace, Zhang, and Renwick (1995) have put these circumstances together and shown that there is a global winter pattern (Fig. 7) with warmth over the northern landmasses and cooling over the northern Atlantic and Pacific that is strongly correlated with the ENSO signal. Although this pattern normally has a month-to-month variability, it has been "stuck" in the polarity that warms the land and cools the ocean and therefore contributes to the mean land hemispheric winter warming that has been apparent since 1976. The extra warmth in winter over that in summer is primarily due to this pattern. This pattern correlated strongly with the ENSO. A fuller account is given in the lectures by Wallace in this volume.

We must conclude that a substantial part of the warming is due to

176

Eastern Eq. Pacific SST Anomalies (C)

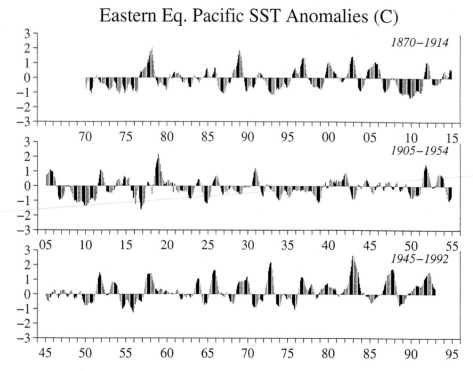

Figure 6: *Time series of monthly mean SST anomalies with respect to the average over the entire record (C) for the cold tongue region (6°S − 6°N, 90°W − 180°W) for 1870-1992 from COADS. Light and dark shading identifies the climatological warm (January to May) and cold (July to November) periods. (Courtesy Todd Mitchell).*

higher frequency natural cycles that seem, for as yet unknown reasons, to have significant modulations of decadal variability. These cycles seem to be influenced by ENSO and it is possible that the decadal variability of these cycles have something to do with the decadal variability of ENSO. It is therefore clear that ENSO is a climate process on decadal time scales as well as on interannual time scales and that models which seek to simulate the mean and variable climate and the response to anthropogenic perturbations must also be able to simulate ENSO, its natural decadal variability, and the response of this variability to anthropogenic changes of the composition of the atmosphere.

Current coupled models that seek to simulate long term changes in

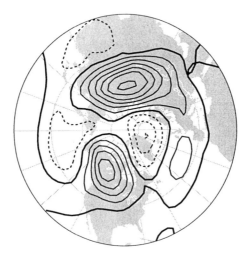

Figure 7: *Dynamical pattern of lower tropospheric winter temperature variability. (From Wallace, Zhang, and Renwick, 1995).*

the climate system, whether natural or anthropogenically influenced, are limited, for practical reasons, to relatively coarse resolution. They therefore lack the resolution near the equator to resolve those processes known to be important for ENSO, especially wave and upwelling processes. Since ENSO and its variability are important climate processes, this remains an important modeling inadequacy in simulating and predicting future climate.

5 Internal Ocean Variability

5.1 Boundary Conditions and the Concept of Internal Ocean Variability

The ocean is forced, primarily at the surface, by fluxes of heat, momentum, and freshwater. These fluxes are in turn determined from the atmosphere in terms of how it interacts with SST. The correct way of dealing with variability in the system is by consistent coupling: when something in the ocean changes and is reflected in surface quantities, the atmosphere responds to adjust the surface heat and momentum fluxes and thereby the ocean and conversely, when something changes in the atmosphere and is reflected in the surface fluxes of heat and momentum, the ocean responds which then adjusts the atmosphere *ad infinitum*. The system is mutually and inextricable coupled.

It has become common, however, for mostly practical reasons, to force the ocean with simplified boundary conditions and examine the consequent variability with the understanding that these simplified boundary conditions do not have the full dynamics of the coupled system. The most common forcing is "restoring boundary conditions", on both temperature and salinity, diagnosing the resulting fresh water fluxes, and then forcing with restoring boundary conditions on temperature and constant flux boundary conditions on salinity, the so called "mixed" boundary conditions introduced by F. Bryan (1986). The argument is given that these mixed boundary conditions are more appropriate than purely restoring boundary conditions since the temperature of the atmosphere does respond to SST while precipitation in the atmosphere does not respond to sea surface salinity.

There have been raging discussions about the suitability of mixed boundary conditions for the correct simulation of the ocean. We will take the point of view that there is no alternative to coupling the ocean to a real atmosphere, but that simplicity is its own reward: to the extent that the simplified boundary conditions gives the correct fluxes, we will consider those boundary conditions reasonable. When variability is found under mixed boundary conditions, it will be taken as smoke: whether or not there is fire will be left to more complex coupled atmosphere-ocean models.

It should be pointed out that in choosing a simple restoring boundary condition and then diagnosing the fresh water flux to get mixed boundary conditions, the restoring time is crucial. As Willebrand (1993) and Marotzke (1994) have pointed out, the restoring time is spatially scale dependent and therefore choosing a single restoring time as an artifice implies choosing a single spatial scale and therefore a single process of interest. We will consider that the most important process is the cooling of water in the Gulf Stream extension to become dense enough to sink in the Greenland-Iceland-Norwegian (GIN) seas and perhaps also in the Labrador sea. The horizontal scale is of order 1000 km and the appropriate restoring time scale is about 50 days. If the time scale is made long enough, the water cannot lose enough heat through surface fluxes and arrives at its final destination with less density: in the extreme case, too light to sink.

A final emphasis about boundary conditions: there is no "correct" boundary condition for an ocean-only model since there is no single boundary condition that can give both the correct fluxes and the correct time change of fluxes: the only correct and consistent boundary condition is given by a correct atmosphere consistently coupled to the ocean.

5.2   General Considerations of Variability

Most studies of ocean decadal thermohaline variability were carried out in idealized coarse resolution OGCMs. These mechanistic explorations mainly investigate internal ocean variability, where the variability arises from internal processes in the ocean without direct help from a changing atmosphere. The coarse resolution is necessitated by practical reasons: the large amounts of time needed to consistently spin up the thermohaline circulation, on the order of 1000 model years.

The first exploration in that direction was the seminal work of F. Bryan (1986). Bryan showed that, in a simplified coarse resolution sector model, a steady circulation, symmetric about the equator, with sinking at high latitudes and upwelling in the rest of the ocean, existed under both restoring (both temperature and salinity are restored) and mixed boundary conditions (surface temperature restored and salinity flux prescribed). When the mixed boundary condition case was perturbed with the addition of fresh water at high southern latitudes, a halocline capped the sinking region and a single cell pole-to-pole circulation, with sinking in the north and rising elsewhere, obtained in only 50 years. This rapid adjustment implied that relatively rapid variability was possible even though the generally agreed-upon time scale for the thermohaline circulation was of order 1000 years. The rapid time scales turned out to be due to the frictional boundary current components of the thermohaline circulation.

Perhaps the simplest idea about thermohaline variability is this: a steady circulation requires the delivery of salt by the circulation itself to balance any freshwater input at high latitudes in the region of the deep sinking. In particular the Gulf Stream Extension in the models must deliver enough salt to counter the freshening by surface freshwater fluxes in the sinking regions of the NADW. Clearly if the circulation is limited in its capacity to deliver salt, there will be some input of freshwater flux to the surface beyond which the circulation cannot counter: the circulation then breaks down and a halocline spreads over the sinking regions. The detailed mechanism by which the circulation revives to give an oscillation will be dealt with below.

5.3   Decadal Variability

Weaver and Sarachik (1991a, 1991b) were among the first to recognize the importance of F. Bryan's halocline catastrophe experiment. They carried out extensive studies of the transient behavior of an ocean-only model upon a switch from the equilibrium solution obtained under restoring boundary conditions to mixed boundary conditions. They found that the model went through an adjustment process during which there is large transient and chaotic variability on interdecadal to centennial time scales. After a detailed analysis of an short time scale (decadal) oscillation cycle, they proposed an advective mechanism: the time scale it takes an anomaly to be advected around a limited portion of the basin is decadal. In resolving the dilemma that transient variability was observed in some models but not in others, they found that the existence of transient variability depends on the characteristics of the forcing, especially the magnitude of the surface fresh water flux (Weaver et al, 1993): variability increased as the braking effect of the fresh water fluxes increased relative to the driving effect of the thermal forcing.

Transient variability of an ocean model, upon switching from restoring to mixed boundary conditions, is usually considered in the context of multiple equilibria: the equilibrium obtained under restoring BC's is unstable upon a switch to mixed BC's, therefore variability arises spontaneously as an instability, after which the model settles either into another equilibrium or stays in a chaotic state. Yin and Sarachik (1995) showed that the ocean model with mixed BC's can also oscillate interdecadally. They analyzed the heat and salt budget in regions where largest variability occurs and proposed an advective and convective mechanism. Horizontal advective heat transports from subtropical regions warm up the subsurface water in the subpolar region and thereby enhance convection. Convection in turn induces surface cyclonic and equatorward flows, which, together with horizontal diffusion and surface fresh water input, advect subpolar fresh water into convective regions to weaken or suppress convection. During an oscillation, convection vertically homogenizes the vertical water column, increases the surface salinity, creates a larger rate of meridionally decreasing surface salinity, and makes surface advective freshening in the convective region more efficient. The periodic strengthening and weakening of convection caused by subsurface advective warming and surface freshening in the subpolar region show up as interdecadal oscillations in the model.

This mechanism focuses on the instability induced by subsurface poleward heat transport through the northeastward currents associated with deep water formation. Because the slow vertical heat transport is small, enough heat accumulates below the surface and leads to instability, a way to release heat, and this happens through convection in the model. An examination of water column T-S properties in regions where largest variations occur show that the magnitude of temperature difference between surface and subsurface increases over the whole oscillation period while salinity differences quickly saturates after convection weakens or stops. This indicates that the oscillation is driven by a thermal instability since short time scale haline process is unlikely the dominant process for the oscillation. In order for an oscillation to occur, there has to be a process acting to restore the system back. In the model, surface freshening process acts to weaken or suppress convective process so that the heat can once again accumulate in the subsurface.

What determines the period of these oscillations? For the advective mechanism, Weaver and Sarachik (1991b) interpreted the period of the oscillation as the time that an anomaly takes to travel a certain distance. With both the advective and convective mechanisms acting, the period of the oscillations can be thought as the time between two consecutive strong phases of convection. This time scale is related to the rate of subsurface warming and surface freshening. If everything else remains the same, it would take less time for the water column to reach a convective critical state when the rate of subsurface warming is larger, or the rate of surface freshening is larger. Since surface freshening is a relatively short time scale process, the oscillation period is mainly determined by the heating rate. In the model, the dominant subsurface warming is due to horizontal advection, which determines an interdecadal time scale (the mechanism is summarized in Fig. 8).

One would still like to put these mechanisms into some conceptual framework. Welander (1982, 1986) proposed simple flip-flop and loop thermohaline oscillation models. The basic idea of flip-flop oscillations is that the system is unable to reach an equilibrium, so it has to oscillate between equilibrium states. The flip-flop model analyzed by Welander (1982) gives

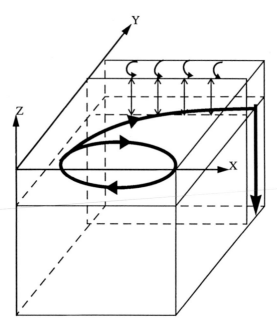

Figure 8: *A schematic diagram for the advective and convective mechanism of ocean interdecadal thermohaline oscillations of Yin and Sarachik, 1995. Horizontal advective heating due to the warm northeastward current (heavy thick arrow lines) associated with deep water formation warms up the subsurface water and induces convection (thin arrow lines). Equatorward cyclonic surface flows (dark arrow lines) advect subpolar fresh water into convective region to end convection. Convective ventilation associated with inter-decadal oscillations release heat to the surface from the subsurface and transport fresh water from the surface to the subsurface.*

a longer time scale for haline process than that for thermal process. Guided by the advective and convective mechanism, Yin (1995) formulated a similar flip-flop model, in which thermal process has a longer time scale than saline process. The model demonstrates that the system oscillates between two unreachable equilibria: one without convection and the other with constant convection.

Results from single buoyancy variable models with fixed surface flux indicate that oscillations still exist with only one buoyancy process (Huang and Chou, 1994; Greatbatch and Zhang, 1995; Winton, 1995c). A "loop" oscillation can occur in a single variable ocean model as has been interpreted by Huang and Chou, 1994. Winton (1995c) has argued that oscillations in a single buoyancy model are actually the prototype of those in models with mixed BC's. Instead of a thermal instability of the heat transport by the northeastward currents, which is associated with deep water formation, he argues that oscillations arise from the instability of this current hitting a no normal flow boundary with subsequent Kelvin-like propagation of anomalies. He points out that sufficient weak damping is essential for the anomalies to propagate along the boundary, and in models with mixed boundary conditions a surface halocline near the northern boundary actually shields subsurface anomaly from being damped by surface fluxes. Since Kelvin-like propagation is the only way to release instability in fixed flux experiment, it is not clear whether this propagation is essential for the existence of oscillations unless feedbacks from this propagation are essential to the instability of the northeastward current associated with deep water formation.

Another type of decadal variability, completely different from that described above, is induced through the nonlinear dynamics of boundary currents. Spall (1995) has recently indicated in a model, how the Gulf Stream, interacting with the southward deep western boundary current fed by the sinking of North Atlantic Deep Water, can dynamically interact to produce decadal oscillations.

5.4 Longer Term Variability

We may be sure that our understanding of the climate system will be on considerably firmer foundation when the task of quantitatively modeling the paleo-record has been completed. At the disposal of the climate theorist are variability mechanisms in a broad range of time scales, from the ten thousand year time scale of insolation changes and ice sheet dynamics down to the interannual time scale of low frequency atmospheric variability. The ocean is quite flexible in this regard, with time scales for relevant processes ranging from seasonal (mixed layer) to thousands of years (vertical diffusion).

We have already seen the record in the Greenland ice cores (Fig. 3). A plausible hypothesis for the temperature signal recorded by the ice cores is that it reflects the variations of surface heat loss associated with variations of intensity of North Atlantic Deep Water production. There is evidence that the low nutrient signal of the NADW diminished during the cold Younger Dryas period (Boyle and Keigwin, 1987) and more complete evidence that a nutrient depleted intermediate water was formed instead of deep water during the Last Glacial Maximum (Duplessy et al, 1988). Observations also show that atmospheric ^{14}C was quite steady during the Younger Dryas period (Broecker, 1994). Since NADW is a major downward pathway for ^{14}C, ^{14}C would be expected to go up in the atmosphere when NADW was shut off, other things being equal, so it remains to explain ^{14}C variability in the NADW hypothesis.

In spite of this ambiguity in the paleo-record, the NADW hypothesis remains attractive because numerous modeling results show that the opposition of thermal and haline forcing involved in NADW formation is capable of producing multiple equilibria and internal variability. The most complete of these studies was performed by Manabe and Stouffer (1988) who showed that a coupled ocean-atmosphere model has two equilibria: one with and one without NADW formation. The difference in the two climates was most pronounced in the North Atlantic region where warmer sea surface and surface air temperatures and reduced sea ice accompany NADW production. Manabe and Stouffer show that the freshening of the North Atlantic in the NADW off mode is due to the increased residence time in the region of freshening rather than changes in the hydrological cycle – in other words, it is due to internal ocean processes rather than a coupled ocean-atmosphere interaction. If we accept that these two climate modes underlie the paleo- record, we are faced with two questions: (1) how do transitions between the two modes occur?, and (2) why have such transitions not been observed since the end of the Younger Dryas, about 11,000 years ago, or in the last warm period 120,000 years ago?

The most obvious place to look for answers to these questions is in the freshwater flux into the North Atlantic. During glacial times, ice sheets enhanced the capacity for freshwater storage on land as well as the potential for pulse-like release. There is evidence for such impulsive freshwater inputs but it is hard to reconcile with direct forcing of mode switches. First consider the period of deglaciation that began 14,000 years ago. The sea level record from Barbados corals shows that the warm periods before and after the Younger Dryas contained meltwater spikes with maximum melting rates near 0.4 Sv (Fairbanks, 1989). The melting was considerably reduced during the cold Younger Dryas period. This suggests a negative feedback: reduced ocean heat transport leads to reduced melting allowing increased overturning. Impulsive freshwater inputs also occur during the heart of the glacial period. Recently, layers of ice-rafted debris associated with enormous calving events off of the Laurentide ice sheet have been discovered in the subpolar North Atlantic (see review in Broecker, 1994). These "Heinrich" events are typically spaced at 10,000 year intervals and occur during the cold part of the Dansgaard-Oeschger (D-O) cycles. The volume of ice involved in these discharges has been estimated to be nearly half of the current Greenland ice sheet or 5% of the Laurentide Ice sheet at its maximum (MacAyeal, 1993). MacAyeal estimates the associated freshwater flux at 0.16 Sv (about 1/2 of the current net freshening of the subpolar North Atlantic). The Heinrich events occur within the D-O cold phases and precede warm phases, sometimes by a substantial interval. The H1 event, for example, preceded the warming at 12,700 ^{14}C years BP by some 1,400 ^{14}C years. Thus these large impulsive salinity forcings are not easily connected to changes in NADW production.

Another possibility is that the Dansgaard-Oeschger events arise as an internal response of the ocean to the changed environment of the glacial period. This hypothesis finds support from some very simple modeling experiments using simplified geometries– we shall describe one here. Winton and Sarachik (1993) produced and analyzed a millennial oscillation in a sector OGCM under steady mixed boundary condition forcing. These oscillations were named "deep-decoupling oscillations" because there are small and large heat fluxes associated with the absence and presence of deep overturning respectively. A variant of this theory (Winton, 1993) uses spherical coordinates in a general circulation model based upon the diagnostic horizontal momentum balance:

$$f\mathbf{k} \times \mathbf{v} = -\nabla p + A\nabla^2 \mathbf{v} + \text{wind stress}.$$

The reason for neglecting the inertia terms is that they are generally small in coarse-resolution runs and, by eliminating them, fast barotropic and internal gravity waves are filtered out allowing long time steps and long simulations. The model basin is a flat bottom, 60°-wide sector extending from 70^oS to 70^oN with a re-entrant channel south of 54^oS. The boundary conditions are zonally uniform and symmetric about the equator (Fig. 9). Mixed boundary conditions are used: surface temperatures are restored to a reference profile and fixed surface salinity fluxes are applied. The net freshening of the subpolar regions is about 0.3 Sv comparable to the observations of Schmitt, Bogden, and Dorman (1988). There are two reference temperature profiles – for the moment we shall use the profile labeled warm. The air-sea thermal coupling coefficient is 40 $Wm^{-2}C^{-1}$

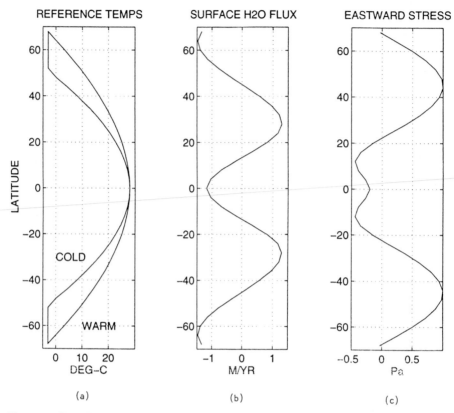

Figure 9: *Boundary conditions for the three dimensional experiment described in section 5.1*

(corresponding to about a 60 day restoring time for a 50 m mixed layer).

With the reference boundary conditions, a steady interhemispheric cell is generated with sinking in the northern North Atlantic and a consequent northward heat transport across the equator of about 0.5×10^{15} W. When the standard freshwater fluxes of Fig. 9b are increased by 25%, a steady interhemispheric cell is also obtained. When the freshwater fluxes are increased by 50%, however, oscillations are produced. Figure 10 shows the heat flux out of the subpolar North Atlantic over the oscillations. This is the quantity that presumably forces the climate changes recorded in the ice cores and the shape is remarkably similar to the ice core ^{18}O record: century/millennial time scale warm and cold phases are separated by decadal

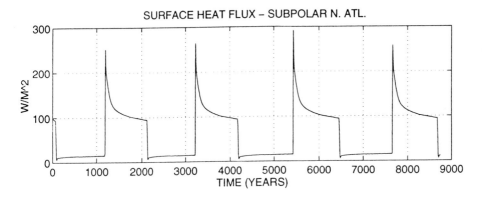

Figure 10: *Upward heat flux out of the subpolar North Atlantic over deep- decoupling oscillations.*

time scale transitions and there is progressive cooling over the warm phase.

The small and large heat flux phases of the model oscillations are associated with the absence and presence of deep overturning respectively. There are two phases involved in deep decoupling oscillations: one with a well developed meridional overturning (a coupled phase), and the other without or with little overturning (a decoupled phase). During a coupled phase, meridional overturning and deep ocean temperature decrease, while low latitude salinity increases. These changes occur progressively until a transition into a decoupled phase occurs suddenly due to the cessation of convection. Once the system is in a decoupled phase, a halocline in the polar region forms and convection weakens; afterward, the deep ocean warms diffusively, while at the same time, heat is mixed poleward beneath the halocline by gyre mode horizontal circulations; the warming eventually destabilizes the halocline to renew convection abruptly, and the model proceeds into a deep coupled phase again: the oscillation continues in a self-sustaining fashion (Winton and Sarachik, 1993).

The abrupt transitions between the phases are due to the rapid change in the convective state at high latitudes as a halocline is formed or breaks down. The longer time scale of the phases themselves comes from the imbalance of the advection and diffusion terms in the heat balance which leads to a trend in basin mean temperature. Heat that accumulates during the deep-decoupled phase is mixed poleward beneath the halocline by gyre mode (horizontal plane) circulations, eventually destabilizing the halocline to renew convection (it may be noted that two-dimensional models lack these gyre mode circulations). The details of this process vary between particular oscillations – in some oscillations convection renews abruptly as the destabilizing effect of the subsurface warming is enhanced by the nonlinearity of the equation of state (Winton, 1993); in other oscillations, an intermediate water forming cell deepens and strengthens breaking down the halocline by advective processes (Winton and Sarachik, 1993). In the subsequent deep- coupled phase, heat is flushed out of the basin and the initial burst of poleward heat transport slowly weakens. Eventually the fixed freshening at high latitudes dominates the reduced upward heat flux to stabilize the water column and reform the halocline. The warming of the basin in the deep- decoupled phase is associated with a storage of potential energy which is later converted to kinetic energy in the deep-coupled phase (Winton, 1995a).

Although diffusive warming of the deep ocean and associated poleward advective mixing of heat is the fundamental cause of the transition to the coupled phase as shown by heat and salt budget calculations, there are three important processes through which this transition can be triggered. The first is the maintenance of shallow convection and intermediate water formation while the deep ocean is warming during the decoupled phase; the convection and overturning strengthen and deepen gradually, importing more salt into the sinking regions and helping the onset of convection. The second is related to the nonlinearity of sea water density with temperature. As the deep ocean warms, the decrease in density grows nonlinearly with temperature, which often leads to an abrupt destabilization of the polar stratification. The third is associated with the convective eddies which depart from a salty western boundary current and propagate eastward through the polar halocline, and deposit salt in the sinking region. After a transition into the coupled phase, heat is flushed out of the basin over high latitudes, and poleward heat transport slowly weakens after an initial burst. Eventually the fixed surface freshening at high latitudes dominates the reduced upward heat flux to stabilize the water column and reform the halocline (Winton and Sarachik, 1993; Winton, 1993).

Note that the transition between the coupled and decoupled phases happens on a decadal time scale. The transition corresponds to rapid changes in convection at high latitudes associated with the formation and destruction of a halocline. Presumably the time scale of transition is the dynamic adjustment time of the polar ocean to a sudden onset of convection, which is decadal according to baroclinic geostrophic adjustment scaling.

In classifying deep-decoupled millennial oscillations, one can use the concepts of low order dynamic systems. Winton (1993) has constructed a three box model to mimic these oscillations. For a weak fresh water input, the model is at an equilibrium resembling the normal thermohaline circulation with steady convection at high latitudes. When the fresh water input is large, a permanent halocline prevents high latitude convection, and the model is at an equilibrium corresponding a state of collapsed thermohaline circulation. For an intermediate fresh water flux, the model oscillates, but unable to settle into either of these two equlibria. The model also has an important characteristics for the oscillations: the adjustment time for the high/deep salinity gradient is shorter than that for the temperature. As shown in the model, the period of these oscillations is determined by vertical diffusive time, which is millennial.

Although the internal oscillations of the sector models are encouraging, three limitations should be borne in mind. Two of these have to do with the way the oscillations work – in particular with the way the halocline is broken down. This is accomplished by the mixing of heat that has entered the basin diffusively during the deep-decoupled phase up to high-latitudes beneath the halocline as discussed above. Thus the oscillatory nature of this model is enhanced by 1) the inhibition of alternative deepwater sources that might import heat making it unavailable to the halocline covered region, and 2) a large value of the vertical diffusivity to quickly provide heat before the halocline becomes too strong to be broken down. The role of the vertical diffusivity is particularly troublesome because large diffusivities also favor thermally direct circulations requiring a larger freshwater forcing to induce oscillations. The vertical diffusivity of 0.5 cm^2/sec used here is several times larger than the value thought to be appropriate for today's ocean.

The third limitation has to do with the way the oscillations are induced: for a given vertical diffusivity and wind stress this is accomplished by increasing the freshwater forcing. This would seem to call upon the cold glacial climate to have a larger atmospheric and riverine meridional transport of water than the warm interglacial climates that exhibit steady thermohaline circulations. Equilibrium climate simulations with various levels of CO_2 by Manabe and Bryan (1985) indicate that the opposite is true: The cold, low CO_2 climates are accompanied by reduced mid-latitude precipitation minus evaporation. By itself this would favor increased thermohaline overturning in glacial times. The Manabe and Bryan low CO_2 experiment was actually accompanied by reduced thermohaline overturning which they attributed to nonlinearity in the equation of state and reduced midlatitude heat loss through the expanded sea ice cover. Another potentially important effect is the influence of ice sheets upon the zonal distribution of precipitation. To our knowledge, the ice age atmospheric simulations have not been analyzed to determine the net change in precipitation minus evaporation over the North Atlantic ocean and its drainage basin.

We may look to a different mechanism to explain these glacial thermohaline oscillations. A more promising correlation seemingly exists between cooling and thermohaline stability than between freshwater forcing and thermohaline instability. The warm climates of the interglacial Holocene and Eemian periods were apparently accompanied by steady thermohaline circulation, while the intervening cold period exhibits instability nearly throughout its duration. Atmospheric models have shown that the ice sheets exert a powerful cooling influence upon the subpolar North Atlantic (Kutzbach and Wright, 1985; Manabe and Broccoli, 1985). Anti-cyclonic low level winds over the ice sheets funnel Arctic air into this region. The strength of this circulation has been reported to be sensitive to the size of the ice sheet (Kutzbach and Ruddiman, 1993). Bond et al (1993) have noted that the D-O events are grouped into long term cooling cycles bracketed by Heinrich events. The reduction in size of the Laurentide ice sheet after a Heinrich event is consistent with the subsequent appearance of the warmest part of these "Bond cycles." It is noteworthy that the interstadials decrease in duration with the cooling in the Bond cycles – suggesting a correlation between surface cooling and thermohaline instability within the glacial period as well as between the glacial and interglacial periods. In Section 7 we will address these problems by using some simple models to investigate the way changes in thermal forcing affect the level of freshening required to break down a steady thermally direct deep overturning.

Let us close this section by noting a distinct type of oscillation associated with a normal thermohaline overturning, which is often called a loop oscillation. A salinity anomaly can be traced as it passes through the deep overturning cell. The overturning is enhanced when it is in the polar region, and retarded when it is in the low latitudes. When the salinity anomaly dominates, it can grow or maintain itself against dissipation through surface fluxes by its effect on the overturning (advective feedback). When the salt (fresh) anomaly is at low (high) latitudes, the circulation slows down and the flux boundary condition becomes more important for determining salinity, thereby enhancing the anomaly. When the polar (low latitude) ocean has the salty (fresh) anomaly, overturning increases, and the flux boundary condition, which is now acting to reduce the anomaly, becomes less effective. This mechanism has also been noted by Mikolajewicz and Maier-Reimer (1990) in their model response to stochastic forcing. Simple model interpretations of loop oscillations have been carried out (Welander, 1986; Winton and Sarachik, 1993), which usually determine the period of loop oscillation as the time it takes the anomaly to travel around the mean circulation loop. Note that the temperature variability involved in loop oscillations are small since the salinity anomaly is the dominant factor in advection feedback, hence variability in the surface heat flux is also small.

6 Coupled Atmosphere-Ocean Modes

The practical problems of coupled atmosphere-ocean modeling make it difficult to simulate long runs of coupled models. Low resolution atmospheric models (R15–approximately $5^o \times 8^o$) have now been used to examine variability in the climate system represented by these models. There have recently been a number of relatively long simulations using such low resolution coupled models and this sections will be based on those results.

In order to have a truly coupled atmosphere-ocean mode, the oceanic SST has to induce atmospheric heat and momentum fluxes that reinforce the SST. The problem is that most higher latitude modes require higher latitude SST to affect the atmosphere and this influence has been difficult to detect. Lau and Nath (1994) and Graham et al. (1994) have indicated that the tropical SST dominates the effects on the atmosphere over the mid-latitude SST.

It should be realized that the definition of a coupled mode, a mode that depends critically on the interaction of the atmosphere and ocean, is far more difficult in mid-latitudes than our favorite example of ENSO. A great deal of analytic work on ENSO has shown that coupled ENSO modes arise solely from the interaction of the atmosphere and the ocean and would not exist without this interaction (e.g. Hirst, 1986). In midlatitudes, we have two possibilities that can masquerade as a coupled mode: the ocean can have decadal variability which then, through its SST expression, drives atmospheric motions on decadal scales, or, more likely, the atmosphere can

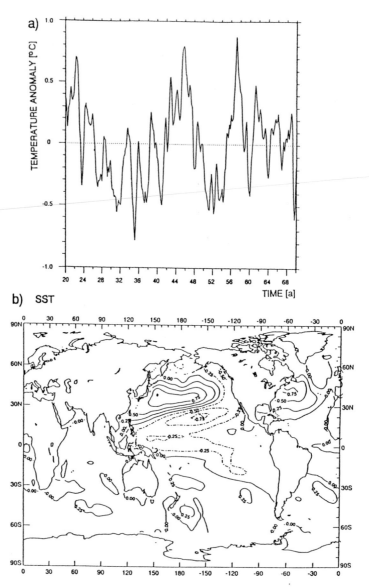

Figure 11: *The interdecadal coupled mode of Latif and Barnett (1994). (a) Time series of model SST in region $150°E - 180°E$ and $25°N - 35°N$ (b) SST regressed on time series in (a).*

have decadal modes which, through the working of the wind on the water, can drive decadal motions in the ocean. It would be very difficult to tell such driven motions from modes.

The only candidate for a truly coupled decadal mode that we know of is Latif and Barnett (1994). Fig. 11 shows the SST in the North Pacific for this mode. In view of our previous discussion of atmospheric forcing of the ocean, the question naturally arises whether or not this (model) mode is in fact a mode or is it the result of forcing the ocean by natural variability in the atmosphere. One way to decide is to examine the natural variability of the atmosphere in the absence of interactive SST, i.e. in response to climatological SST and seeing whether or not the coupling significantly enhances the atmospheric variability. Another way would be to see if the prediction skill exceeds the persistence skill. Since neither of these has yet been done, we will simply remark the claim of a coupled mode and reserve judgment.

In a coupled model, Delworth et al. (1994) analyzed a 600 year simulation with interdecadal irregular oscillations in meridional overturning. One would have expected this to be a coupled mode yet the dynamics turns out to be very much like ocean internal variability. The power spectrum of SST shows a peak at a broad time scale around 50 years. A regression analysis reveals that the oscillation is driven by density anomalies in the confined sinking region of the thermohaline circulation combined with much smaller density anomalies of opposite sign in the more extensive rising regions. They proposed the following mechanism: beginning with a weak THC, the reduced heat transport associated the this weak THC results in the development of a cold, dense pool through the top 1 km in the middle of the North Atlantic; the cold, dense pool has an associated cyclonic circulation; the anomalous flow across the mean salinity contours associated with this gyre circulation results in an enhanced salt transport into the sinking region, thereby increasing the salinity and density of the sinking region; this processes is associated with a strengthening of the THC until it reaches its maximum. The result of this strengthening of the THC is that the northward advection of heat is enhanced, thereby generating a pool of warm, less dense water throughout the top 1 km in the middle of the North Atlantic; as this warm pool develops, an anomalous anticyclonic gyre circulation (i.e. a weakened cyclonic circulation) is created, which acts to decrease the salt transport into the sinking region and weakening the THC.

The essence of this mechanism is the 10 year lead of saline to heat transports and associated "gyre mode" circulation (deviation from the zonal mean). It is not clear why there should be an anomalous anticyclonic gyre circulation associated with a warm pool. It should be noted that this is a consistent picture with the advective and convective mechanism (Yin and Sarachik, 1995), in which at the weak phase of convection, there is an anomalous warm water, anomalous high overturning, and anomalous high surface pressure related to anomalous surface divergence, so an anticyclonic horizontal circulation.

7 Sudden Transitions

Perhaps the simplest model of the counteracting effects of thermal and haline forcing on the control of the buoyant overturning is the Stommel two–box model (Stommel, 1961; Marotzke, 1989). Here we use a variant of this model that, after non-dimensionalizing, has two free parameters: F, the magnitude of the salt flux from the high-latitude to the low latitude box, and P, a coefficient for restoring the thermal difference between the boxes to a reference value (Fig. 12a). This model is identical to that of Marotzke except that we allow the temperature difference to deviate from its reference value. The equations for the differences in salinity, ΔS, and temperature, ΔT, between the low and high latitude boxes are:

$$\frac{d\Delta S}{dt} = -2\|\Delta T - \Delta S\| \cdot \Delta S + 2F$$

and

$$\frac{d\Delta T}{dt} = -2\|\Delta T - \Delta S\| \cdot \Delta T + P(1 - \Delta T).$$

Fig. 12b shows the magnitude of the overturning, $\Delta T - \Delta S$, normalized by its magnitude with the same thermal restoring, P, but with no freshwater forcing (F=0) in the region of the parameter space with a stable, thermally dominated equilibrium. The portion of the solution space without such an equilibrium is shaded. Two aspects of this figure are noteworthy: (1) With stronger thermal restoring, more freshening can be sustained by a thermally dominated equilibrium; (2) A substantial reduction of the overturning (more than 40%) occurs before the thermally dominated solution becomes untenable. Both of these properties are evidence of an advective instability: the salinity gradient retards the overturning, increasing the influence of the freshwater flux boundary condition relative to internal mixing, thereby further increasing the salinity gradient. Thus the advective instability works by countering the effect of the thermal torque upon the overturning. With stronger thermal restoring, a larger freshwater flux is required to retard the overturning and obtain significant positive feedback from the salinity boundary condition.

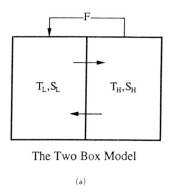

The Two Box Model

(a)

OVERTURNING/OVERTURNING(F=0)

Figure 12: *The two–box model: (a) configuration, and (b) results (contours are over-turning magnitude normalized by its value without freshwater forcing (F=0); the shaded region has no thermally direct steady state.*

Now consider a non-rotating, two-dimensional, viscous model with convective adjustment. Models of this sort make a considerable step in the direction of realism by forming their own thermocline through advective-diffusive processes and modeling the important nonlinear dependence of vertical mixing upon vertical density gradient (convection). A version of this model without salinity effects was used by Winton (1995b). Here we restore surface temperatures to a half-cosine reference profile, with coefficient P, and apply a half-cosine shaped salt flux pattern with magnitude, F. Fig.13 shows the result of a series of experiments designed to determine the maximum freshwater forcing that could be sustained by a steady thermally direct cell over a range of thermal restoring coefficients. Although, P in this model is not entirely analogous to P in the two–box model since only the surface temperature gradient is directly under its influence, it is rather surprising to see that this model becomes more susceptible to freshwater forcing with stronger clamping to the reference temperature profile over most of the range of P. For each experiment we have noted the percentage reduction in overturning magnitude from $F = 0$ to $F = F_{crit}$, the point where the cell breaks down. In the low P range where F^{crit} is increasing with P, there is a substantial reduction indicating a role for advective instability of the type characterizing the two-box model. In the region where F^{crit} is decreasing with P, the reduction goes to zero. In this region, another kind of instability must be responsible for the breakdown of steady, thermally direct overturning. We will argue that this is a convective instability.

One piece of evidence for convective instability comes from applying a different values of P over the two halves of the basin. When this is done, it is found that the critical level of freshwater forcing is not sensitive to the value of P in the warm, "low-latitude", half, of the basin. This is consistent with the time scales involved in the problem. In the low-latitude half, heat penetrates diffusively through the thermocline. The time scale for this process is the flushing time by the thermal overturning (Winton, 1995b). This is long compared to the time scales of the processes that control surface, or equivalently, mixed layer temperature: months for air-sea exchange processes to several years for the top of atmosphere radiative balance. Notice that, starting near mid basin, shallow convective layers are found; these deepen moving toward the cold end, finally striking the bottom at the boundary (Fig. 14a). Convection is modeled as an instantaneous process; so in this convecting region, the heat balance is affected by the strength of the surface temperature restoring.

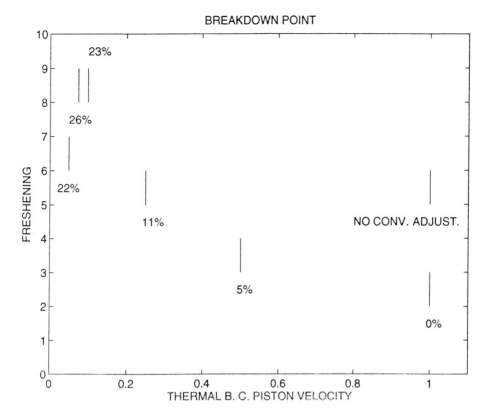

Figure 13: *Two-dimensional model sensitivity to surface thermal restoring. The maximum freshening sustained by a steady thermally direct circulation falls upon the vertical lines. The percentages refer to the reduction in overturning from an experiment without freshwater forcing to the experiment with just subcritical freshwater forcing (the bottoms of the vertical lines).*

Fig. 13 also shows the result of a series of experiments performed with a large surface restoring coefficient ($P = 1$), but without convective adjustment. In this case, the cell was able to sustain nearly twice as much freshening as in the convective adjustment case. This sensitivity is surprising in light of the fact that the deep water is substantially colder in the convective adjustment case, and so the no-convection circulation has forgone a substantial portion of the meridional temperature gradient available from the boundary condition. The cold deep water turns out to be exactly the source of instability for the convectively adjusted circulation. Since the deep water has had the benefit of convective cooling, the circu-

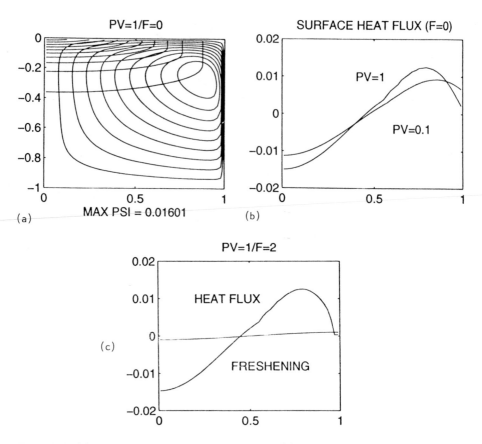

Figure 14: *(a), Two-dimensional model circulation, (b) surface heat fluxes with P=1 and P=0.1, and (c) surface heat and freshwater flux in the just sub-critical experiment with P=1 (F=2).*

lation is committed to maintaining steady convection. Otherwise, warmer candidate deep water will not sink rapidly and runs risk of taking on so much freshwater that it may never sink. Thus salinity effects can break down a thermal circulation in two ways: by an adverse meridional torque (as in the two-box model) and by inhibiting high-latitude cooling through stratification.

Now let us examine the details behind the sensitivity to P in Fig. 13. A rough prerequisite for halocline formation is that the downward buoyancy flux due to surface freshening exceed the upward flux due to surface heat loss. This condition is only approximate because it does not consider horizontal effects that might come into play when the state of the stratification changes. Note from Fig. 14a that thermal advection is dominantly vertical at low latitude and horizontal at high latitude. Fig. 14b shows the upward heat flux with $F = 0$ for steady states with $P = 0.1$ and $P = 1$. In the $P = 1$ case the temperature of the convecting layers is effectively pinned at the overlying reference temperature. Since the reference profile is a half-cosine curve, its meridional derivative goes to zero at $y = 1$ (the "northern" boundary). Near $y = 1$, the flow is predominantly meridional, and so the convergence of the heat transport and the surface heat flux go to zero. With the smaller surface restoring coefficient, heat is released through the surface more "frugally", and the water sinks with a temperature above the coldest reference temperature. This excess heat may be thought of as left over convecting capacity and these considerations highlight the importance of accurately modeling the temperature of the deepwater when the issue of thermohaline stability is being addressed.

Fig. 14c shows the surface heat flux for the $P = 1$ case with $F = 2$, just below the critical level of freshening. The small amount of freshening is able to dominate the upward heat flux and stratify the water column at the $y = 1$ boundary. With slightly larger freshening $(F = 3)$, freshwater pools at the surface and continually interferes with the sinking branch of the circulation to the south. The sinking moves to lower latitudes and ventilates shallower depths as the halocline expands. This process has been termed a "halocline catastrophe". It should be pointed out that the two-dimensional model, particularly this one with isotropic diffusion, exaggerates the sensitivity to freshening because of the neglect of horizontal plane motions that can provide an alternative outlet for freshening aside from incorporation into deepwater [see Winton and Sarachik (1993) for a comparison of the sensitivity of similarly formulated two- and three-dimensional models].

The sensitivity described above is associated with the flattening of the reference temperature profile as the y=1 boundary is approached. This might be thought a rather artificial aspect of the forcing. However, it might also be argued that the thermal boundary condition seen by the actual ocean flattens near the poles due to the fact that seawater does not go below freezing and the thermal buoyancy flux associated with a given heat flux decreases at low temperatures due to the temperature dependence of the thermal expansion coefficient. Beyond the latitude where the water column, or perhaps a substantial portion thereof, is brought to near

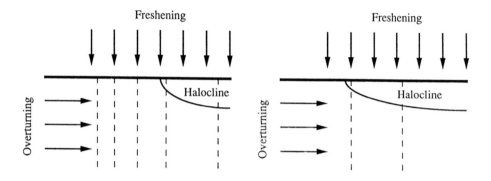

Figure 15: *Cooling induced convective instability. Horizontal mixing with the larger halocline covered portion of the net surface freshening region represents a larger threat to steady deep water formation in the cold climate case where horizontal advection is less effective at supplying heat to drive convection.*

freezing temperature, horizontal advection is incapable of supplying a surface heat flux to balance the stratifying effect of a downward freshwater flux and a halocline is formed. Fig. 15 depicts a hypothetical scenario for thermohaline instability based upon this effect. The situation at top corresponds to today's warm climate where a substantial portion of the region of net freshening in the North Atlantic (north of roughly 40^oN) has temperatures above freezing. Convection in this region mixes down the surface freshening and the bits of halocline that are imported from the north by Ekman drift, gyre circulations, coastal currents, etc. In the cold climate case below, a larger fraction of the surface freshening region has near freezing temperatures, and is stratified. Now horizontal mixing of halocline into the convecting region becomes a larger threat to deepwater formation.

Now we explore the hypothesis developed in the last section with the three- dimensional rotating model presented previously. Recall that the warm reference temperature profile (Fig. 9a) gave a steady state with an interhemispheric circulation sinking in the North Atlantic. This reference experiment was not particularly close to the critical level of freshening – between 25% and 50% increased freshwater forcing was required to break the steady circulation down into deep-decoupling oscillations. Now we perform a second experiment, identical to the reference experiment except that the

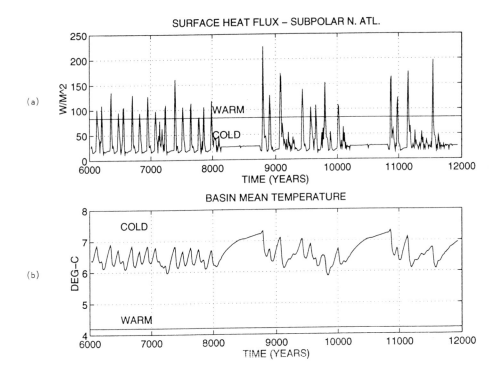

Figure 16: *Three-dimensional model results from warm and cold boundary condition experiments: (a) upward heat flux out of the subpolar North Atlantic, and (b) basin mean temperature.*

cold reference temperature profile shown in Fig. 9a is used. The temperature range is identical to the warm experiment, but the area averaged reference temperature is colder. The result is the irregular variability depicted in Fig. 16. Halocline formation continually interferes with deepwater formation. As a result the basin mean temperature is actually warmer with the cold boundary condition. This result indicates that the same fresh water input can have a large effect at colder temperatures: cooling is potentially a powerful mechanism for inducing thermohaline circulation instability.

The kind of instability that has been induced is not entirely of the deep- decoupling variety discussed above. There is basin warming during two extended decoupled phases, and sporadic warming in the intervening periods. These fluctuations in basin mean temperature are symptomatic of the deep- decoupling type of variability. An examination of the details of the circulation changes, however, shows that there are elements of the pure decadal variability present as well. This kind of variability will be discussed next.

We can "tune in" the pure decadal variability in the cold boundary condition experiment described above by halving the surface temperature restoring coefficient. Recall that this moves the model in the direction of stability. The result is a regular oscillation with about a 20 year period (Fig. 17a). Figure 17b shows the details of the oscillation. The velocity vectors at 175 m depth are overlain by contours of vertically averaged density. The first panel shows a zonally oriented baroclinic jet impinging upon the eastern boundary between 30^oN and 50^oN. A portion of this jet splits off and propagates in the Kelvin wave direction along the northern boundary. Light water is left on the northern boundary in the wake of this disturbance, reversing the direction of the zonal flow along the northern boundary.

Similar propagating features can be found in the mixed boundary condition decadal variability studied by Weaver and Sarachik (1991) and Yin and Sarachik (1995). The mechanism for the variability also appears to be the same as that for internal decadal variability found under a fixed flux boundary condition for a single buoyancy variable (Winton, 1995c). Apparently, the mixed boundary condition variability does not involve the interplay of heat and salt nor either of the advective and convective instabilities described above. It is rather a product of the adjustment of baroclinic currents impinging on weakly stratified coasts to the no normal flow boundary condition. This adjustment spins up deep warm and cold features on the boundary that, upon reaching sufficient amplitude, propagate slowly in the Kelvin wave direction (Winton, 1995c). The property shared by the mixed boundary condition and fixed flux boundary condition experiments that permits this variability to occur is weak damping. In the mixed boundary condition case, a halocline shuts off convection and the timescale for modification of the water column reverts from the short timescale for the surface boundary condition (about 10 years) to a much longer diffusive timescale. The propagation is particularly clear in the case presented here because of the uniform coverage of the high-latitude ocean by the halocline.

8 Mechanism and Predictability

We can define three types of predictability. Deterministic predictability is best known from our experience with weather prediction and is limited to

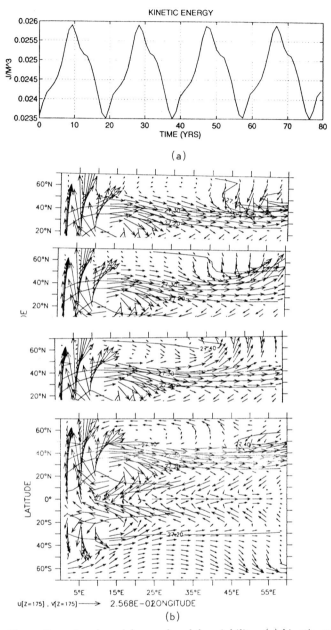

Figure 17: *Three-dimensional model pure decadal variability: (a) kinetic energy, and (b) horizontal circulation at 175 m at four year intervals.*

about two weeks, by inevitable growing errors in the atmosphere–there is no way to deterministically predict the motions of the atmosphere beyond this limit of deterministic predictability no matter how small the initial error.

The second type of predictability is governed by the slowly varying boundary conditions to which the atmosphere responds. If we can predict SST, or soil moisture, or ice and snow cover, then we will be able to know the statistics of the atmosphere in equilibrium with these boundary conditions. The system is initialized by initializing those factors that affect the boundary condition: in practice, this is only possible for the SST boundary conditions and those parts of the ocean that affect the surface in a given prediction time must be initialized. This is the type of prediction that has been so successful for SST in the tropical Pacific characteristic of ENSO (e.g. Latif et al., 1994): only the upper ocean is initialized since only the near-surface ocean will affect SST on predictions times of order one year. Because the maritime tropical atmosphere is so tightly coupled to SST, skillful forecasts of rainfall around the tropical Pacific basin can be made in terms of the SST predictions.

The third type of predictability is that typified by the predictions made for greenhouse warming. No aspect of the climate system is initialized but scenarios for the radiatively active constituents of the atmosphere are specified as function of time and the statistics of the entire atmosphere-ocean- land system are predicted in response to the changing atmospheric constituents.

Thus for prediction of the first type, we predict what the precise state of the atmosphere is for, say, next Wednesday. For prediction of the second type, we might predict the March averaged rainfall over Peru nine months in advance. For the third type of prediction, all we can say is that summer temperatures over the US will generally be warmer, but we can't offer a prediction for the specific month, August 2050, say. In this type of prediction, the magnitude and phase of the natural cycles are not predicted for a specific time. Only in the second type of prediction can we capture the magnitude and phase of a decadal-to-centennial cycle and we will therefore confine ourselves to that type of prediction.

In order for the coupled atmosphere-ocean system to be predictable we must be able to predict the slowly varying boundary conditions and the nature of the mechanism determines the possible range of predictability. When the mechanism for low frequency variability is due to random forcing of the boundary conditions by the atmosphere, we can expect predictability due only to the slowness of the variation: we would therefore expect the limit on the prediction time to be given by persistence, i.e. by the autocorrelation time of the boundary condition. Clearly only those mechanisms we have discussed that intrinsically involve the ocean in a deterministic way will lead to predictability beyond the persistence time. Thus if most decadal-to- centennial variability is due to random forcing of the ocean by the atmosphere, we have no hope in ever predicting the future SST beyond persistence. If however the mechanism of variability involves the ocean either as part of a coupled mode or by internal ocean variability as described in Sections 4 and 5, then there is some predictability of the future ocean dynamics and we have some possibilities of predicting the future SST for times longer than the persistence time. One way of unambiguously identifying a coupled atmosphere-ocean mode, for example, is this enhanced predictability beyond the persistence time.

We have to know those aspects of the ocean that will affect the SST on decadal-to-centennial time scales and initialize those aspects. Unfortunately, we don't know how to answer to this basic question. We have been able to study, by means of tracer experiments (e.g. with Clorofluorocarbons) the inverse problem, how water proceeds from the surface into the interior. We have not been able to answer the basic prediction question of which water reaches the surface in a given time and how this water reaches the surface. If the ocean can be initialized, and if the future evolution of the coupled system is at least partially determined by this dynamical evolution of this initial state, then we have hope of some skill in climate prediction years into the future.

Acknowledgements

This work was supported by a grant from the NOAA/Office of Global Programs to the Stanley P. Hayes Center of the University of Washington. MW was supported by a UCAR ocean modeling post-doctoral fellowship while 50% of F.L.Y.'s time was supported by the US Department of Energy's NIGEC through the NIGEC Regional Center at Univ. of California, Davis (DOE Cooperative Agreement No. DE-FC03-90ER61010): financial support does not constitute endorsement by DOE of the views expressed in this article. The invaluable assistance of the Margaret Black Lab is gratefully acknowledged. Conversations with Bob Charlson, Todd

Mitchell, Gregor Nitche, Jim Renwick, Mike Wallace and Yuan Zhang have helped the authors in writing this paper. This paper is JISAO contribution 335.

References

ALLEY, R. B., D.A. MEESE, C.A. SHUMAN, A.J. GOW, K.C. TAYLOR, P.M. GROOTES, J.W.C. WHITE, M. RAM, E.D. WADDINGTON, P.A. MAYEWSKI, AND G.A. ZIELINSKI, 1993: Abrupt increase in Greenland snow accumulation at the end of the Younger Dryas event, *Nature*, **362**, 527-529.

BATTISTI, D.S., AND A.C. HIRST, 1989: Interannual variability in the tropical atmosphere/ocean system: influence of the basic state and ocean geometry. *J. Atmos. Sci.*, **46**, 1687-1712.

BOND, G., W. BROECKER, S. JOHNSEN, J. MCMANUS, L. LABEYRIE, J. JOUZEL, AND G. BONANI, 1993: Correlations between climate records from North Atlantic sediments and Greenland ice. *Nature*, **365**, 143-147.

BOYLE, E. A., AND KEIGWIN, L. D., 1987: North Atlantic thermohaline circulation during the past 20,000 years linked to high latitude surface temperature, *Nature*, **330**, 35-40.

BRADLEY, R.S., AND P.D. JONES, EDITORS, 1992: Climate Since A.D. 1500., Routledge, *679 pp.*

BROECKER, W. S., 1994: Massive iceberg discharges as triggers for global climate change, *Nature*, **372**, 421-424.

BRYAN, F., 1986: High latitude salinity effects and inter-hemispheric thermohaline circulations. *Nature*, **323**, 301-304.

CHAPPELLAZ, J., AND COAUTHORS, 1993: Synchronous changes in atmospheric CH4 and Greenland climate between 40 and 8 kyr bp. *Nature*, **366**, 443-445.

CHARLSON, R.J., J. LANGNER, H. ROHDE, C.B. LEOVY AND S.G. WARREN, 1991: Perturbation of the northern hemisphere radiative balance by backscattering from anthropogenic sulfate aerolsols. *Tellus*, **43AB**, 152-163.

CHARLSON, R.J., AND T.M.L. WIGLEY, 1994: Sulfate aerosol and climatic change. *Sci. Am.*, **270**, 48-57.

CLIVAR, 1995: CLIVAR Science Plan. World Climate Research Program, *WCRP-89, 189pp*

DANSGAARD, W., S.J. JOHNSEN, H.B. CLAUSEN, D. DAHL-JENSEN, N.S. GUNDESTRUP, C. U. HAMMER, C.S. HVIDBERG, J. P. STEFFENSEN, A.E. SVEINBJRNSDOTTIR, J. JOUZEL, AND G. BOND, 1993: Evidence for general instability of past climate from a 250-kyr ice-core record, *Nature*, **364**, 218- 220.

DELWORTH, T., S. MANABE, AND R. J. STOUFFER, 1993: Interdecadal variability of the thermohaline circulation in a coupled ocean-atmosphere model. *J. Climate*, **6**, 1993-2011.

DESER, C., AND M. L. BLACKMON, 1993: Surface climate variations over the North Atlantic Ocean during winter: 1900-1989. *J. Climate*, **6**, 1743-1753.

DUPLESSY, J. C., N. J. SHACKLETON, R. G. FAIRBANKS, L. LABEYRIE, D. OPPO, AND N. KALLEL, 1988: Deepwater source variations during the last climatic cycle and their impact on the global deepwater circulation. *Paleoceanography,* **3,** 343-360.

EDDY, J.A., 1976: The Maunder Minimum. *Science,* **192,** 1189-1202.

FAIRBANKS, R. G., 1989: A 17,000 year glacio-eustatic sea level record: Influence of glacial melting rates on the Younger Dryas event and deep ocean circulation, *Nature,* **342,** 637-642.

FELLER, W., 1968: An Introduction to Probability Theory and its Applications. Vol 1, Third Edition, John Wiley, *509 pp.*

GRAHAM, N. E., T. P. BARNETT, R. WILDE, M. PONATER, S. SCHUBERT, 1994: On the role of tropical and midlattitude SSTs in forcing interannual to interdecadal variability in the winter northern hemisphere circulation. *J. Climate,* **7,** 1416-1441.

GREATBATCH, R. J., AND S. ZHANG, 1995: An interdecadal oscillation in an Idealized ocean basin forced by constant heat flux. *J. Climate,* **8,** 81-91.

HANSEN, J., A. LACIS, R. RUEDY, AND M. SATO, 1992: Potential climate impact of Mount Pinatubo eruption. *Geophys. Res. Lett.,* **19,** 215-218.

HANSEN, J., A. LACIS, R. RUEDY, M. SATO, AND H. WILSON, 1993: How sensitive is the world's climate. *Nat. Geograph. Res. & Exploration,* **9,** 142-158.

HANSEN, J., M. SATO, R. RUEDY, A. LACIS, K. ASAMOAH, S. BORENSTEIN, E. BROWN, B. CAIRNS, G. CALRI, M. CAMPBELL, B. CURRAN, S. DE CASTRO, L. DRUYAN, M. FOX, C. JOHNSON, J. LERNER, M.P. MCCORMICK, R. MILLER, P. MINNIS, A. MORRISON, L. PANDOLFO, I. RAMBERRAN, F. ZAUKER, M. ROBINSON, P. RUSSELL, K. SHAH, P. STONE, I. TEGEN. L. THOMASON, J. WILDER, AND H. WILSON, 1995: A Pinatubo climate modeling investigation. In The Effects of Mt. Pinatubo Eruption on the Atmosphere and Climate., NATO ASI Series, *Editors G. Fiocco, D. Fua, and G. Visconti, Springer-Verlag, in press.*

HASSELMANN, K., 1976: Stochastic climate models. I, Theory. *Tellus,* **28,** 473- 485.

HIRST, A.C., 1986: Unstable and damped equatorial modes in simple coupled ocean-atmosphere models. *J. Atmos. Sci.,* **43,** 606-630.

HUANG, R. X., AND R. L. CHOU, 1994: Parameter Sensitivity Study of the Saline Circulation. *Climate Dynamics,* **9,** 391-409.

HURRELL, J. W., 1995: Decadal trends in the North Atlantic Oscillation: Regional temperature and precipitation. *Science,* **269,** 676-679.

INTERGOVERNMENTAL OCEANOGRAPHIC COMMISSION, 1992: Oceanic Interdecadal Variability, IOC Technical Series No. 40, *UNESCO, 37pp.*

IPCC, 1990: Climate Change: The IPCC Scientific Assessment. J. T. Houghton, G. J. Jenkins, J. J. Ephraums, Editors, *Cambridge University Press, 403 pp.*

IPCC, 1995: Climate Change 1994: Radiative forcing of Climate Change and An Evaluation of the IPCC 1992 IS92 Emission Scenarios. J.T. Houghton, L.G. Meira Filho, J. Bruce, H. Lee, B.A. Callender, E. Haites, N. Harris, and K. Maskell, Editors. Cambridge University Press, *325pp.*

JAMES, I.N. AND P.M. JAMES, 1989: Ultra-low frequency variability in a simple atmosphereic circulation model. *Nature,* **342,** 53-55.

JOHNSEN, S.J., H. B. CLAUSEN, W. DANSGAARD, K. FUHRER, N. GUNDESTRUP, C. U. HAMMER, P. IVERSON, J. JOUZEL, B. STAUFFER, AND J.P. STEFFEMSEN, 1992: Irregular glacial interstadials recorded in a new Greenland ice core. *Nature,* **359,** 311-313.

KARL, T.R., R.W. KNIGHT, G. KUKLA, AND J. GAVIN, 1995: Evidence for the radiative effects of anthropogenic sulfate aerosols in the observed climate record. In Aerosol Forcing of Climate, eds., R. Charlson and J. Heintzenberg, John Wiley & Sons .

KEIGWIN, L. D., W. B. CURRY, S. J. LEHMAN, AND S. JOHNSON, 1994: The role of the deep ocean in North Atlantic climate change between 70 and 130 kyr ago, *Nature,* **371,** 323-326.

KUSHNIR, Y., 1994: Interdecadal variations in North Atlantic sea surface temperature and associated atmosphereic conditions. *J. Climate,* **7,** 141-157.

KUTZBACH, J. E., AND W. F. RUDDIMAN, 1993: Model description, external forcing, and surface boundary conditions, In: Global Climates Since the Last Glacial Maximum, H. E. Wright, ed., University of Minnesota Press .

KUTZBACH, J. E., AND H. E. WRIGHT, 1985: Simulation of the climate of 18,000 years bp: Results for the North American/North Atlantic/European sector and comparison with the geologic record of North America, *Quat. Sci. Rev.,* **4,** 147-187.

LATIF, M., AND T.P. BARNETT, 1994: Causes of decadal climate variability over the North Pacific/North American sector. *Science,* **266,** 634-637.

LATIF, M., T.P. BARNETT, M.A. CANE, M. FLUGEL, N.E. GRAHAM, H. VON STORCH, J.-S. XU, AND S.E. ZEBIAK, 1994: A review of ENSO prediction studies. *Climate Dynamics,* **9,** 167-179.

LAU, N.-C., AND M.J. NATH, 1994: A modeling study of the realtive roles of tropical and extratropical SST anomalies in the variability of the global atmosphere-ocean system. *J. Climate,* **7,** 1184-1207.

LEAN, J., 1991: Variations in the sun's radiative output. *Revs. Geophys.,* **29,** 505-535.

LEAN, J., A. SKUMANICH, AND O. WHITE, 1992: Estimating the sun's radiative output during the Maunder minimum. *Geophys. Res. Lett.,* **19,** 1591-1594.

LINDZEN, 1994: Climate dynamics and global change. *Ann. Rev. Fluid Mech.,* **26,** 353-378.

LOCKWOOD, G.W., B.A. SKIFF, S.L. BALIUNIS, AND R.A. RADICK, 1992: Long term solar brightness changes estimated from a survey of Sun-like stars. *Nature,* **360,** 653-655.

MACAYEAL, D. R., 1993: Binge-purge oscillations of the Laurentide ice sheet as a cause of the North Atlantic's Heinrich events. *Paleoceanography,* **8,** 775-784.

MCMANUS, J. F., G.C. BOND, W.S. BROECKER, S. JOHNSEN, L. LABEYRIE, AND S. HIGGINS, 1994: High-resolution climate records from the North Atlantic during the last interglacial, *Nature,* **371,** 326-329.

MANABE, S., AND A. J. BROCCOLI, 1985: The influence of continental ice sheets on the climate of an ice age, *J. Geophys. Res.,* **90,** 2167-2190.

MANABE, S., AND K. BRYAN, 1985: CO2-induced change in a coupled ocean- atmosphere model and its paleoclimatic implications, *J. Geophys. Res.*, **90**, 11,689-11,707.

MANABE, S., AND R. J. STOUFFER, 1988: Two stable equilibria of a coupled ocean-atmosphere model, *J. Climate.*, **1**, 841-866.

MANABE, S., AND R. J. STOUFFER, 1995: Low frequency variation of surface air temperature in a 1,000 year integration of a coupled ocean-atmosphere model. *J. Climate, submitted.*

MAROTZKE, J., 1989: Instabilities and multiple steady states of the thermohaline circulation. In: Oceanic Circulation Models: Combining Data and Dynamics, D. L. T. Anderson and J. Willebrand, Eds., NATO ASI series, Kluwer, *501-511.*

MAROTZKE, J., 1994: Ocean models in climate problems. In Ocean Processes in Climate dynamics: Global and Mediterranean Examples. *P. Malanotte and A.R. Robinson, eds., Kluwer, 79-109.*

MARTINSON, D.G., K. BRYAN, M. GHIL, M.M. HALL, T.R. KARL, E.S. SARACHIK, S. SOROOSH IAN, AND L.D. TALLEY, EDS, 1995: The Natural Variability of the Climate System on Decade-to-Century Time Scales, *National Academy Press, Washington D.C., in press.*

MIKOLAJEWICZ, U. AND E. MAIER-REIMER, 1990: Internal secular variability in an ocean general circulation model. *Climate Dyn.*, **4**, 145-156.

NATIONAL RESEARCH COUNCIL, 1994: Solar Influences on Global Change. *National Academy Press, Washington, DC, 163pp.*

RIND, D., AND J. OVERPECK, 1993: Hypothesized causes of decadal-to-century- scale climate variability: climate model results. *Quat. Sci. Revs.*, **12**, 357-374.

ROPELEWSKI, C.F., AND M.S. HALPERT, 1987: Global and regional scale precipitation patterns associated with the El Nino/Southern Oscillation. *Mon. Wea. Rev.*, **114**, 2352-2362.

SCHMITT, R. W., P. S. BOGDEN, AND C. E. DORMAN, 1989: Evaporation minus precipitation and density fluxes in the North Atlantic, *J. Phys. Ocean.*, **19**, 1208-1221.

SCHNEIDER, E.K., AND J.L. KINTER, III, 1994: An examination of internally generated variability in long climate simulations. *Climate Dyn.*, **10**, 181-204.

SPALL, M.A., 1995: Dynamics of the Gulf Stream/deep western boundary current crossover. Part II: Low frequency internal oscillations. *J. Phys. Ocean., in press.*

STOMMEL, H., 1961: Thermohaline convection with two stable regimes of flow. *Tellus*, **13**, 224-230.

TAYLOR, G.I., 1921: Diffusion by continuous movements. *Proc. Lond. Math. Soc.*, **20**, 196-211.

TRENBERTH, K.E., AND J.W. HURRELL, 1994: Decadal atmosphere-ocean variations in the Pacific. *Climate Dyn.*, **9**, 303-319.

WALLACE, J.M, Y. ZHANG, AND L. BAJUK, 1995: Interpretation of interdecadal trends in Northern Hemisphere surface air temperature. *J. Climate*, **270** , 780-780.

WALLACE, J.M, Y. ZHANG, AND J. RENWICK, 1995: Dynamical contribution to hemispheric temperature trends. *Science, in press.*

WEAVER, A. J., AND E. S. SARACHIK, 1991A: The role of mixed boundary conditions in numerical models of the ocean's climate. *J. Phys. Oceanogr.,* **21**, 1470-1493

WEAVER, A. J., AND E. S. SARACHIK, 1991B: Evidence for decadal variability in an ocean general circulation model: An advective mechanism. *Atmos.-Ocean,* **29**, 197-231

WEAVER, A. J., J. MAROTZKE, P. F. CUMMINS, AND E. S. SARACHIK, 1993: Stability and Variability of the Thermohaline Circulation. *J. Phys. Oceanogr.,* **22**, 39-60.

WEBB, T., 1989: The spectrum of temporal climatic variability: current estimates and the need for global and regional time series. In Global Changes of the Past, R.S. Bradley, editor, UCAR Office of Interdisciplinary Studies, *61-82.*

WEISSE, R., U. MIKOLAJEWICZ, AND E. MAILER-REIMER, 1994: Decadal Variability of the North Atlantic in an Ocean General Circulation Model. *J. Geophys. Research,* **99**, 12411-1242.

WELANDER, P., 1982: A simple heat-salt oscillator. *Dyn. Atmos. and Oceans,* **6**, 233-242.

WELANDER, P., 1986: Thermohaline effects in the ocean circulation and related simple models. *In Large-Scale Transport Processes in Oceans and Atmospheres, J. Willebrand and D. L. T. Anderson, eds, 163-200.*

WILLEBRAND, J., 1993: Forcing the ocean by heat and freshwater fluxes. In Energy and Water Cycles in the Climate System. *E. Raschke and D. Jacob, eds., Springer-Verlag, 215-233.*

WILLIAMS, M., 1994: Forests and Tree Cover. In Changes in Land Use and Land Cover: A Global Perspective. *Edited by W.B. Meyer and B.L. Turner II. Cambridge University Press, 97-124.*

WILLSON R., AND H. HUDSON, 1988: A solar cycle of measured and modeled solar irradiance. *Nature,* **332**, 810.

WINTON, M., AND E. S. SARACHIK, 1993: Thermohaline oscillations induced by strong steady salinity forcing of ocean general circulation models. *J. Phys. Oceanogr.,* **23**, 1389-1410.

WINTON, M., 1993: Deep decoupling oscillations of the oceanic thermohaline circulation. In: Ice in the Climate System, *W. R. Peltier, Ed., NATO ASI Series, Springer-Verlag, 417-432.*

WINTON, M., 1995A: Energetics of deep-decoupling oscillations, *J. Phys. Oceanogr.,* **25**, 420-427.

WINTON, M., 1995B: Why is the deep sinking narrow? *J. Phys. Oceanogr.,* **25**, 997-1005.

WINTON, M., 1995C: On the role of horizontal boundaries in parameter sensivity and decadal-scale variability of coarse-resolution ocean general circulation models. *J. Phys. Oceanogr., submitted.*

WUNSCH, C., 1992: Decade-to-century changes in the ocean circulation. *Oceanography,* **5**, 99-106.

YIN, F. L., AND E. S. SARACHIK, 1995: On interdecadal thermohaline oscillations in a sector ocean general circulation model: Advective and convective processes. *J. Phys.Oceanogr,* **25**, 2465-2484.

YIN, F. L., 1995: A mechanistic model of ocean interdecadal thermohaline oscillations. *J. Phys. Oceanogr.,* **25**, 3239-3246.

LONG-TERM COORDINATED CHANGES IN THE CONVECTIVE ACTIVITY OF THE NORTH ATLANTIC

ROBERT DICKSON, JOHN LAZIER[1], JENS MEINCKE[2] and PETER RHINES[3]
MAFF Fisheries Laboratory
Lowestoft, Suffolk, England

Contents

1   Indroduction

The North Atlantic is a peculiarly convective ocean. The convective renewal of intermediate and deep waters in the Labrador Sea and Greenland/Iceland Sea both contribute significantly to the production and export of North Atlantic Deep Water, thus helping to drive the global thermohaline circulation, while the formation and spreading of 18-degree water at

---

[1] Bedford Institute of Oceanography, Dartmouth, Canada
[2] Institut für Meereskunde, Hamburg, Germany
[3] School of Oceanography, University of Washington, Seattle, USA

NATO ASI Series, Vol. I 44
Decadal Climate Variability
Dynamics and Predictability
Edited by David L. T. Anderson and Jürgen Willebrand
© Springer-Verlag Berlin Heidelberg 1996

shallow-to-intermediate depths off the US eastern seaboard is a major element in the circulation and hydrographic character of the west Atlantic. For as long as time-series of adequate precision have been available to us, it has been apparent that the intensity of convection at each of these sites, and the hydrographic character of their products have been subject to major interannual change, as shown by Aagaard (1968), Clarke et al (1990), and Meincke et al (1992) for the Greenland Sea, in the OWS BRAVO record from the Labrador Sea, (eg Lazier,1980 et seq.), and at the PANULIRUS / Hydrostation "S" site in the Northern Sargasso off Bermuda (eg Jenkins, 1982, Talley and Raymer, 1982). This paper reviews the recent history of these changes showing that the major convective centres of the Greenland- and Labrador Seas are currently at opposite convective extrema in our postwar record, with vertical exchange at the former site limited to 1000 m or so, but with Labrador Sea convection reaching deeper than previously observed, to over 2300 m. As a result, Greenland Sea Deep Water has become progressively warmer and more saline since the early '70's due to increased horizontal exchange with the Arctic Ocean through Fram Strait, while the Labrador Sea Water has become progressively colder and fresher over the same period through increased vertical exchange; most recently, convection has become deep enough there to reach into the more saline NADW which underlies it, so that cooler, but now saltier and denser LSW has resulted. The horizontal spreading of these changing watermasses in the northern gyre is described from the hydrographic record. The theory is advanced that the scales of atmospheric forcing have imposed a degree of synchrony on convective behaviour at all three sites over the present century, with ventilation at the Sargasso and Greenland Sea sites undergoing a parallel multi- decadal evolution to reach their long term maxima in the 1960's, driven by the twin cells of the North Atlantic Oscillation (NAO). During the NAO minimum of the 1960's, with an extreme Greenland ridge feeding record amounts of fresh water into the northern gyre in the form of the Great Salinity Anomaly, and its partner cell over the Southeast USA causing a southwestward retraction of storm activity (Dickson and Namias, 1976), the surface freshening and postwar minimum in storm activity in the intervening area of the Labrador Sea also brought a progressive reduction, and ultimately a cessation, of wintertime convection there during the 1960's. In other words, the evolution of winter convective activity during the century was in phase but of different sign at the three sites. In these events, we see strong evidence of a direct impact of the shifting atmospheric circulation on the ocean; while this certainly does not rule out either feedbacks from anomalous ice and SST conditions on the atmosphere, or autonomous oscillations of the ocean's overturning circulation, it does tend to minimise them.

2 Variability of deep convection in the Greenland Sea, and its controls

2.1 General description

The deep water of the Greenland Sea [GSDW] is renewed from two sources, either by horizontal exchange with the deep waters of the Arctic Ocean through Fram Strait [sill depth 2600 m], or by vertical exchange as a result of local, deep-reaching, open-ocean convection.

Inputs from these two sources have very different effects on the hydrographic character of the GSDW (Aagaard, 1981; Swift, Takahashi and Livingston, 1983). Arctic Ocean Deep Water is largely formed by "slope convection", in which dense water from the broad Arctic shelves (the result of brine rejection during seasonal ice-formation; Nansen, 1906, Midttun, 1985) spills over the shelf-edge and descends the Slope, encountering and entraining a warm intermediate layer of Atlantic water as it does so.The product then penetrates to a depth determined by its initial salinity and by the subsequent alteration of its $\theta - S$ characteristics during descent. As Rudels and Quadfasel point out [1991], the large volumes of the Eurasian and Canadian Basins effectively buffer the Arctic Ocean Deep Water against change so that its $\theta - S$ properties are thought to be relatively stable in time, despite obvious interannual changes in the production of sea-ice, and presumably of brine.

However, in the context of the present discussion, the key point is that, as pointed out by Meincke, Jonsson and Swift, [1992], the downslope penetration of high salinity shelf water at freezing temperatures and the net entrainment of warm Atlantic waters produce a net downward flux of heat and salt for the Arctic Basins, in stark contrast to the situation in the Greenland Sea where, in the stepwise convective process, the entrainment of freshwater in haline plumes at freezing temperatures, and the penetrative character of the plumes in the intermediate warm layer results in an upward flux of heat and salt [Figure 1, from Meincke, Jonsson and Swift, op cit]. As a direct result, the AODW is the warmest and most saline

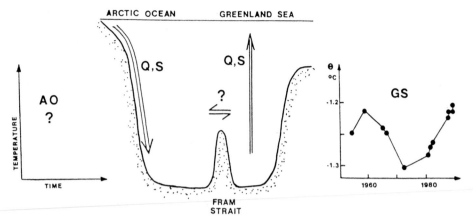

Figure 1: *Schematic representation of the Arctic Ocean/Nordic Seas system as a two-basin system with different deep water formation processes. Fluctuations are known only for deep Greenland Sea. From Meincke, Jonsson and Swift, 1992.*

deepwater component in the Arctic Ocean/ Nordic Seas system, while the GSDW is the coldest and freshest. Other local deep waters are a mixture of the two.

Since the deep waters filling the Greenland Sea Basin are very much more homogeneous in space than their observed variation in time [see below], the implication is clear. Cooling of GSDW can only be carried out from above, by convection, while warming can only be effected from outside the basin, by lateral exchange. In addition the relatively small volume of the Greenland Sea Basin should make for clear inter-annual changes in its deep water characteristics.

2.2   Time-dependence of GSDW characteristics

Figure 2 shows a recent update [Meincke and Rudels, 1995] of the well-known variation in potential temperature [solid line] and salinity [dots] for waters deeper than 2000 m in the central Greenland Sea, using all records of sufficient precision. [original by Clarke, Swift, Reid and Koltermann, 1990, later updated by Meincke, Jonsson and Swift, 1992.]. While CTD temperatures have been adequate for this purpose since the mid-1950's, salinities determined to 0.002 psu have only been routinely available since the late '70's, so that for much of this period, we must infer the convective history of the Basin largely from its temperature, supported however by the annual estimates of the depth of convective overturning provided by

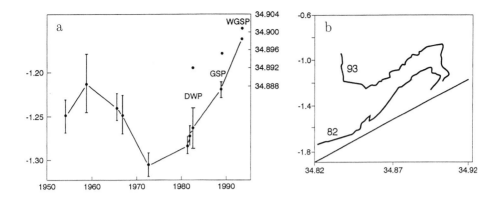

Figure 2: (a) The time variation of the mean temperature and salinity below 2000 m in the central Greenland sea. DWP (Deep Water Project), GSP (Greenland Sea Project). (b) The $\theta - S$ curves from the Hudson Station 58 (1982) and the Valdivia Station 8 (1993). The straight line indicates a σ_3 isopycnal. Both from Meincke & Rudels (in press).

the Deep Water Project and Greenland Sea Project [DWP and GSP].

In Figure 2a, the long cooling of the deepwater from the late '50's to the early '70's is interpreted as the result of intensifying convection, culminating [though the record is gappy] in the winter of 1971 when Malmberg [1983] reports convection reaching to 3500 m. Since then, the evidence is of a progressive reduction in the intensity and penetration of winter convection, with no convective renewal of waters deeper than 1600 m during the 1980's [GSP Group, 1990; Meincke, Jonsson and Swift, 1992], or deeper than 1000 m in the most recent years [Meincke and Rudels, op cit]. In Figure 2b, Meincke and Rudels underline the resulting dramatic change in the $\theta - S$ characteristics of GSDW over the period 1982 to 1993, where, with vertical exchange restricted, an increasing horizontal exchange with the Arctic Ocean has produced the warmest and most saline conditions ever observed in the deep waters of this Basin.

Schlosser, Bonisch, Rhein and Bayer, [1991] have quantified the extent of this convective capping from the change in mean tracer concentrations of GSDW [depths > 1500 m] as a function of time. They show that the decay of tritium, ingrowth of ^3He, and increase of $^3H/^3He$ age in the GSDW is not significantly different from that of a stagnant water-body, [Figure 3 a-c from Schlosser et al, 1991], conclude that the weakening of convection began sometime between 1978 and 1982, and estimate an

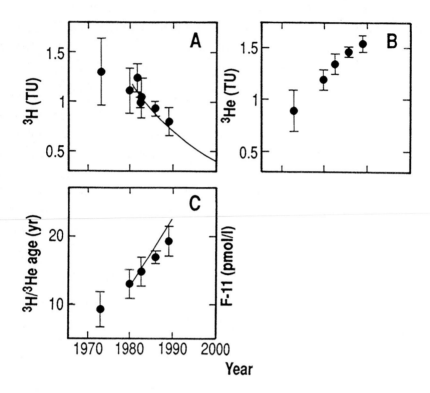

Figure 3: *Mean tracer concentrations in Greenland Sea Deep Water as a function of time. (A) tritium; (B) ^3He; (C) $^3H/^3He$-age. The line in (A) is the radioactive decay curve for a stagnant water body, the line in (C) is the corresponding theoretical increase of the $^3H/^3He$-age. The error bars indicate 1 SD of the mean value of all data from stations in the central Greenland Sea (depths $\geq$ 1500m). From Schlosser et al., 1991.*

80% ± 10% reduction in deep water formation from the ≈ 0.47 Sv of the 1960's and '70's to the 0.1 Sv of the 1980's, corresponding to an increase in the mean turnover time of GSDW, by renewal from the surface, from 34 to 170 years. Their model however cannot comment on cause.

2.3 Causes of change in Greenland Sea Convection

Various candidate mechanisms are proposed for the shutdown in Greenland Sea convection.

(i) Jonsson [1991] suggests that its proximate cause is **a reduction in the windstress curl** over the Greenland Sea. The mean windstress curl pattern there causes upper-layer divergence through Ekman pumping and drives a single-gyre 35 Sv cyclonic circulation, [figure 4], resulting in the doming of isopycnal structures across the Greenland Sea, [e.g.Kiilerich, 1945]. Both the cyclonic circulation and the doming have important implications for convection. It is at the centre of this dome, where stratification is weakest and where dense water is brought closest to the surface that there is the greatest potential for convective instability, since we have to remove less buoyancy there to promote overturn. In turn, the upper-layer divergence has the effect of sharpening the fronts which surround the dome, thus isolating the convective centre against the lateral spread of fresh water from the East Greenland Current.

Figure 5, from Meincke, Jonsson and Swift [1992] certainly confirms a dramatic reduction in windstress curl over the Greenland Sea between the late 1970's and the 1980's, implying that conditions favourable to convection would have been replaced in the 1980's by weaker doming, a smaller, leakier gyre circulation and weak convection.

Meincke and Rudels [1995] provide the first direct evidence of the collapsing dome, pointing out that the Eurasian Basin Deep Water layer [the warmest, most saline and deepest of the Arctic Ocean Deep Waters which pass south through Fram strait (sill depth=2600 m)] had recently been observed to increase its depth in the Greenland Sea Basin by several hundred metres, *but without changing its density*. They conclude that without a convective input, the dome collapses and spreads towards the rim, causing a compensating influx of Canadian Basin Deep Water [CBDW] and Eurasian Basin Deep Water [EBDW] into the centre of the Basin. A succession of hydrographic transects extending towards the centre of the Greenland Sea from its eastern boundary in 1991-94 appear to show the spread of a salinity maximum layer across the basin at 2000-2500 m depth [not shown; Blindheim, pers comm.]

[Note that the increasing CFC concentrations at depth in the Greenland Sea, reported in a recent informal note by Schlosser, Wallace, Bullister, Boenisch and Blindheim (1994) may be just a further reflection of the collapsing dome rather than any sign of renewed convective mixing].

(ii) Aside from the change in wind stress curl, it has been supposed since the time of Nansen [1906] that changes in deep water temperatures might be some response to **changes in winter air temperature**, acting through the cooling and sinking of surface water. Aagaard [1968] constructs a winter cooling index to confirm the point.

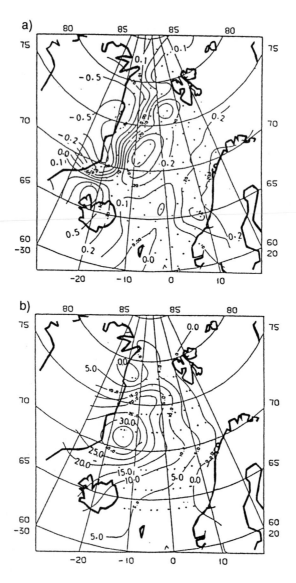

Figure 4: *(a) Mean wind stress curl over the Nordic seas. The unit is pascals per 1000 km, and the contour interval is 0.1 Pa per 1000 km. (b) The integrated Sverdrup transport corresponding to the mean wind stress curl in Figure 2. The unit is sverdrups, and the contour interval is 5 Sv. Both from Jonsson, 1991.*

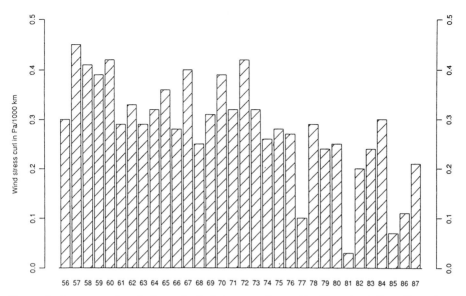

Figure 5: *Interannual variations of wind stress curl over the Greenland Sea. the values are averages over an area of* 200 000 km^2 *in the central part of the Greenland Sea and over a one-year period from summer of year-1 to the following summer. Note the decrease in the late 1970s and 1980s. From Meincke and Swift, 1992.*

Direct winter chilling of the surface layers does appear to be a mechanism of some considerable significance in explaining the time-dependence of Greenland sea convection, and hence the cold-warm cycling in its deeper layers. Over almost the whole of this century, the mean winter pressure at Greenland has shown a spectacularly-steady tendency towards increase, [Figure 5, from Rodgers, 1984] so that by the time of its maximum in the late 1960's, mean winter pressures were 12.8 mb above normal. [Dickson, Lamb, Malmberg and Colebrook, 1975]. As a result, an increasingly strong, direct-northerly airflow along the eastern flank of this Ridge was sweeping the Greenland/Iceland Sea in every month of the year on average, but especially in winter. *Decade-mean* air temperatures at Franz-Josef's Land dropped precipitously by 4.5°C, [Figure 6, from Scherhag, 1970], while Rodewald shows that a more- extensive mean cooling of between 2 and 4°C characterised the Nordic seas between winters [December- March] of the 1950's and 1960's [Figure 7, from Rodewald, 1972]. Comparing the cumulative sum of monthly mean air temperature anomalies at Jan Mayen with deep water temperatures leaves little doubt of the causal connection [Figure 8]. The cooling of the GSDW layer in the 1960's corresponded to a long run of colder-than-normal air temperatures, while the sustained

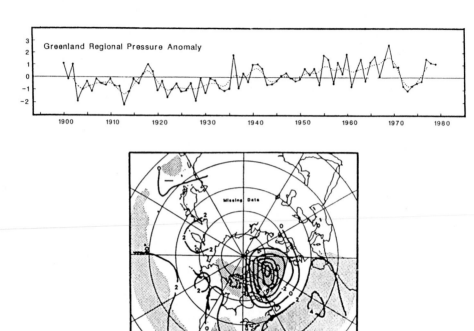

Figure 6: *(a) Normalized winter mean pressure anomaly for the region 60- 70° N, 30-65° W from 1900-79. (b) Difference in winter mean SLP between 1904-25 and 1955-71. (Stippled areas significant at the 95 % level.) Both from Rogers (1984).*

warming since then has occurred while air temperatures were generally increasing.

(iii), (iv) Aagaard and Carmack (1989) suggest that the present-day Greenland Sea is "rather delicately poised" in its ability to sustain convection. They calculate that if the freshwater spreading from the East Greenland Current to the Greenland Sea increased by only a few percent (from 3% to 6% of the $3950km^3$ of ice and freshwater that are carried south in the EGC each year), it would be sufficient to shut down convection there completely. This is not inconsistent with the fact that Greenland Sea convection was at or near maximal intensity when the augmented freshwater signal of the developing Great Salinity Anomaly passed south through the western Greenland Sea in the 1960's, since at that time, the intense wind stress curl and strong Ekman divergence had sharpened the Polar Front between the East Greenland Current and the gyre interior into a highly effective barrier against freshwater intrusion. Conversely, the much

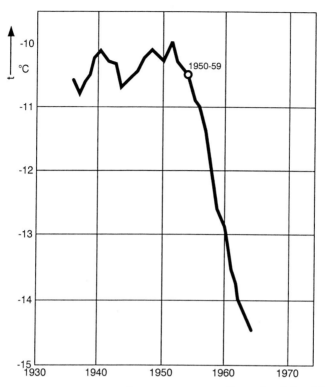

Figure 7: *Running ten-year means of air temperature at Franz Josefs Land from Scherhag, 1970.*

lesser **freshening associated with the return of the Great Salinity Anomaly** to the region in 1981-82 [Dickson, Meincke, Malmberg and Lee, 1988] might well have caused some local suppression of convection since the interannual minimum in windstress curl at that time (see Figure 5) could support only a small-scale, leaky gyre circulation. The **relatively light ice conditions** which characterised the area in the the 1980's cf. the 1970's, with less brine rejection to initiate convection, is perhaps a more plausible supplementary mechanism, as Meincke, Jonsson and Swift, [1992] point out.

2.4 Remote Effects on the Deep Waters of the Norwegian Sea

Since Norwegian Sea Deep Water [NSDW] is partly made up of GSDW, — most probably in a 2:1 mixture of AODW and GSDW according to Swift and Koltermann [1988], who discuss the recipe, — it is not surprising that

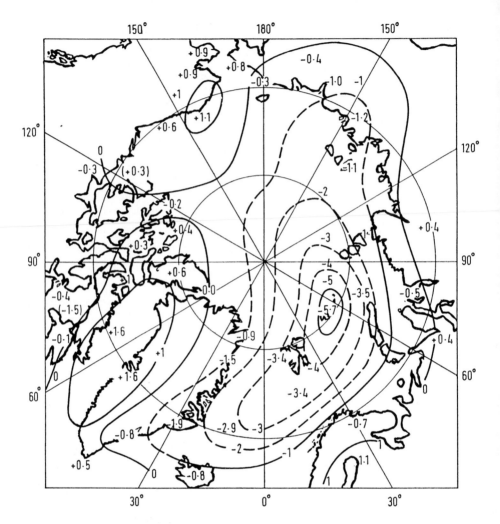

Figure 8: *Mean change of Arctic air temperature (deg C) in the four-month period December-March between the decades of 1951-60 and 1961-70. From Rodewald 1967.*

the temperature of the NSDW should exhibit a similar time-dependence. The route taken by the spreading GSDW to the only available monitoring-point at OWS MIKE is a circuitous one, [Figure 9; from Osterhus and Gammelsrod, in press], passing through the Jan Mayen Channel and around the western and southern margins of the Norwegian Sea Basin before reaching OWS M, so it is not surprising that the temperature change observed there is a lagged, lower-amplitude version of that from the Greenland Sea.

[Figure 10, from Osterhus and Gammelsrod, in press]. Since the maximum T and S change in the deep Greenland Sea was centred on 2000-2500 m depth, [see Aagaard, Fahrbach, Meincke and Swift, 1991, inset to their Figure 8], and since the Jan Mayen Channel has a sill depth [$\approx$ 2000m] which permits exchange at this level, it is perhaps also unsurprising that this was the depth first affected in the Norwegian Sea. Figure 11, also from Osterhus and Gammelsrod, neatly captures the depth-lag of the ascending signal in showing that the end of the cooling and beginning of the warming at MIKE took place in the early-to-mid 1980's at 2000 m, but in the late '80's at 1500 m and around 1990 at 1200 m. There seems little doubt from the precipitous nature of the change that this is the same feature at all these depths or that it represents a remote response to the equally radical warming and salinification of deep waters in the central Greenland Sea. [The salinity data-set from MIKE is not adequate to comment on whether a similar salinification affected the NSDW].

What is *unexpected* in the deep record from OWS M is the levelling-off of warming which appears to have taken place at 2000 m since 1990, [not yet evident at shallower depths], despite the continuation of warming in its supposed source in the Greenland Sea. Though the data are far from conclusive, the working hypothesis put forward by Osterhus and Gammelsrod [Osterhus, pers comm] is that the virtual cessation of GSDW production may have been reflected in a reversal of flow in the Jan Mayen Channel since that date, effectively cutting-off the deep Norwegian Sea from the influence of GSDW and its changes. The available evidence is summarised in Figure 12 a,b showing that long-term, near-bottom current measurements close to the sill of the Jan Mayen Channel in April-November 1981 [Saelen, 1988] and from September 1983-July 1984 were all strongly and constantly eastward, while measurements at the same depth and site in November 1992-July 1993 showed a sluggish westward flow [Osterhus and Gammelsrod, op cit]. Further measurements to confirm this change are clearly necessary and are planned.

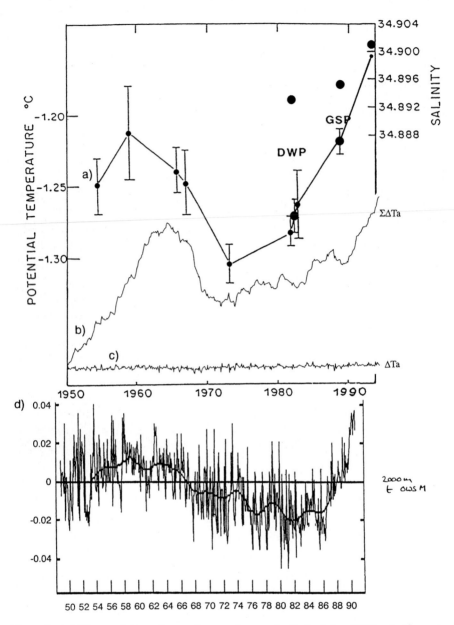

Figure 9: *(a) Time variation of mean temperature and salinity below 2000 m in the central Greenland Sea. (Meincke and Rudels, in press) (b,c) Monthly mean air temperature anomaly and its integral at Jan Mayen 1921- 94 (normal period 1961-90). (d) Temperature variation at 2000 m, ocean weather station 'MIKE', 1948-91. From Gammelsrod, Osterhus and Godoy, 1992.*

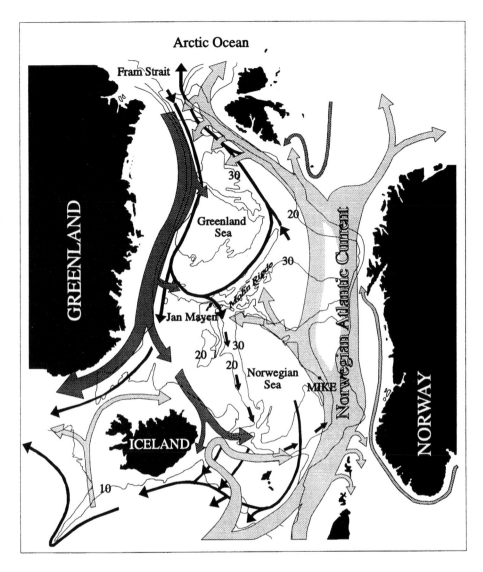

Figure 10: *Schematic of the principal currents of the Nordic Seas with the position of OWS 'MIKE' indicated. Open hatched arrows indicate surface currents while black arrows indicate the deep/bottom current pattern. From Østerhus and Gammelsrød, in press.*

3 Ventilation of the West Atlantic and its causes

3.1 Changes in the North Atlantic Oscillation, and the Storm Climate of the West Atlantic

Though there is no argument but that the second major centre of deep convection in the North Atlantic is that of the Labrador Sea, we switch

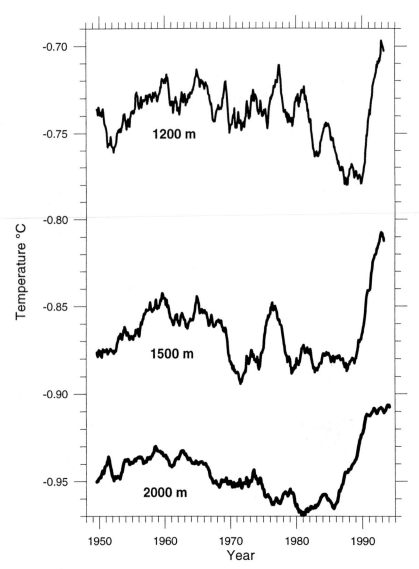

Figure 11: *Time series of smoothed monthly mean temperatures (°C) at depths of 1200 m, 1500 m and 2000 m at OWS MIKE. Monthly mean values are smoothed over 21 month by a running mean filter. From Østerhus and Gammelsrød, in press.*

first to a brief discussion of ventilation in the West Atlantic. The switch is deliberate. The argument we develop below, in the remainder of this paper, is that the time-dependence of deep ventilation at all three of these sites — Greenland Sea, Labrador Sea and Sargasso — is semi-synchronous, [though of differing sign], and that the reason for this concerted behaviour has to do

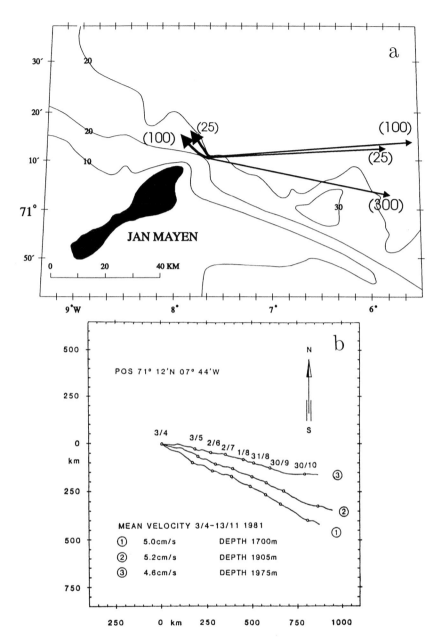

Figure 12: (a) Mean current vectors in the Jan Mayen Channel from September 1983 -
July 1984 (thin arrows), and from November 1992 - July 1993 (thick arrows). Depths of
current meters above bottom are indicated in brackets. From Østerhus and Gammelsrød,
in press. (b) Progressive vector diagrams (based on 25 hourly means) in the Jan Mayen
Channel between 3 April and 13 November, 1981. From Saelen, 1988.

with the scales of atmospheric forcing responsible. More specifically, it has to do with the North Atlantic Oscillation [NAO] and *its* time-dependence, and as such, we will argue that the convective history of the Labrador Sea is determined by the combined *indirect* effects of the two cells which make up the NAO pattern. The *direct* effects of these two cells must therefore be described first.

A version of the NAO pattern — a pressure-anomaly centre over Greenland with a band of opposite sign to its south — is shown in Figure 13. Classically, the NAO index is simply a measure of the pressure difference between Iceland and the Azores so that a high-index pattern [indicating strong midlatitude westerlies], is one of a strong Iceland Low with a strong Azores Ridge to its south, while in the low-index pattern [as Figure 13] the signs of these anomaly- cells are reversed.

Among its rich blend of frequencies, the NAO index has exhibited a considerable long-term variability [Figure 14; see also Rodgers 1984], with interdecadal minima in the 1880's and 1960's. We have already seen in Figure 5 the progressive evolution of the high-pressure anomaly cell at Greenland over much of this century to its maximum in the 1960's and in fact the "version of the NAO pattern" shown in Figure 13 is that which rather frequently occurred during winters of the '60's, when the high pressure anomaly cell over Greenland was partnered by a cell of abnormally low pressure centred over the Southeastern USA.

Dickson and Namias [1976] describe the accompanying change in the storm climate off the US eastern seaboard. Essentially, from the late 1950's to 1970, a regime of low winter air temperatures [Figure 15] and the refrigerating effect of an extreme snowcover [see Dickson and Namias, 1976, their Figure 5] acted to steepen the strong winter land:sea temperature gradient [Figure 16] with the effect of spinning-up more storms than normal in a narrow band following the main coastal baroclinic zone [Figure 17]. It is important in the present context that this increased storm activity was also accompanied by a faster development rate of storms to occlusion so that storms reached their maximum development by Newfoundland rather than [as more-normally] at Iceland and thereafter tended to track east on a southerly trajectory towards Europe or northwest into Arctic Canada. The net effect was that the area of maximum storm frequency was withdrawn > 700 n mi to the southwest [Dickson and Namias op cit, their Figure 8], *so that storm activity over the Labrador Sea during the 1960's was minimal as a result* [See Figure 17].

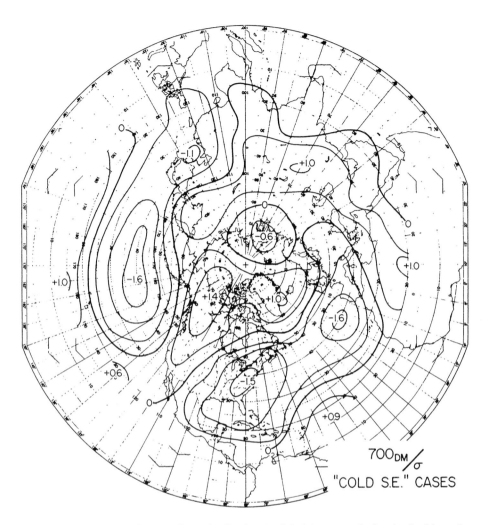

Figure 13: *Mean distribution of standardised 700 mb height anomaly for the "cold south-east" group of winter months described by Dickson and Namias, 1976.*

3.2   Effect on ventilation of the west Atlantic

Before describing the dramatic effect on convection in the Labrador Sea, we can first deal briefly with the resulting changes in the West Atlantic. There, coastal storm activity was not only at an interannual- but at an interdecadal-maximum in the 1960's, illustrated in Figure 18 from the time-variation of the 1st eigenvector of annual cyclone frequencies [i.e., the EOF pattern that reflects east coast cyclogenesis; Hayden, 1981]. As we might expect from

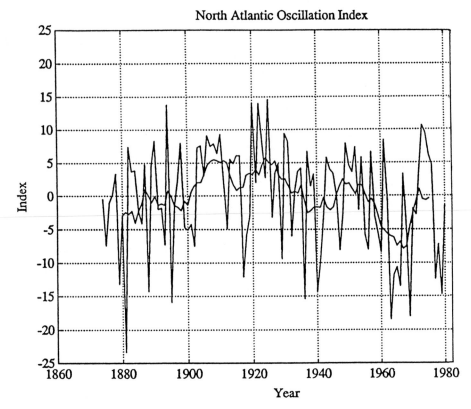

North Atlantic Oscillation Index

Figure 14: *Time dependence of the North Atlantic Oscillation index 1875-1980. Data from Dr. Mick Kelly, Climate Research Unit, University of East Anglia, 1995, pers. comm.*

two cells that are so closely coupled as those of the NAO, the century-long secular variation in coastal storminess is essentially parallel to that shown by the positive pressure-anomaly cell over Greenland [Figure 5]. The twin cells of the NAO are, as normal, acting in concert.

The ocean's response to this long-term maximum of coastal storm activity in the 1960's can be demonstrated in a range of ocean-climate indices. As we might expect, cold air outbreaks meant that the shelf waters off the US and Canadian Atlantic coasts were chilled to a deep and sustained postwar minimum [e.g., Figure 19, from Drinkwater pers comm.]. Further offshore, at the *Panulirus* station off Bermuda [32°10'N, 64°30'W], the storminess shows up as a long- term increase in winter mixed-layer depth during this decade, with a brief *reprise* during winter 1976-77 [Figure 20b from Michaels et al 1994], while Talley and Raymer [1982] use the same *Panulirus* data set to show the development of a long-term maximum in

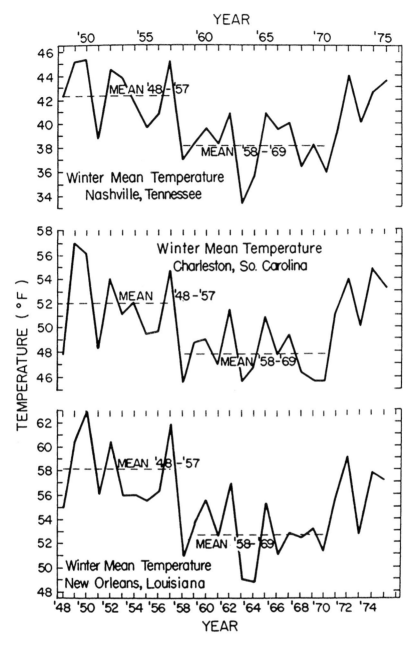

Figure 15: *Winter mean air temperatures at Nashville, Charleston and New Orleans from 1947-48 (labelled "48") to 1974-75 (labelled "75"). From Dickson and Namias, 1976.*

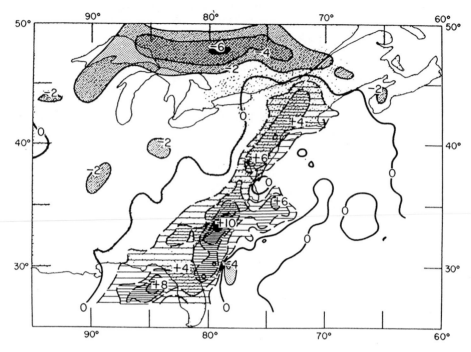

Figure 16: *Magnitude of the mean change in surface temperature gradient [°F (100 km)] over the eastern United States and west Atlantic between the warm SE and cold SE groups of winter months. The surface temperature fields used were a composite of surface air temperature data over land and sea surface temperature data over the ocean. For full explanation see Dickson and Namias, 1976.*

density and minimum in potential vorticity at $\sigma\theta = 26.4 - 26.6 mg/cm^3$, indicating strong formation and ventilation of the 18-Degree Water [an almost homogeneous pycnostad or mode water at 250-400 m depth] during the '60's and again around winter 1976-77 [Figure 20c, based on the analysis by Talley and Raymer op cit, kindly updated by Lynn Talley, 1995, pers comm]. Since a major portion of the heat transfer in this area is by latent heat flux, Jenkins [1982] is able to use the isopycnal-salinity change at *Panulirus* as well as oxygen to confirm directly that watermass renewal to $\sigma\theta = 26.7$ was at a postwar maximum in the late 60's [Figure 21]. As he points out, change to the depth of this isopycnal [550 m] is not all accomplished locally but reflects local ventilation to 350 m or so in the west of the Basin spreading east along deepening isopycnals. [Isopycnal salinity change shows an apparently-increasing lag with depth from $\sigma\theta = 26.1$ to 26.5]. As he also points out however, the relatively rapid horizontal circu-

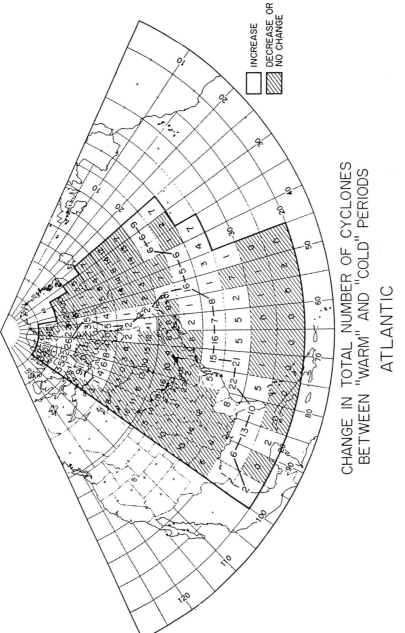

CHANGE IN TOTAL NUMBER OF CYCLONES
BETWEEN "WARM" AND "COLD" PERIODS

ATLANTIC

INCREASE

DECREASE OR
NO CHANGE

Figure 17: *Change in the total number of cyclones between the "warm SE" and the "cold SE" groups of winter months in the south-eastern USA. From Dickson and Namias, 1976.*

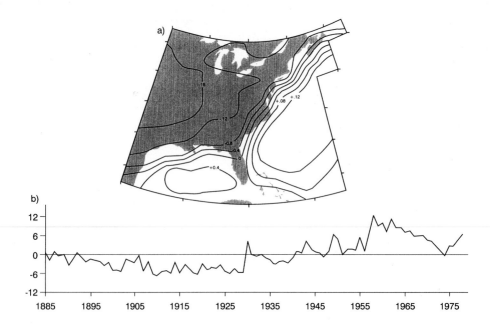

Figure 18: *(a) The first eigenvector of annual cyclone frequencies. (b) Time variation of the annual weightings of the first eigenvector (E_1) of annual cyclone frequencies, N America and W Atlantic. From Hayden 1981.*

lation in the Subtropical Gyre compared with ventilation timescales means that the *Panulirus* data are representative of watermass conditions in the gyre as a whole.

Thus, acting in response to the twin coupled cells of the NAO, each at its long term extreme state, deep convection in both the Greenland Sea and the [shallower] ventilation of the Sargasso were both at maximum intensity during the 1960's, and have since largely shut down.

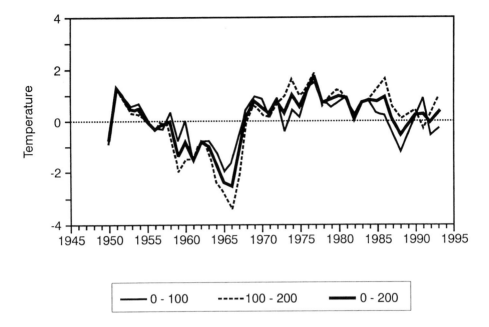

Figure 19: *Annual temperature anomaly (°C) for the Emerald Basin, Scotian Shelf at depths from 100-200 m. K F Drinkwater, Bedford Institute of Oceanography, 1995 pers. comm.*

4 Variability of Deep Convection in the Labrador Sea, and its Controls

4.1 The remote effects of the NAO minimum

We are now in a position to explain the changing convective history of the Labrador Sea in similar terms. As with the Greenland Sea and Sargasso, the key changes appear linked to the North Atlantic Oscillation and *its* changes, but in this case we suggest that the changing production of Labrador Sea Water can be explained as a *remote* response to the two cells which make up the NAO pattern, each of which was at record intensity during the 1960's.

The Greenland Ridge, and the northerlies which it generated over the Greenland Sea were the main factors responsible for bringing south vast quantities of fresh Polar water in a swollen East Greenland Current at that time, and for exporting it into the Northern Gyre as the Great Salinity Anomaly [Dickson, Meincke, Malmberg and Lee, 1988]. As a result, an estimated $2000km^3$ of extra fresh water [Aagaard and Carmack, 1989] was circuiting the margins of the Labrador Sea in the 1960's, finally passing east into the open Atlantic in 1971-72.

236

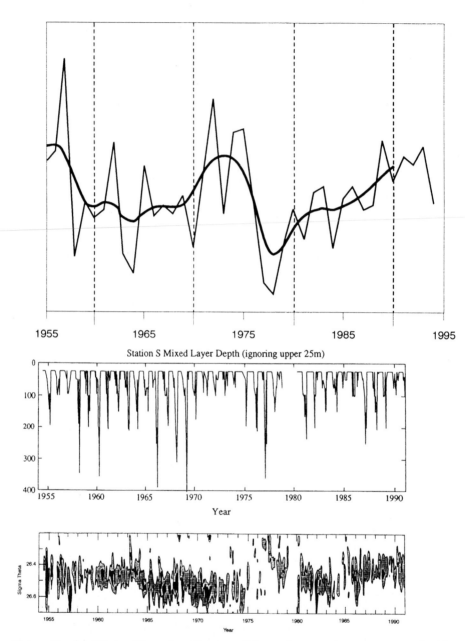

Figure 20: (a) Winter (December, January, February) mean air temperatures for New Orleans, 1955-94. (Data from National Climatic Data Centre, Divisional Data Set). (b) Mixed layer depths, calculated from 38 year time series at Hydrostation S, off Bermuda. From Michaels et al., 1992. (c) Potential vorticity $(f/\rho)(\partial\rho/\partial z)$ as a function of potential density σ_θ for 1954-1994 based on monthly changes of hydrographic data at Hydrostation S. Contours are 1.0 and $1.3 \times 10^{-12} cm^{-1} s^{-1}$. An approximate update of Talley and Raymer, 1982, (their Figure 3), kindly provided by Dr. Lynne Talley, SIO, 1995.

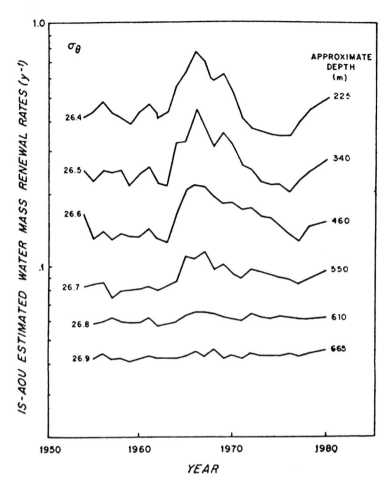

Figure 21: *Watermass renewal rate* (y^{-1}) *for Eighteen Degree Water in the West Atlantic as a function of time for* $26.4 < \sigma_\theta < 26.8$. *From Jenkins 1982.*

At the same time, an intense trough over the SE USA, as we have seen, was setting up changes in the storm climate at the US eastern seaboard which had the effect of withdrawing the offshore centre of winter storm activity far to the southwest of normal, and away from the Labrador Sea. As a result, the winter windstress and more-specifically its chilling north-west wind component were both at a postwar minimum over the Labrador Sea in the 1960's, [Figure 22a and b respectively from Rhines, 1994 and Drinkwater, 1994; see also Drinkwater and Pettipas, 1993] with the effect of suppressing winter heat loss, while Figure 22c, [kindly provided by Gilles

Reverdin, LDEO, pers. comm.] confirms that the windstress curl over the Labrador Sea was also at its postwar minimum at this time. By reducing the Ekman divergence, this reduced curl would have allowed fresh surface water to spread into the convective centre of the Labrador Sea.

The separate effects of the two main pressure-anomaly centres at the time of the NAO minimum were therefore to promote an increased accession of fresh surface water to the Labrador Sea while minimising its winter storminess and cyclonic circulation, and the long hydrographic record at OWS BRAVO captures the expected net response — reduced winter heat-loss, suppressed convection rarely reaching deeper than 1000m [see ERIKA DAN 1962, Hudson 1966], increasing stratification and decreasing salinity [by about 0.7 psu at 10 m] in the surface layers during the 1960's, with a corresponding increase in salinity at depth. [Figure 23, from Lazier,1980].

4.2   The restoration of deep convection in the Labrador Sea

This tight capping of convection ended abruptly in the winter of 1971-72, as the freshening influence of the GSA passed eastward out of the area [Dickson et al 1988], as the cold regime of winter air temperatures along the US eastern seaboard came to an abrupt end [see Figure 15] and as severe winter storminess with chilling northwesterly winds and intense windstress curl were restored over the Labrador Sea [see Figure 22]. The continued monitoring of subsurface hydrographic conditions at OWS BRAVO, close to the centre of the cyclonic gyre, provides evidence of increased and deepening ventilation of the Labrador Sea and increasing production of Labrador Sea Water since then.

The restoration of deep convection was not regular but tended to intensify in two main steps,- — the first from 1972-76, and the second from 1988 [approx.] onward. [Figure 24]. This intensifying convection, and the progressive cooling and freshening in the 1000-1500 db layer of the Labrador Sea that has taken place since the early '70's are clearly in total contrast to the contemporaneous tendencies for suppressed convection, warming and salinification in the deep Greenland Sea, already described. Overall, Lazier [1995] shows that over the whole 3500 m watercolumn of the Labrador Sea, a freshening by 0.059 psu, equivalent to the removal of 200 kg of salt per m^2, [or the addition of $\approx$ 6 m of freshwater per m^2 !] and a cooling by 0.46°C, equal to the additional heat loss of $8W/m^2$ continuously for 26 years, has taken place between 1966 and 1992. The freshening was not

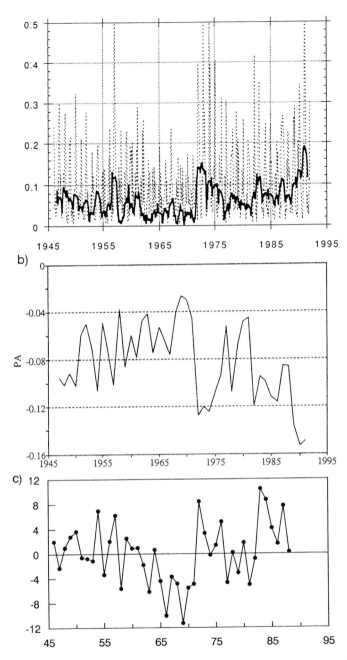

Figure 22: (a) Time variation of winter wind stress over the central Labrador Sea, smoothed and unsmoothed, from data of Drinkwater; see Rhines, 1994. (b) The annual vector-averaged NW wind-stresses for the Labrador Sea (average of 6 sites). Negative stresses indicate winds from NW. From Drinkwater, 1994. (c) Mean wind stress curl at 55N, 50W and 55N, 45W, based on surface daily NMC pressure fields. From Giles Reverdin, Lamont-Doherty Earth Observatory, pers. comm., 1995.

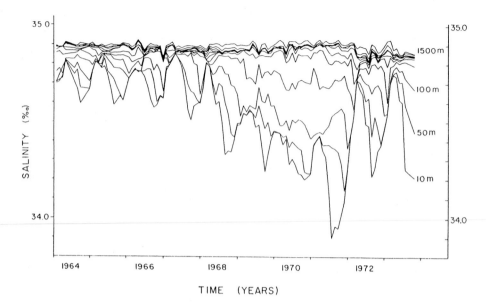

Figure 23: *Monthly averages of salinity at eleven depths at OWS Bravo from 1964 to 1973. From Lazier, 1980.*

uniform with depth however but was maximal in the convectively-formed LSW layer where, at 2000 db, the salinity decreased by 0.09 psu over the same period, presumed due to the downward mixing of the freshwater accumulation from the surface [Lazier, 1995]. At the same time, the base of the LSW layer has eroded from < 2000 db, [often < 1000 m] t= 3.4°C during the mid-sixties when isolated from the surface to 2300 db, t= 2.8°C in 1992, and even deeper in 1993-94 [see below].

4.3   The link with air temperature

Analagous to the example for the Greenland Sea, [Figure 8], Figure 24 b compares the time- variation in θ at intermediate depths in the Labrador Sea with monthly air temperature anomalies and their integral at Iqaluit [Frobisher Bay].To a first approximation, the curves correspond, with declining LSW temperatures following protracted periods of negative air temperature anomaly, most notably after the early '70's and mid-to-late '80's. The link with the *persistence* of cooling, not merely its intensity per se, is thought to be important; a series of cold months or cold winters will create a more homogeneous watermass into which deep convection will be initiated with less heat-loss than after a series of mild winter months.

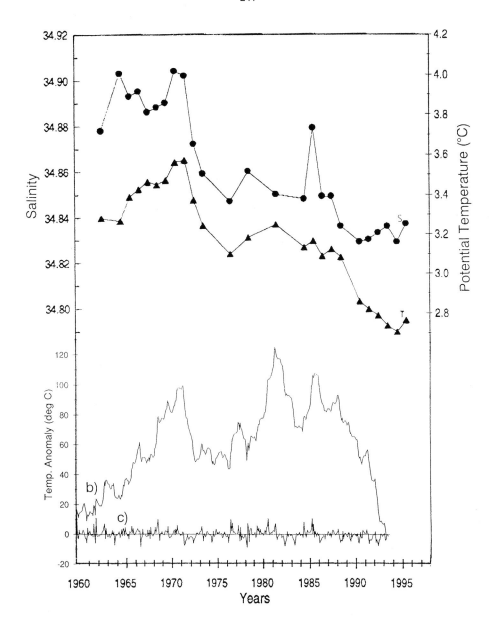

Figure 24: *(a) Average potential temperature and salinity at OWS BRAVO over the 1000 and 1500 dbar pressure interval for each year when data are available. Lazier, unpublished. (b,c) Monthly mean air temperature anomaly and its integral, 1960-94 at Iqaluit (Frobisher Bay, Canada).*

In detail, the instantaneous heat budget in the central Labrador Sea is not easy to reconcile. The air-sea heat flux and the heat storage below (assuming a one-dimensional heat balance over short times) show the right "events" yet disagree in magnitude [Smith and Dobson, 1984] ; yet after averaging over a year, the anomalies in surface flux and storage were in better agreement. Monthly mean heat flux can, according to Smith and Dobson, exceed 300 W/m^2 in severe winters. Their annual average heat loss for the central Labrador Sea is about 85 W/m^2 in the severe winter of 1972, yet over the more-placid period 1964-73, averages 28 W/m^2. This is far less than the values obtained using Bunker's flux formulas, which yield 98 W/m^2. As Clarke and Gascard [1983] emphasise, the air temperature is much lower close to the rim of the Labrador Sea and the oceanic heat loss is likely to be greater there.

These small annual heat flux numbers highlight the delicate balance of the high latitude ocean: while we assume and believe some connection between heat loss and convection, the net cooling is difficult to establish in detail, and this leads to widely-varying estimates of the rate of production of Labrador Sea Water.

More recent modelling results by Wallace and Lazier [1988] suggest that cooling may not be the sole factor at work. In their experiments, most of the vertical profiles were not prone to convection below 1000 m under normal conditions, and deep convection to 1400 m could only be achieved patchily or with unrealistic heat fluxes. They conclude that some form of pre- conditioning of the watercolumn,— a cyclonic circulation for example,— may also be required for deep convection to occur, but as already described, (Figure 22), the observed changes in air temperature and windstress curl *have* tended to act in the same sense during the postwar period .

4.4   Changes in LSW density

It is plain enough from Figure 24 that deep $\theta - S$ changes occur in the core of the Labrador Sea Water layer. Figure 25 shows equally clearly that in most years for which we have records, the changes in θ and S at BRAVO have been mutually-compensating in density, so that annual changes in the $\theta - S$ characteristics of LSW have tended to occur along isopycnals. In fact, Clarke and Gascard [1993] proposed that the density of Labrador Sea Water should remain constant with time because of this effect.

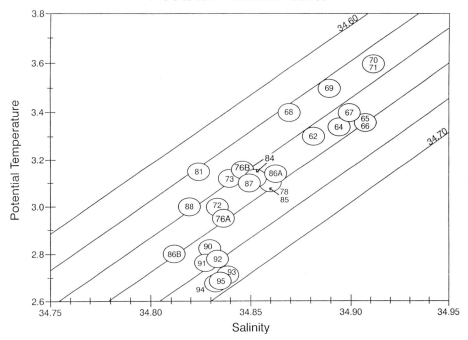

Figure 25: *Annual potential temperature: salinity characteristics at the Labrador Sea water temperature minimum, 1962-95.*

This self-compensating behaviour broke down after 1992, however. By then, the long intensification and deepening of convection at OWS BRAVO had proceeded to the point where it exceeded the normal limit of convective overturn [2000 m or so] and began to excavate the underlying layers of North Atlantic Deep Water, which are characteristically colder but more saline than LSW. By 1992, Lazier's profiles of θ, S and $\sigma\theta$ referenced to 1500 db [Figure 26, from Lazier, 1995] show an almost-homogeneous layer extending down to > 2300 m at BRAVO, and a Hudson cruise in June 1993 from Cartwright, Labrador to Cape Desolation, Greenland shows even deeper mixing to 2400 m in the centre of the basin. This depth of overturn, never previously observed, is reflected in the time-series of Figure 24a where, in the 1990's, the long parallel trend towards cooling and freshening in the LSW is replaced by a divergence of trend, continuing to cool but now increasing in salinity as the more-stratified sublayer of NADW becomes incorporated. The result, [see Figure 25] is that LSW became steadily more dense from 1990 to 1993, and has remained dense until 1995.

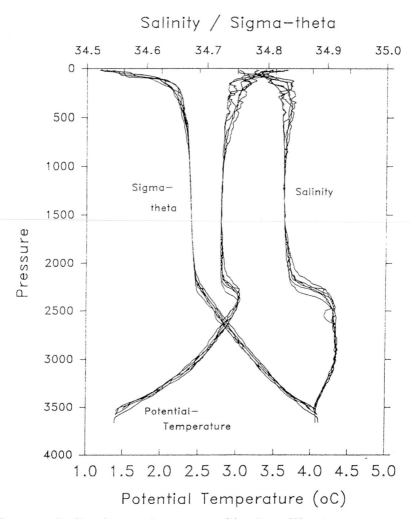

Figure 26: *Profiles of potential temperature (θ), salinity (S) and $\sigma_{1.5}$ versus pressure at four stations in the middle of the AR7/W line, Labrador Sea, in July 1990. From Lazier, in press.*

In fact, something similar may have happened previously. By using Ice Patrol hydrography to extend the OWS BRAVO record back beyond 1945, Hallberg, Lazier and Rhines, pers. comm. have compiled a discontinuous record of annual $\theta - S$ characteristics for LSW back to 1928. Though it occurred at a higher base-temperature and salinity than today [these were the years when a considerable wave of warming and salinification was

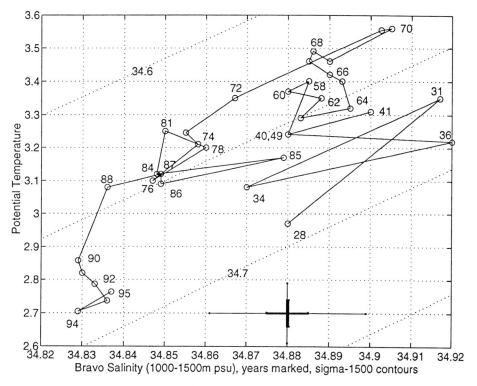

Figure 27: *History of the annual average potential temperature and salinity of Labrador Sea Water between 1000 and 1500 m depth at Ocean Weather Station Bravo. Years are indicated and $\sigma_{1.5}$ contours are shown. The bold cross shows the $\pm$ standard deviation within individual years of the Weathership Bravo hydro data, in this depth range; the double-bold cross shows the actual annual range of the Bravo mooring data. Hallberg, Lazier and Rhines, 1995, pers. comm.*

passing through the Northern Gyre; see Dickson and Brander, 1993, their Figure 1], Figure 27 provides evidence of an earlier densification-trend in the LSW during the late '20's and early '30's.

If, using this Figure, we plot deviations from the central isopycnal [$\sigma_{1.5} = 34.65$] as a function of time, [Figure 28b], we find not only that the density of Labrador Sea Water is subtly [perhaps even continuously] variable with time but also that there is a sufficient correspondence with indices of local wind-strength [Figure 28 a,c] to suggest that this change is in some way [heat flux? wind mixing?] dependent on storminess, as we had earlier assumed.

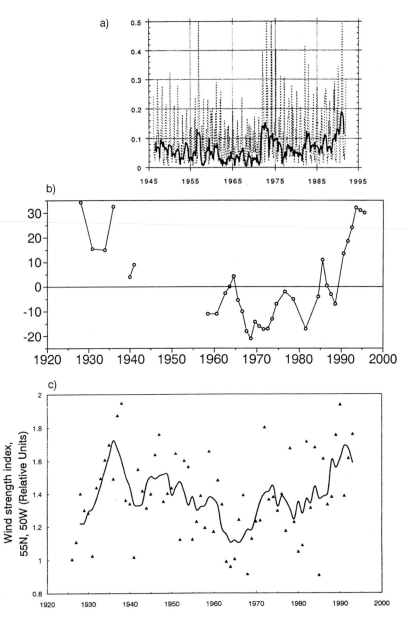

Figure 28: *(a) Post-war variation of winter wind stress over the Labrador Sea. From Rhines, 1994. (b) Annual departure of $\sigma_{1.5}$ from the central density of 34.65, based on the annual $\theta - S$ characteristics for LSW shown in Figure 27. (c) Wind strength index (relative units) for the approximate positions of OWS BRAVO, computed by Dr. Mick Kelly, CRU/UEA Norwich. from the zonal and meridional gradients in surface pressure at 55N, 50W.*

4.5   Time dependence in the advective sublayer at OWS BRAVO

Lazier [1988] and Lazier and Gershey, [1991] address the issue of time-dependence in the bottom-most layers of the Labrador Sea and its causes. Below the depth of even the recent record convection this sublayer of North Atlantic Deep Water must owe the largest part of its T and S variability to changes in its source areas, transferred by lateral advection. Though NADW has various constituents, the most likely source-water suggested by Lazier is Arctic Intermediate Water from the Iceland-Greenland Sea which overspills the Greenland-Scotland Ridge via the Denmark Strait. Dickson and Brown [1994], however, show that this water entrains an equal volume of resident water from the upper layers of the Irminger Sea as it descends along the Greenland Slope, making this another possible source of change. Lazier estimates the transit time from overflow to BRAVO to be around 100 days. Except as a recent source of change in the LSW itself, the deepest layers have little relevance here, but it is perhaps worth noting [Figure 29] that the variability in the deepest layers at BRAVO [θ at $\sigma_2 = 37.16kg/m^3$; from Lazier and Gershey, 1991] bears a closer similarity to the temperature anomaly at 0-200 m in the Irminger component of the Labrador Current, [Borovkov and Tevs, 1991] than anything in the superjacent watercolumn.

4.6   The spreading of Labrador Sea Water

As was the case with the GSA signal, the varying characteristics of LSW provide us with a valuable [perhaps the only] direct means of assessing spreading rates and pathways at the intermediate depths occupied by this watermass. Spreading rates may of course vary from one part of the North Atlantic to another:

(i) The F/S METEOR cruise in November-December 1994 provided spectacular information on **spreading rates within the Labrador-Irminger Basin** itself. In Figure 30, Lazier, Rhein, Sy and Meincke [pers. comm.] demonstrate what is unambiguously the same anomalous increase in density at intermediate depths in the Irminger Sea as Lazier showed for the LSW core at OWS BRAVO in 1991-94. Strictly, it is not known whether the dense water found in the Irminger Sea in November 1994 is the counterpart of the anomalously dense water encountered in the Labrador Sea in the surveys of March 1993 or March 1994. Lazier et al conclude that it is "very likely" the latter, but in either case the spreading rate indicated

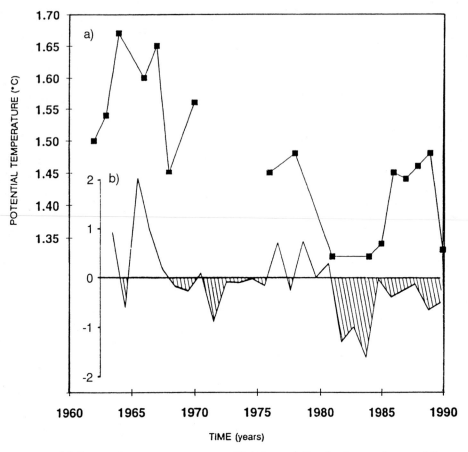

Figure 29: *(a) Potential temperature at $\sigma_2 = 37.16\,kg\,m^{-3}$ for the deepest layers of the Labrador Sea. From Lazier and Gershey, 1991. (b) Temperature anomaly (°C) at 0-200 m in the Irminger component of the Labrador Current. From Borovkov and Tevs, 1991.*

— 700 km in 8 or 20 months at mean speeds of 1-3 cm/s — is an order of magnitude faster than the published spreading rates, as these authors point outThey suggest that the two areas might even be regarded as the same entity from the viewpoint of LSW formation and spreading.

(ii) Evidence on **the trans-Atlantic spreading rate** of LSW is provided from two main studies.

In his carefully-compiled salinity series from the Rockall Channel, Ellett describes an abrupt, 7- standard deviation drop in salinity in 1990-91 in the depth layer [1600-1800 m] occupied by LSW in the East Atlantic, [Figure 31, from Ellett, 1993]. He interprets this extreme event as indicating the sudden arrival at the eastern boundary of the fresh LSW vintage

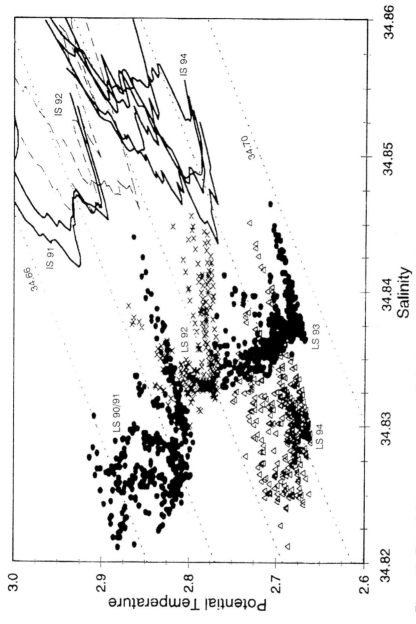

Figure 30: *Potential temperature versus salinity diagram for data collected in the central Labrador Sea in 1990-94 and in the central part of the Irminger Sea in 1991-94. Contours of constant σ are referenced to 1500 db. See Lazier et al., 1995.*

formed in the winter of 1971-72, when the restoration of intense convection at BRAVO first redistributed the accumulated fresh water from the surface layers to intermediate depths [see Figure 24a]. This interpretation is not without its puzzling aspects, not least the questions of why the freshening at Rockall was so large [at 0.1 psu, not far short of the entire freshening observed at BRAVO; see Figure 24], and still so abrupt after an elapsed spreading-time of 18 years. However, the spreading rate of 0.4 km/d, [0.47 cm/s], about half the lower of recent estimates from the Western Basin, seems not inappropriate, and what other North Atlantic site than the convective centre of the Labrador Sea could produce such rapid freshening to 1600 m?

Cunningham and Haine [1995 a,b] comment usefully on the fate of the LSW vintage formed during the second period of intense convection at BRAVO in the late-'80's. Since a potential vorticity minimum is a characteristic and conserved property of any convectively-formed watermass, they follow Talley and McCartney [1982] in employing it as an LSW tracer in their analysis of results from the 1991 VIVALDI surveys of the Eastern Atlantic. Specifically, they find that the range of potential vorticity values from the VIVALDI area in 1991 [i.e. $1.5 - 7 \times 10^{-14}/cm/s$] showed no overlap with values from the same area in 1957-64 [$8 - 16 \times 10^{-14}/cm/s$] when Talley and McCartney compiled their map [Figure 32]. The LSW in that area had evidently been completely renewed since the early '60's. Noting that the minimum salinity at the LSW core [$\sigma_{1.5} = 34.64$] on any of their zonal VIVALDI sections was the value of 34.866 at 51 °N — the latitude where the p.v. minimum crosses the Mid-Atlantic Ridge according to both studies — Cunningham and Haine suggest from the "source function" at BRAVO [Figure 24] that salinities of this value must have been generated in the Labrador Sea either between 1975-80 or after 1985. In line with Top, Clarke and Jenkins [1987] who propose upper limits of 1.4 cm/s and 0.7cm/s for the southward and eastward spreading of LSW, respectively, Cunningham and Haine discount the spreading rates of mm/s that the former dates would imply, and therefore conclude that the freshest water they observe on VIVALDI was of the 1986 vintage. Thus at the time of the 1991 VIVALDI survey, the products of the late-'80's convective event were east of the Mid-Atlantic Ridge but had not then reached the eastern boundary. [A similar conclusion emerged from CFC analyses of trans-ocean sections worked during the CONVEX program of 1991; Denise Smythe-Wright pers comm, 1995].

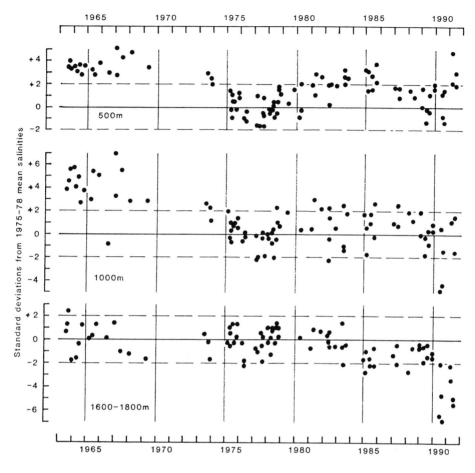

Figure 31: *Salinity variations in the central Rockall Channel at three depths, 1963-92. (Standard deviations from mean salinity values for 1975-78.) From Ellett, 1993.*

A subsequent zonal section through the eastern basin by F/S METEOR in November- December 1994 shows colder, deeper, denser LSW characteristics in the Southern Rockall Trough at that time [Sy, Koltermann and Paul, 1994], and these conditions were again encountered in even more extreme form in this area during F/S VALDIVIA Cruise 154 in May-June 1995 [Manfred Bersch, 1995, pers comm]. It is therefore suggested by Sy et al [op cit] that the products of the second [late-80's] period of convective intensification in the Labrador Sea are now beginning to arrive at the

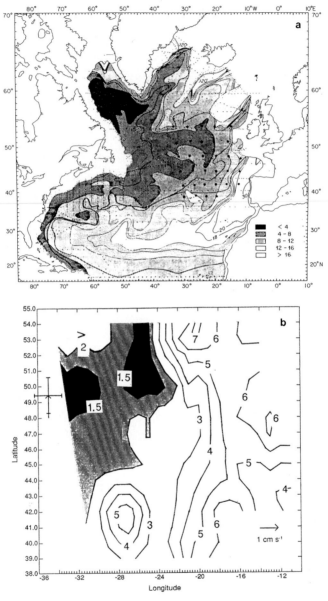

p.v. distribution (×10⁻¹⁴ cm⁻¹ s⁻¹)

Figure 32: (a) Potential vorticity distribution in units of $10^{-14}\,cm^{-1}\,s^{-1}$ at the Labrador Sea Water potential vorrticty minimum. Compilation by Talley and McCartney, 1982, based on data from years 1957-1964. Dots indicate the location of the 1991 VIVALDI survey. (b) VIVALDI potential vorticity distribution at the potential vorticity minimum of the Labrador Sea Water in the eastern North Atlantic expressed in units of $10^{-14}\,cm^{-1}\,s^{-1}$. Location shown in Figure 32a. From Cunningham and Haine, 1995.

eastern boundary. If so, the implied trans- Atlantic spreading rate would be very much [×3] faster than that assumed by Read and Gould [1992] or Ellett [1993], as Sy et al point out.

(iii) As Labrador Sea Water characteristics have changed, **the southward spread of these anomalous conditions into the western Atlantic** has also been recognised in the hydrographic record. McCartney, Worthington and Raymer [1980] compare meridional sections along 55W to highlight the arrival, in July 1977, of small cells of pure LSW, 800 km from their nearest possible source. Localised salinity anomalies of up to -0.106 were encountered in the Slope Water, under the Gulf Stream and beneath the centre of the recirculating gyre to its south, and were clearly new features compared with the KNORR 60 section of the previous October. Though McCartney et al were evidently unaware that convection in the Labrador Sea had resumed since winter 1971-2, we would now probably assign salinities as low as these [S=34.86 at t=3.52 °C] to the vintage of LSW formed in about 1973 during that first intense period of renewal [cf source function, Figure 24], implying a spreading rate of 0.56 cm/s. This is very similar to the estimated spreading rate to the eastern boundary, described above.

More recently, Koltermann and Sy [1994] give evidence of the continued increase in LSW production during the 1980's in showing a deepening of the LSW east of Newfoundland from 1400 to 2100 m in July 1993, and a more-than-doubling of LSW volume between 1982 and 1993 at 48 °N in the West Atlantic.As we have seen, this dramatic increase of southward spreading has coincided with the succession of exceptionally-harsh and worsening winters in the Labrador Sea during the 1980's and 1990's, with postwar maxima developing in windstress, northwesterly wind component and windstress curl (Figure 22 a,b,c).

5 Summary and Conclusions

This review suggests the following conclusions:

1. The convective activity of the North Atlantic is subject to radical interannual and interdecadal changes.

2. In the case of the Greenland Sea convection-centre, intensifying deep convection from the 1950's to the early 1970's, when overturn reached

254

3500 m, was replaced by a progressive reduction of deep convection since then, with an $80 \pm 10\%$ reduction in GSDW formation during the 1980's and with convective exchange restricted to < 1000 m by 1993. The dwindling depth and intensity of vertical exchange and an increasing deep lateral exchange with the Arctic Ocean through Fram Strait brought a progressive warming, salinification, deoxygenation and [weak] stratification of Greenland Sea Deep Water.

3. This waxing and waning of Greenland Sea deep convection is attributed two main factors; first to a reduction of local wind stress curl during the 1980's, so that pre-existing conditions favourable to convection [a strong cyclonic gyre, with surface layer divergence through Ekman pumping, and a doming of isopycnals] were replaced by weaker doming, a smaller-scale, leaky gyre circulation and weak convection. Reinforcing this trend, the century-long build-up of wintertime pressure over Greenland, which had resulted in an extreme northerly airflow and record winter cooling over the Nordic seas by the 1960's, showed an abrupt weakening in the late 1970's and steadily-warming air temperatures since then have continued the suppression of convective overturn. A reduced sea-ice extent during the 1980's, hence lesser brine-production, may have been a third contributory factor.

4. A recent deepening of the Eurasian Basin Deep Water layer in the Greenland Sea, without any associated change in density, is the first direct evidence that the suppression of convection is being accompanied by some relaxation or collapse of the Greenland Sea dome. This, rather than any resumption of deep mixing, is the likely cause of the reported increase in CFC concentrations at all depths to 3000 m between 1981 and 1991.

5. The Deep Water of the Norwegian Sea is composed partly of GSDW and, until the early 1990's, exhibited a low-amplitude and lagged version of the temperature changes shown by GSDW itself. There was also an apparent lag with decreasing depth in the Norwegian Sea, so that the recent abrupt warming appeared first at 2000 m where the greatest change in the GSDW had been recorded. Since 1990, as GSDW production dwindled to near-extinction, the NSDW has ceased to warm further at depth, and there are some signs that the connecting flow through the Jan Mayen Channel may have reversed.

6. The ventilation history of the west Atlantic is in synchrony with that of the Greenland Sea since both are driven directly by the twin coupled cells of the North Atlantic Oscillation [NAO]. The NAO index reached its extreme low-index state — with ridging over Greenland and low pressure to its south — during the 1880's and 1960's, with the opposite high index state peaking during the 1920's. Thus during the 1960's, the Greenland high pressure- anomaly cell and the trough over the southeastern United States were each the result of a slow amplification over decades.

7. From the late 1950's to 1970, under the more-southerly of these two cells, a regime of cold winter air temperatures and extreme snowcover greatly enhanced the land-sea temperature gradient at the US eastern seaboard, spinning-up more storms than normal offshore, and causing them to develop more rapidly to occlusion in a narrow band following the main coastal baroclinic gradient. The local effect was to draw the centre of maximum storm activity far to the southwest of normal, while the remote effect was to reduce storminess over the Labrador Sea to a postwar minimum. Along the US and Canadian seaboard, shelf-waters chilled to their long-term minimum values in the '60's, while offshore the cold, stormy conditions caused maximum formation and ventilation of the 18-Degree Water pycnostad.

8. The changing convective history of the Labrador Sea is explained as the combined, *remote* response to the coupled cells of the North Atlantic Oscillation. During the NAO minimum of the 1960's, while an intense Greenland Ridge directed an increased accession of freshwater to the Labrador Sea in the form of the Great Salinity Anomaly, and its partner-cell over the SE USA caused the withdrawal of its winter storm activity, convection in the Labrador Sea became increasingly suppressed. LSW production resumed abruptly in winter 1971-72 as the GSA passed eastwards out of the area and as the cold winter regime ended at the US east coast. Since then the tendency has been towards increasing and deepening ventilation of the Labrador Sea [to > 2300 m in the most recent years] and a general cooling and freshening of LSW, in marked contrast to the contemporaneous tendencies in the deep Greenland Sea.

9. Labrador Sea convection intensified in two main steps, during the early '70's and from the late '80's onwards.Though the annual changes in the $\theta - S$ characteristics of LSW appear to be mutually-compensating in density over most of the period of record, the excavation of the cold but saline NADW sublayer during the deepest-reaching convection since 1992 has caused a clear recent increase in LSW density. A similar increase in density is indicated in the hydrographic record from the late '20's/early '30's, although at a higher base-temperature and salinity. Closer inspection of the full record suggests that LSW density may in fact be a subtly- and continuously-varying function of windspeed and heat exchange, contrary to the established view that it is stable in time.

10. Changes in the $\theta - S$ characteristics of the NADW sublayer of the Labrador Sea appear to be advectively transferred from source, reflecting changes in the Arctic Intermediate Water that overflows the Denmark Strait, and/or the near-surface layers of the Irminger Sea that are entrained as it does so.

11. Using the new dense LSW as a tracer, spreading rates at intermediate depths in the Labrador-Irminger Basin have been estimated at 1-3 cm/s, an order of magnitude greater than published values. The arrival of LSW vintages from the early '70's in the Rockall Trough in 1990-91 and in the Gulf Stream Recirculation by 1977 suggest lower long-range spreading rates of around 0.5 cm/s. Salinity, CFC and potential vorticity characteristics suggest that the products of the late-'80's convective event have crossed the Mid-Atlantic Ridge and have reached the eastern boundary in 1994.

12. Thus, coordinated by the scales and configuration of the forcing, we suggest that convective activity at all three sites evolved over decades to a long-term extreme state — in phase but of differing sign — during the NAO minimum of the 1960's, at which time the ventilation of the Greenland Sea and Sargasso was at a maximum and that of the Labrador Sea was tightly capped. Since then all three centres have evolved rather more rapidly towards their opposite extreme states, in which convection in the Greenland Sea and Sargasso is suppressed but vertical exchange in the Labrador Sea is reaching deeper than previously observed. The local mechanisms by which the atmospheric

Area	Controls	Convective Status	
		1960's	Now
Greenland Sea	1,2,3,4b,(4a)	1:Max	1:Min
		3:Max	3:Min
		4b:Min	4b:Max
		(4a:Min)	(4a:?(GSA))
Labrador Sea	1,2,3,4a,4b	1:Min	1:Max
		2:Min	2:Max
		4a:Max (GSA)	4a:Min
		4b:Max	4b:Min
Sargasso Sea	2	2:Max	2:Min

Key to dominant processes:

1: Cyclonic wind stress curl for doming and intnsity of fronts around convective gyre
2: Cold air outbreaks for thermal density increase
3: Brine release during freezing for haline density increase
4: Cross-frontal advection of non-local waters into gyre
   a. Capping of gyre by surface freshwater influx
   b. Stratifying the gyres interior

Table 1: The changing dominance of open-ocean convective processes, by area

forcing is translated into convective change may differ at all three sites. [Table 1]. We can expect some correspondence in the pattern of convective activity between the NAO minima of the 1960's and the early 1880's.

13. In these events, we see strong evidence of a direct impact of the shifting atmospheric circulation on the ocean; while this certainly does not rule out either feedbacks from anomalous ice and SST conditions on the atmosphere, or autonomous oscillations of the ocean's overturning circulation, it does tend to minimise them.

Acknowledgement This paper was coordinated and compiled during a visit by RRD to Scripps Institution of Oceanography as Visiting Research Fellow, funded by the Joint Institute for Marine Observations and at the invitation of Professor Russ Davis, SIO Physical Oceanography Research Division. Their support is gratefully acknowledged. The contribution by JM was supported by Sonderforschungbereich 318. PBR acknowledges support from the Office of Naval Research and NOAA.

258

References

AAGAARD, K. 1968. Temperature variations in the Greenland Sea Deep Water. *Deep-Sea Res.*, **15**, 281-296.

AAGAARD, K. 1981. On the deep circulation in the Arctic Ocean. *Deep Sea Res.*, **28**, 251-268.

AAGAARD, K. AND E.C. CARMACK. 1989. The role of sea ice and other fresh water in the Arctic circulation. *J. Geophys. Res.*, **94**, 14485-14498.

AAGAARD, K., E. FAHRBACH, J. MEINCKE AND J.H. SWIFT. 1991. Saline outflow from the Arctic Ocean: Its contribution to the deep waters of the Greenland, Norwegian and Iceland seas. *J. Geophys. Res.*, **96**, 20433-20441.

BOROVKOV V.A.AND I.I.TEVS 1991 Oceanographic Conditions in NAFO Subareas 0,1,2 and 3 in 1990. *NAFO SCR Doc. 91/11, 20 pp (mimeo)*

CLARKE, R.A., J.H.SWIFT, J.A.REID AND K.P.KOLTERMANN 1990. The formation of Greenland Sea Deep Water:double diffusion or deep convection? *Deep-Sea Res.*, **37 (9)**, 1385-1424.

CLARKE, R.A. AND J.C. GASCARD. 1983. The formation of Labrador Sea Water. Part I: Large- scale processes. *J. Phys. Oceanogr.* **13**, 1764-1778.

CUNNINGHAM, S.A. AND T.W.N HAINE 1995A Labrador Sea Water in the Eastern North Atlantic. Part I.: A Synoptic Circulation Inferred from a Minimum in Potential Vorticity. *J. Phys. Oceanogr.*, **25 (4)**, 649-665.

CUNNINGHAM S.A. AND T.W.N.HAINE, 1995B Labrador Sea Water in the Eastern North Atlantic. Part II: Mixing dynamics and the Advective-Diffusive Balance. *J. Phys. Oceanogr.*, **25 (4)**, 666-678.

DICKSON R.R. AND J. BROWN., 1994. The production of North Atlantic Deep Water: Sources, rates and pathways. *J. Geophys. Res.*, **99, C6,** 12319-12341.

DICKSON, R.R. AND K.M. BRANDER, 1993. Effects of a changing windfield on cod stocks of the North Atlantic. *Fish. Oceanog.2, 124-153. ICES mar. Sci. Symp., 198: 271-279.*

DICKSON, R.R. AND J. NAMIAS. 1976. North American Influences on the Circulation and Climate of the North Atlantic Sector. *Mon.Wea. Rev.* **104 (10)**, 1256-1265.

DICKSON, R.R., J. MEINCKE, S.-A. MALMBERG AND A. J. LEE. 1988. The "Great Salinity Anomaly" in the northern North Atlantic 1968-1982. *Prog. Oceanogr.*, **20**, 103-151.

DICKSON, R.R., H.H. LAMB, S.A. MALMBERG AND J.M. COLEBROOK. 1975. Climatic reversal in northern North Atlantic. *Nature, Lond.*, **256**, 479-482.

DRINKWATER, K.F. 1994. Environmental Changes in the Labrador Sea and some Effects on Fish Stocks. *ICES CM 1994/MINI:4, 19 pp, (mimeo)*

DRINKWATER, K.F. AND E.G. PETTIPAS, 1993. Climate Data for the Northwest Atlantic: Surface Windstresses off Eastern Canada, 1946-1991. *Can. Data Rept. of Hydrog. and Ocean Sci., 123, 130 pp*

ELLETT, D.J. 1993. Transit times to the NE Atlantic of Labrador Sea water signals. *ICES CM 1993/C:25 11pp + 6 figs (mimeo)*

GAMMELSROD, T., S. OSTERHUS, AND O. GODOY, 1992. Decadal variations of ocean climate in the Norwegian Sea observed at Ocean Station MIKE (66N 2E). *ICES mar. Sci. Symp.*, *195:68-75*

GSP GROUP 1990. The Greenland Sea Project - a venture toward improved understanding of the ocean's role in climate. *EOS*, **71: 750-751**, 754-755.

HAYDEN, B.P. 1981. Secular variation in the Atlantic coast extratropical cyclones. *Mon. Wea. Rev.*, **109**, 159-167.

JENKINS, W.J. 1982. On the climate of a subtropical gyre: decade timescale variations in water mass renewal in the Sargasso Sea. *J. Mar. Res.*, **40. (suppl.)**, 265-290.

JÓNSSON, S. 1991. Seasonal and interannual variability of wind stress curl over the Nordic Seas. *J. Geophys. Res.*, **96: C2**, 2649-2659.

KOLTERMANN, K.P.AND A.SY 1994. Western North Atlantic cools at intermediate depths. *WOCE Newsletter*, **15**, 5-6.

KILLERICH, A.B. 1945. On the hydrography of the Greenland Sea. *Medd. Grönl.* **144**, 1-63.

LAZIER, J., M. RHEIN, A. SY AND J. MEINCKE, in press. Surprisingly Rapid Renewal of Labrador Sea Water in the Irminger Sea.

LAZIER, J.R.N. 1995. The Salinity Decrease in the Labrador Sea over the Past Thirty Years. In: Natural Climate Variability on Decade-to-Century Time Scales. *D. G. Martinson, K. Bryan, M. Ghil, M. M. Hall, T. M. Karl, E.S Sarachik, S. Sorooshian, and L. D. Talley, (eds.). National Academy Press, Washington, D.C.*, pp. m-n.

LAZIER, J.R.N. 1980. Oceanographic conditions at Ocean Weather Ship Bravo, 1964-1974. *Atmos. Ocean*, **18**, 227-238.

LAZIER, J.R.N. 1988. Temperature and salinity changes in the deep Labrador Sea, 1962-1986. *Deep-Sea Res.*, **35**, 1247-1253.

LAZIER, J.R.N. AND R.M. GERSHEY. 1991. AR7W: Labrador Sea Line - July 1990. *WOCE Newsletter*, **11**, 5-7.

MALMBERG, SV.-A. 1983. Hydrographic investigations in the Iceland and Greenland Seas in late winter 1971 - "Deep Water Project.", *Jökull, 33:133-140.*

MCCARTNEY, M.S., L.V. WORTHINGON AND M.E. RAYMER. 1980. Anomalous water mass distributions at 55W in the North Atlantic in 1977. *J. Mar. Res.*, **38 (1)**, 147-172.

MEINCKE, J. 1990. The Greenland Sea interannual variability. *ICES CM 1990/C:17, 6pp +8 figs. (mimeo).*

MEINCKE, J. AND B. RUDELS. 1995. Greenland Sea Deep Water: A balance between convection and advection. *Nordic Seas Symposium, Hamburg March 1995. Extended Abstr. Vol.,U Hamburg, 143-148.*

MEINCKE, J., S. JONSSON AND J.H. SWIFT. 1992. Variability of convective conditions in the Greenland Sea. *ICES mar. SCI. Symp.* **195**, 32-39

MICHAELS, F.A., A.H. KNAP, R.L. DOW, K. GUNDERSEN, R.J. JOHNSON, J. SORENSEN, A. CLOSE, G.A. KNAUER, S.E. LOHRENZ, V.A. ASPER, M. TUEL AND R BIDIGARE. (1992) Seasonal patterns of ocean biogeochemistry at the UK JGOFS Bermuda Atlantic Time-series Study site. pp 1013- 1038. *In: T.M. Powell & J.H. Steele (Eds.). Ecological Time Series. Chapman and Hall, N.Y. 491 pp*

MIDTTUN, L. 1985. Formation of dense bottom water in the Barents Sea. *Deep-Sea Res.,* **23** 1233-1241.

NANSEN, F. 1906. Northern Waters: Captain Roald Amundsen's oceanographic observations in the Arctic seas in 1901. *Videnskabs-Selskabets Skrifter, I. Mathematisk-Naturv. Klasse,* **No. 3,** 145 pp.

ØSTERHUS. S. & T. GAMMELSROD, in press. The Abyss of the Nordic Seas is Warming. Nature . Lond.

READ J.F. AND W.J. GOULD, 1992. Cooling and freshening of the subpolar North Atlantic Ocean since the 1960's. *Nature Lond.,* 360 55-57

RHINES, P. 1994. Climate change in the Labrador Sea, its Convection and Circulation. *Pp 85-96 In: Atlantic Climate Change Program: Proceedings of the PI's meeting, Princeton, May 9-11, 1994.*

RODEWALD, M. 1972. Temperature conditions in the NorthAtlantic during the decade 1960-70. *ICNAF Spec. Publ. No 8, Symp on Environmental Condits in NW Atlantic, 1960-69,* 9-34.

ROGERS, J.C. 1984. The Association between the North Atlantic Oscillation and the Southern Oscillation in the Northern Hemisphere. *Mon Wea. Rev.,* **112,** 1999-2015.

RUDELS, B. AND D. QUADFASEL. 1991. Convection and deep water formation in the Arctic Ocean- Greenland Sea system. *J. Mar. Sys.* **2 (3/4),** 435-450.

SAELEN, O. H. 1988. On the Exchange of Bottom Water Between the Greenland and Norwegian Seas. *Geophys. Inst. Univ. Bergen Rept. 67, 14 pp + 8 figs.*

SCHERHAG, R. 1970. Die gegenwartige Abkuhlung der Arktis. *Beil z. Berliner Wetterkt. 105/70 so 31/70.*

SCHLOSSER, P., D. WALLACE, J. BULLISTER, G. BOENISCH AND J. BLINDHEIM. 1994 . New results from long-term tracer observations in the Greenland/Norwegian and Labrador seas. *Pp 129-133 In: Atlantic Climate Change Program: Proceedings of P.I.'s meeting, Princeton, May 9-11, 1994*

SCHLOSSER, P., G. BONISCH, M. RHEIN AND R. BAYER. 1991. Reduction of deepwater formation in the Greenland Sea during the 1980's: Evidence from tracer data. *Science,* **251,** 1054-1056.

SMITH, S.D.AND F.W. DOBSON, 1984. The heat budget at Ocean Weather Station BRAVO. *Atmos-ocean,* **22,** 1-22.

SWIFT, J.H., AND K.P.KOLTERMANN, 1988. The origin of Norwegian Sea Deep Water. *J. Geophys. Res.,* **93,** 3563-3569

SWIFT, J.H., T. TAKAHASHI AND H.D. LIVINGSTON. 1983. The contribution of the Greenland and Barents Seas to the deep water of the Arctic Ocean, *J. Geophys. Res.,* **88,** 5981-5986.

SY, A., K.P. KOLTERMANN, AND U.PAUL. 1994. Ausbreitung und Veranderlichkeit des Labradorseewassers. *pp 44-50 In: Anon, WOCE Status Rept., Germany, 1994*

TALLEY, L.D. AND M.E. RAYMER. 1982. Eighteen Degree Water variability. *J. Mar. Res.,* **40 (Supp.),** 757-775.

TALLEY, L.D., AND M.S. McCARTNEY. 1982. Distribution and circulation of Labrador Sea Water. *J. Phys. Oceanogr.,* **12,** 1189-1205.

TOP, Z., W.B. CLARKE AND W.J.JENKINS, 1987. Tritium and primordial 3-He in the North Atlantic: a study in the region of Charlie-Gibbs fracture Zone. *Deep-Sea Res.,* **34 (2)** 287-298.

WALLACE, D.W.R. AND J.R.N LAZIER 1988. Anthropogenic Chlorofluoromethanes in newly formed Labrador Sea Water. *Nature, Lond.,* **332, 61590,** 61-63.

A MECHANISM FOR DECADAL CLIMATE VARIABILITY

M. Latif, A. Grötzner, M. Münnich, E. Maier-Reimer, S. Venzke, T. P. Barnett[1]

Max-Planck-Institut für Meteorologie
Hamburg, FRG

Contents

Abstract

We describe in this paper a mechanism for decadal climate variability that can lead to decadal climate cycles in the North Pacific and North Atlantic Oceans. A hierarchy of numerical models and observations are used to understand the fundamental dynamics of these decadal cycles. They are generated by large-scale ocean-atmosphere interactions in mid-latitudes and must be regarded as inherently coupled modes. The memory of the coupled system, however, resides in the ocean and is associated with slow changes in the subtropical ocean gyres.

When, for instance, the subtropical ocean gyre is anomalously strong, more warm tropical waters are transported poleward by the western boundary current and its extension, leading to a positive SST anomaly in mid-latitudes. The atmospheric response to this SST anomaly involves a weakened storm track and the associated

---

[1] Scripps Institution of Oceanography, La Jolla, Ca. , U. S. A.

NATO ASI Series, Vol. I 44
Decadal Climate Variability
Dynamics and Predictability
Edited by David L. T. Anderson and Jürgen Willebrand
© Springer-Verlag Berlin Heidelberg 1996

changes at the air-sea interface reinforce the initial SST anomaly, so that ocean and atmosphere act as a positive feedback system. The atmospheric response, however, consists also of a wind stress curl anomaly which spins down the subtropical ocean gyre, thereby reducing the poleward heat transport and the initial SST anomaly. The ocean adjusts with some time lag to the change in the wind stress curl, and it is this transient ocean response that allows continuous oscillations. The existence of such decadal cycles provides the basis of long-range climate forecasting at decadal time scales.

1 Introduction

The study of decadal climate variability is not only of scientific but also of enormous social and economical interest. Civilization can cope with extreme climate variations that last briefly, say a particularly severe winter. Climate variations with time scales of centuries are beyond a single human's or even a single political system's lifetime and so elicit little but intellectual interest. But the impact of variations in a key climate variable, say precipitation, that last a decade can generally not be avoided by society, e.g. the extended droughts in California, Australia, or the Sahel.

We concentrate in this paper on the decadal variability in the North Pacific and North Atlantic Oceans and describe one particular mechanism that explains many of the aspects of the decadal variability observed in these regions. Different competing hypotheses to explain the decadal variability observed in the North Pacific were put forward. Both Trenberth and Hurrell (1994) and Graham (1994) argue that the decadal variations in the North Pacific are forced by changes in the equatorial Pacific sea surface temperature (SST) and subsequent changes in the atmospheric circulation over the North Pacific. However, both studies disagree about the nature of the decadal variability. While Graham (1994) and some other studies (e. g. Nitta and Yamada (1989)) explain the changes in the mid-1970s as a discrete "shift", Trenberth and Hurrel (1994) view the decadal variability as more oscillatory.

A completely different hypothesis was offered by Jacobs et al. (1994). They argue that the recent decadal variation in the North Pacific is caused by planetary waves which were excited during the strong 1982/1983 El Niño/Southern Oscillation (ENSO) warm extreme (see Philander (1990) for a comprehensive review on ENSO). The planetary waves propagate slowly westward across the Pacific basin at off-equatorial latitudes and interact eventually with the Kuroshio current, leading lo large-scale SST anomalies in the North Pacific and associated changes in the overlying atmosphere. Jiang et al. (1995) attribute low-frequency climate variability in mid-latitudes to the chaotic nature of the ocean that allows, for instance, multiple equilibria of the wind-driven ocean circulation.

The observations show pronounced decadal variability in the North Atlantic also (e. g. Kushnir (1994)). Most studies attribute the decadal variability in the Atlantic to variations of the thermohaline circulation (e. g. Yang and Neelin (1993), Delworth et al. (1993), Weisse et al. (1994)). Such variations are forced by density anomalies at the sea surface which modify the pattern and strength of oceanic convection, which in turn leads to large-scale variations in the oceanic general circulation.

Here, we propose an alternative hypothesis for the generation of decadal variability in both the North Pacific and North Atlantic Oceans which involves variations in the wind-driven ocean circulation. Our hypothesis is based on unstable large-scale ocean-atmosphere interactions in mid-latitudes leading to climate cycles in these regions. The possibility of unstable ocean-atmosphere interactions in mid-latitudes on seasonal and longer time scales was originally hypothesized by Namias (for the North Pacific) in a series of papers (e.g. Namias, 1959, and 1969) and by Bjerknes (1964) (for the North Atlantic). Namias argued that SST anomalies in the North Pacific can change the transient activity in the atmosphere, which in turn changes the mean westerly flow reinforcing the initial SST anomalies. The paper by Bjerknes is interesting, since it postulates a climate cycle with a period of about 10 years in the Atlantic which involves interactions of the westwind regime and the subtropical ocean gyre. As we shall see below, the "early" ideas of both Namias' and Bjerknes' apply remarkably well to the decadal climate variability observed in both oceans.

The decadal variability is studied by means of a hierarchy of numerical models and a mechanism is postulated for the generation of decadal variability in mid-latitudes. We describe in section 3 a decadal mode in the North Pacific from observations and as simulated in a multi-decadal integration with a coupled ocean-atmosphere general circulation model which is described in section 2. A preliminary description of this mode and the physics responsible for it were given by Latif and Barnett (1994). Suffice to say, it has many of the features of the decadal variability observed in the North Pacific. The results of uncoupled experiments with the individual model components are also shown in section 3. These uncoupled experiments were performed, in order to get more insight into the nature of the

266

atmospheric response to mid-latitudinal SST anomalies and the ocean adjustment to decadal wind stress variations, which are both crucial parts of our mechanism. We describe the results of a simple coupled ocean-atmosphere model which retains the essential physics of the decadal cycles in section 4. Section 5 deals with the decadal variability in the North Atlantic, as simulated by our coupled ocean-atmosphere general circulation model. The paper is concluded with a summary and discussion of our major findings in section 6.

2 Coupled ocean-atmosphere general circulation model and observational data

2.1 Coupled model

The coupled general circulation model (CGCM) and its behaviour with respect to climate drift and interannual variability during the first twenty years of a 125-year integration are described in Latif et al. (1994). Overall, the model drift is relatively small and the CGCM simulates well the decadal variability in the North Pacific, as shown in Latif and Barnett (1994) and Latif and Barnett (1996). As will be shown below, the CGCM succesfully reproduces also some fundamental aspects of the decadal variability in the North Atlantic.

The atmospheric component of our coupled GCM "ECHO" is ECHAM-3, the Hamburg version of the European Centre operational weather forecasting model. The model is described in detail in two reports (Roeckner et al., 1992; DKRZ, 1992). ECHAM-3 is a global low-order spectral model with a triangular truncation at wavenumber 42 (T42). The nonlinear terms and the parameterized physical processes are calculated on a 128 x 64 Gaussian grid which yields a horizontal resolution of about $2.8^o \times 2.8^o$. There are 19 levels in the vertical which are defined on σ-surfaces in the lower troposphere and on p-surfaces in the upper troposphere and in the stratosphere.

The ocean model is "HOPE" (Hamburg Ocean Model in Primitive Equations) which is based on primitive equations (see Latif et al., 1994 and references therein). Its domain is global and we use realistic bottom topography. The meridional resolution is variable, with 0.5^o within the region 10° N to 10° S. The resolution decreases poleward to match the T42-resolution of the atmosphere model. The zonal resolution is constant and also matches the

a)

Observed

Index region: 170E-160W,25N-40N; 5 year running mean; detrended

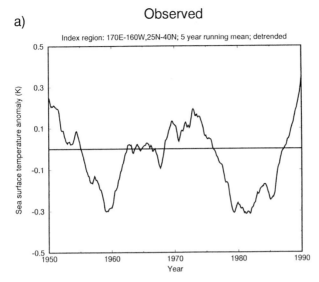

b)

Observed

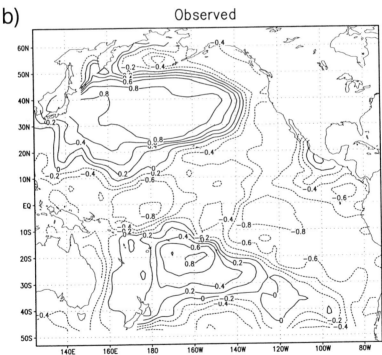

Figure 1: *Observed low-pass filtered (retaining variability on time scales longer than 5 years) SST anomalies averaged over the North Pacific ($25^o N - 45^o N$, $170^o E - 160^o W$) for the period 1950-1990, as derived from the GISST data set. b) Spatial distribution of correlation coefficients of the index time series shown in (a) and SST anomalies in the Pacific.*

c)

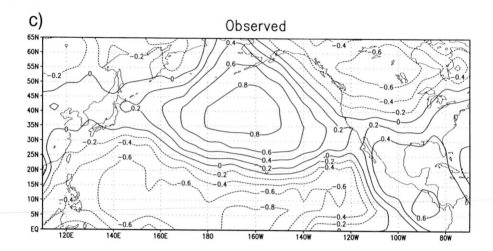

Figure 1: *continued: c) Spatial distribution of correlation coefficients of the index time series shown in (a) and 700 hPa height anomalies over the Pacific and North America.*

atmospheric model resolution. Vertically, there are 20 irregularly spaced levels, with ten levels within the upper 300 m. Since we have not yet included a sea-ice model in HOPE, the SSTs and sea surface salinities are relaxed to Levitus (1982) climatology poleward of 60° , using Newtonian formulations with time constants of about 2 and 40 days, respectively, for the upper layer thickness of 20 meters. The vertical mixing is based on a Richardson-number dependent formulation and a simple mixed layer scheme to represent the effects of wind stirring (see Latif et al. (1994) for details).

The two models were coupled without flux correction. They interact over all three oceans in the region 60° N - 60° S. The ocean model is forced by the surface wind stress, the heat flux, and the freshwater flux simulated by the atmosphere model, which in turn is forced by the SST simulated by the ocean model. The coupling is synchronous, with an exchange of information every two hours. The coupled GCM is forced by seasonally varying insolation. The coupled integration started at 1 January and continued for 125 years.

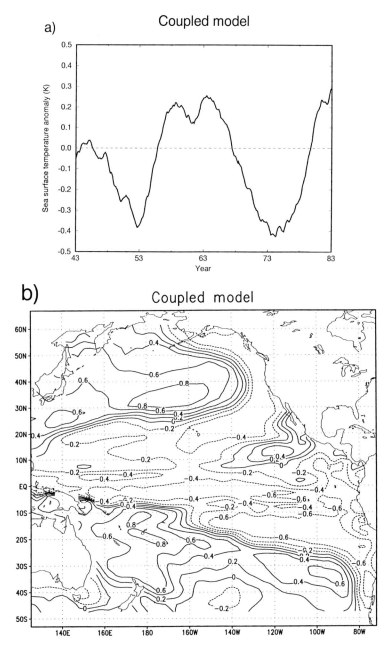

Figure 2: a) With the CGCM simulated low-pass filtered (retaining variability with time scales longer than 5 years) SST anomalies averaged over the North Pacific ($25^o N - 45^o N$, $170^o E - 160^o W$) for the model years 43-83. b) Spatial distribution of correlation coefficients of the index time series shown in (a) and SST anomalies in the Pacific.

c) **Coupled model**

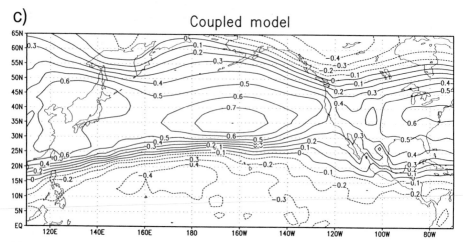

Figure 2: *continued: c) Spatial distribution of correlation coefficients of the index time series shown in (a) and 500 hPa height anomalies over the Pacific and North America.*

2.2 Observational data

Various observational data sets will be used to describe the decadal variability observed. A short description of these data is given below. The data were available at monthly intervals. Sea surface temperature (SST) data were obtained from the SST data set of the British Meteorological Office (UKMO) which is referred to as the "GISST" data set, covering the period 1949-1991, and 700 hPa height data were provided by the Climate Analysis Center. The latter were available for the Northern Hemisphere and the period 1950-1991. Near surface air temperature over North America was obtained from an update of the Climate Research Unit (CRU) product (Jones et al. (1986)). This record will be used back to 1866. Inspection of the individual monthly maps for the era showed a surprising amount of information in the regions wherein the North Pacific decadal mode, that we shall discuss in the next section, has maximum variance. So we have some confidence in the results for the whole period 1866-1992. Rainfall, in gridded field format, was also provided courtesy of the CRU (cf Hulme and Jones (1993)). Comparison of these data with the US divisional data and individual data suggest they are valid in the main areas of interest back to 1900. We show additionally subsurface temperature measurements of the North Atlantic which are described by Levitus et al. (1994).

3   The decadal variability in the North Pacific

3.1   Observations

In order to describe the decadal variability in the North Pacific, we performed a correlation analysis of low-pass filtered SST and 700 hPa height anomalies observed during the period 1950 to 1990. The correlation analysis was performed as follows: First, the observations were detrended and smoothed with a five year running mean filter. We then defined a SST index by averaging the SST anomalies over the region $25^o N - 40^o N$ and $170^o E - 160^o W$ (Fig. 1a). This particular region was chosen because it is located in the region of maximum decadal variability. Finally, we computed locally the correlation coefficients of the SST index with the low-pass filtered SST and 700 hPa anomalies.

The decadal variability is clearly seen in the North Pacific SST index time series, with minima around the years 1960 and 1980 and maxima around the years 1950, 1970, and 1990. Although the time period considered is definitely too short to estimate a reliable period for the decadal variability, there is, however, some evidence of a 20-year time scale. The results of the correlation analysis are shown in Figs. 1b, and 1c. The dominant SST anomaly in the North Pacific is positive (by definition) and surrounded by negative anomalies (Fig. 1b), a result that was found also by Trenberth and Hurrell (1994). These features are statistically highly significant. The SST anomaly pattern shows a remarkable symmetry about the equator, with equatorial Pacific SST anomalies of opposite sign and SST anomalies of the same sign in the Southern Hemisphere relative to the main anomaly in the North Pacific.

The anomalous atmospheric circulation, as expressed by the 700 hPa height anomaly field (Fig. 1c), is dominated by the reverse of the "Pacific-North-American" (PNA) pattern which is an eigenmode of the atmosphere and also excited during extremes of the El Niño/Southern Oscillation (ENSO) phenomenon (Horel and Wallace (1981)). Anomalous high pressure is found over the North Pacific, which was also noted by Trenberth and Hurrell (1994). A negative height anomaly is centered over Canada and a positive height anomaly over the southern part of the United States

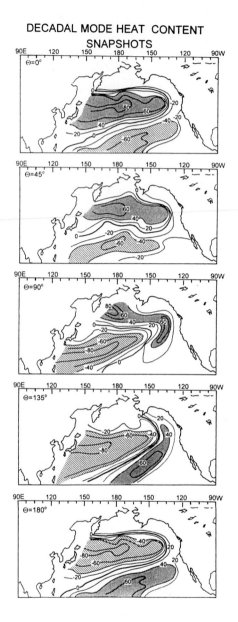

Figure 3: *Reconstruction of anomalous heat content (^{o}Cm) simulated by the CGCM from the leading CEOF mode. The individual panels show the heat content anomalies at different stages of the decadal cycle, approximately 2.5 years apart from each other.*

and Central America. The anomalies associated with the PNA pattern are statistically highly significant, with explained variances relative to the low-pass filtered height anomalies of about 90% in the center of the Aleutian anomaly and about 60% in the centers further downstream.

3.2 Coupled general circulation model simulation

The corresponding results for the coupled general circulation model are shown in Fig. 2. We have chosen a model period that begins in year 43 and extends to year 83 for the correlation analysis. This time period shows a very similar time behaviour in North Pacific SST as the observations during the period 1950-1990 (compare Fig. 2a with Fig. 1a). Overall, the coupled model results are consistent with the observations. The associated SST correlation pattern as derived from the CGCM (Fig. 2b) shows the main positive anomaly in the North Pacific which is surrounded by negative anomalies. The symmetry about the equator is also found in the coupled model simulation. The negative equatorial anomaly, however, is less pronounced than in the observations.

We use the 500-hPa field to describe the changes in the large-scale atmospheric circulation in the coupled simulation. Since the changes in the height field are equivalent barotropic, the model results (Fig. 2c) can be compared directly to the observations (Fig. 1c), although they are presented at a different vertical level. The atmospheric response in the coupled model is also characterized by the reverse of the PNA pattern, but the changes are systematically too zonal relative to the observations. The fundamental spatial phase relationship between the North Pacific SST and height anomaly fields, however, are very similar: anomalously warm SSTs are accompanied by anomalously high pressure over the North Pacific and thus a weakened Aleutian low.

Thus, we believe that our CGCM simulates reasonably well the decadal variations observed in the North Pacific, so that we can use it to investigate the dynamics of the decadal variability in more detail. This is not possible with the available observations which are too sparse and not homogeneous. To elucidate the mechanism producing the decadal variations in SST, we investigated the characteristic evolution of upper ocean heat content anomalies, as defined by the vertically averaged temperatures over the upper 500 meters of the water column via a complex empirical orthogonal function (CEOF) analysis (Barnett (1983)). Before the CEOF analysis, the heat content data were smoothed with a low-pass filter that retained variability at time scales longer than 3 years (further details can

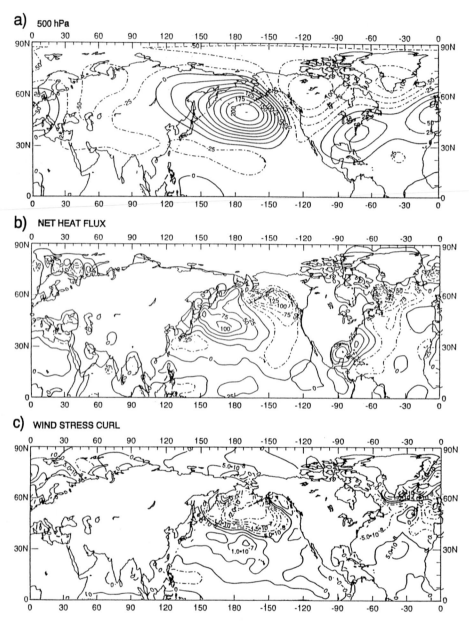

Figure 4: *Atmospheric response to a mid-latitudinal SST anomaly. The upper panel shows the response in the 500 hPa field (gpm), the middle panel shows that of the net surface heat flux (W/m²), while the lower panel shows that of the wind stress curl (Pa/m). The mean fields shown in the panels were obtained by averaging the results of a 12 member ensemble of 30-day perpetual January integrations using the same SST forcing. Please note that the response weakens considerably when the ensemble size is increased.*

be found in Latif and Barnett (1994)).

The leading CEOF mode, accounting for about one third of the variance in the filtered heat content data, has a period of about 20 years. Anomalies in upper ocean heat content reconstructed from this leading CEOF mode (Fig. 3) are displayed at intervals of about two and a half years. When the SST anomalies are fully developed and in a stage corresponding to that shown in Fig. 2b ($\Theta=0$) the main heat content anomaly is positive and covers the majority of the western and central Pacific. A negative anomaly extends to the southwest from North America and increases in area and strength as it approaches the tropics. With time, through one-half of a cycle, the large anomalies rotate around the Pacific in a clockwise fashion reminiscent of the general gyral circulation. Thereafter, the whole sequence of events is repeated, but with reversed signs, and that completes one full cycle.

This evolution is characteristic of the transient response of a midlatitude ocean to a variable wind stress, as described in many theoretical and modeling papers (e. g. Anderson and Gill (1975), Anderson et al. (1979), Gill (1983)). The response is mostly baroclinic at climate time scales longer than several months and involves the propagation of long, relatively fast planetary waves with westward group velocity and their reflection into short, relatively slow planetary waves with eastward group velocity. However, the mean horizontal currents will affect the wave propagation. The net effect of this wave propagation is to modify the strength of the subtropical gyre circulation. In particular, resultant fluctuations in poleward transport of warm tropical waters by the western boundary current lead to the generation of SST anomalies along the path of the Kuroshio and its extension. The spin-up time of the subtropical gyre is several years to a decade or even longer, which accounts for the decadal time scale of the mode under discussion.

3.3 Atmospheric response experiments

The remaining task is to explain the oscillatory nature of this mode. Our hypothesis is that it arises from an instability of the coupled ocean-atmosphere system in the North Pacific. The characteristic SST anomaly pattern exhibits a strong meridional gradient which either reduces or enhances the meridional SST gradient normally found in the central Pacific. Suppose the coupled system is in its reduced meridional SST gradient state

Wind Stress Anomaly

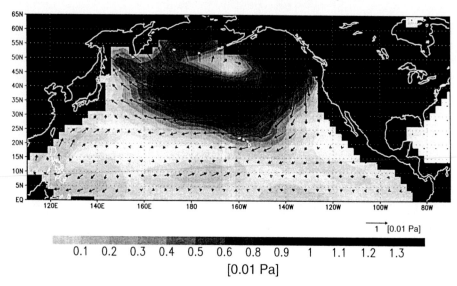

Figure 5: *Wind stress anomaly (Pa) used in the periodic forcing experiments with our OGCM.*

shown in Fig. 2b. It has been suggested by Palmer and Sun (1985) that such a distribution of SST would result in a northward shift of the baroclinic eddy activity in the atmosphere, leading to a weakened Aleutian Low, and subsequently reduced westerly winds over the mid-latitudes ocean. Indeed, the model results show that a reduced meridional SST gradient goes along with anomalous high pressure over the entire North Pacific (Figs. 2b and 2c).

To further investigate the nature of the atmospheric response to the anomalous SST pattern, we forced the atmospheric component of our coupled model in a stand-alone integration by a SST pattern similar to that shown in Fig. 2b. The integration was done in a perpetual January mode and an ensemble of 12 January integrations was performed. The atmospheric response is highly significant and shows the expected result: anomalous high pressure over the North Pacific (Fig. 4a). We note, however, that the mean response becomes much weaker when the ensemble size is increased, but the response pattern remains virtually unchanged (Latif and Barnett (1996)). The response pattern is the weak Aleutian-low extreme of the "Pacific North American" (PNA) mode. The response shown in Fig. 4 arose from an atmospheric model simulation in which tropical SSTs were near their climatological norms, that is the tropics played a minor role in the result. This was confirmed by an additional integration in which SST anomalies south of 25° N were entirely neglected.

The associated changes in the net surface heat flux (Fig. 4b) are such that they tend to reinforce the SST anomalies over the western North Pacific. Heat is anomalously pumped into the ocean in most of the region where it is already warm and vice versa (compare Fig. 2b with Fig. 4b). Changes in the latent and sensible heat fluxes contribute most to the net surface heat flux anomaly. Furthermore, because the westerlies are weakened over the warm SST anomaly, the mean wind speed is reduced, leading to reduced mixing in the ocean, which tends also to strengthen the initial SST anomaly (Miller et al. (1994)). Thus, ocean and atmosphere form a positive feedback system capable of amplifying an initial disturbance, giving rise to instability of the coupled system. The growth is eventually equilibrated by nonlinear processes and the phase switching mechanism described below.

The changes in the wind stress curl (Fig. 4c) force characteristic changes in the ocean, eventually reducing the strength of the subtropical gyre and enhancing meridional SST gradients. The ocean has a memory to past changes in the wind stress and is not in equilibrium with the atmosphere, as described below. It is this transient response of the ocean to imposed wind stress, expressed in terms of planetary wave propagation illustrated in Fig. 3, that provides a mechanism to switch from one phase of the decadal mode to another and, hence, enables the coupled system to oscillate. The oscillation can become easily irregular in the presence of high frequency weather fluctuations. Elements of these ideas were put forward some decades ago (Namias (1959 and 1969), (Bjerknes (1964), White and Barnett (1972)).

3.4 Oceanic response experiments

As described above, a crucial part of the mechanism for the generation of decadal climate variability in the North Pacific is the transient response of the ocean circulation to low-frequency wind stress variations. In order to obtain more insight into the oceanic adjustment, we conducted a series of "ocean-only" integrations in which we forced a coarse-resolution version ($3.5° \times 3.5°$) of our oceanic general circulation model (OGCM) HOPE by prescribed low-frequency periodic wind stress variations. The spatial wind stress pattern by which we forced our OGCM was derived from the stand-alone integrations with our atmosphere model (section 3.3) and is shown in Fig. 5, and the time evolution was assumed to be sinusoidal.

We discuss here two cases in which the forcing period amounts to five years and twenty years, respectively. As can be seen clearly in Fig. 6 showing the time evolution of the sea level anomalies along two particular latitudes bands, the ocean is not in equilibrium with the wind stress forcing at a period of five years. Consistent with planetary wave propagation, slow westward phase propagation can be seen between $20°$ N and $30°$ N (Fig. 6a), while the response further to the north is more stationary in character (Fig. 6b). Similar results are obtained for a forcing period of ten years (not shown). The ocean approaches an equilibrium response for forcing periods longer than about twenty years. This is demonstrated in Fig. 7 which shows the case for a forcing period of exactly twenty years. No phase propagation is seen at this period, neither in the subtropics (Fig. 7a) nor in the mid-latitudes (Fig. 7b). Thus, our ocean-only experiments with low-frequency periodic wind stress forcing support fully the picture that the ocean has a sufficiently long memory to enable decadal oscillations. The precise period of the decadal oscillations, however, will not only depend on the wave transit times but also on the nature of the ocean-atmosphere interactions.

4   Simple coupled model

In summary, both our CGCM results and the uncoupled atmospheric and oceanic experiments support the picture that the decadal mode arises from unstable ocean-atmosphere interactions over the North Pacific, so that the mode is best described as an inherently coupled air-sea mode. The memory of the coupled system resides in the ocean, while the atmosphere responds passively to the slowly varying boundary conditions. It is interesting to note that this paradigm for the generation of the decadal variability over the North Pacific is similar to that for the ENSO phenomenon in the tropical Pacific (e.g. Schopf and Suarez (1988), Neelin et al. (1994)).

It is therefore tempting to construct a simple ocean-atmosphere model

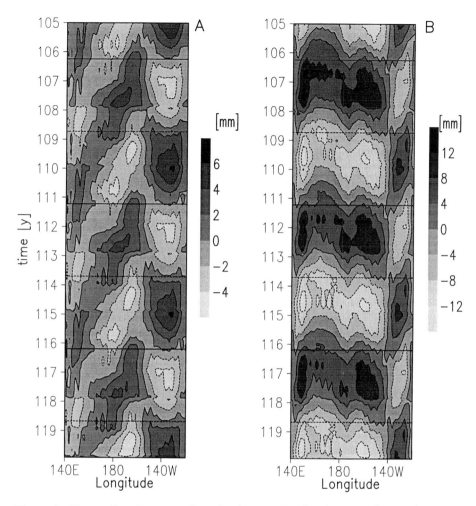

Figure 6: *Hovmoeller diagrams of sea level anomalies (mm) averaged over the region* 20°N − 30°N *(left) and* 35°N − 45°N *(right) for a forcing period of 5 years. Rigdes (troughs) of the wind stress anomaly are indicated by solid (dashed) horizontal lines.*

that retains the essential physics of the decadal variability, as described above. The ocean dynamics is described according to Anderson and Gill (1975) who studied the spin-up problem of a stratified ocean. The relevant equations are the linearized shallow water equations for a particular vertical mode in the quasi-geostrophic approximation, combined into a single

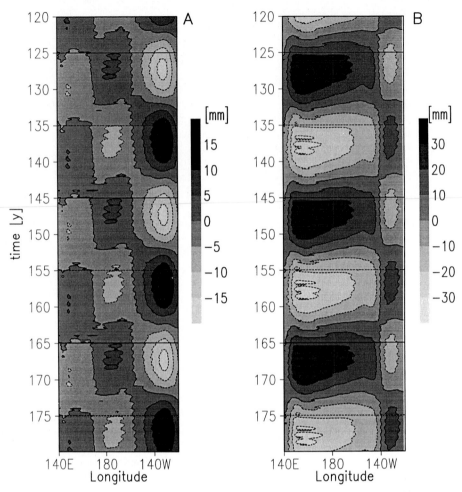

Figure 7: *Hovmoeller diagrams of sea level anomalies (mm) averaged over the region* $20^\circ N - 30^\circ N$ *(left) and* $35^\circ N - 45^\circ N$ *(right) for a forcing period of 20 years. Rigdes (troughs) of the wind stress anomaly are indicated by solid (dashed) horizontal lines.*

equation for the streamfunction ψ:

$$\left(\psi_{xx} + \psi_{yy} - \frac{f^2}{c^2} \cdot \psi\right)_t + \beta \cdot \psi_x = -\frac{1}{H} \cdot \nabla \times \boldsymbol{\tau} + \nu \cdot \psi_{xxxx} \qquad (1)$$

Subscripts denote partial derivatives. Some friction is added in (1) for numerical reasons. The special form of the friction is chosen to account for the existence of western boundary currents. Boundary conditions are ψ=const.=0 along boundaries. The parameters $f = 2\omega \sin \phi, \beta, c, H$, and

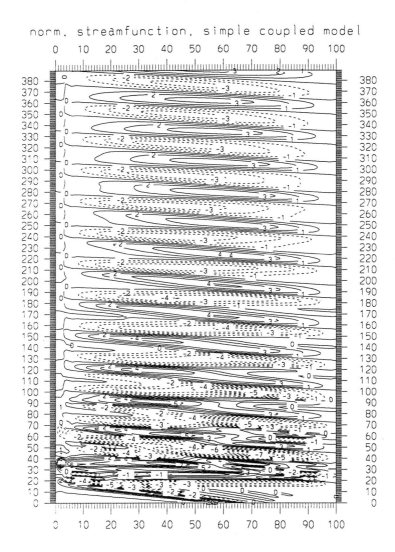

Figure 8: *Hovmoeller diagram of normalized pressure anomalies in the center of the subtropical gyre in a multi-century integration with the simple coupled ocean-atmosphere model. Parameters are:* $c = 2.6\,m/s$, $\nu = 3 \cdot 10^3\,m^2/s$, $\gamma = 0$, $\alpha = 40$. *The basin size is $100°$ in the zonal direction and $50°$ in the meridional direction. A Gaussian disturbance in the eastern quarter of the basin was introduced at time t=0.*

ν denote the Coriolis parameter and its variation with latitude, the speed of the vertical mode considered and a measure of its degree of forcing, and the eddy viscosity, respectively. The surface wind stress vector devided by the reference density is denoted by $\underline{\tau}$, but we consider its zonal component

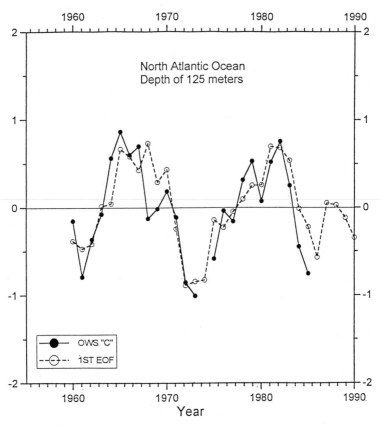

Figure 9: *Time series of subsurface temperature anomalies at 125 m measured at weather ship "C" (35.5°W, 52.5°N) and time series of the first EOF (multiplied by -1) of North Atlantic temperature anomalies at the same depth. Redrawn after Levitus et al. (1994).*

$\underline{\tau}^x$ only in our simple coupled model. Thus, the wind stress curl is given by the meridional derivative of $\underline{\tau}^x$:

$$\nabla \times \underline{\tau} = -\frac{\partial \underline{\tau}^x}{\partial y} \tag{2}$$

The (mean) forcing curl $\overline{\underline{\tau}}$ for our ocean model is assumed to be of the form $A \cdot \sin(2\pi \frac{y}{B})$, with B the meridional extent of the basin. This is a reasonable approximation of the real westwind regime in mid-latitudes, leading to a double-gyre circulation which can be identified with the subtropical and subpolar gyres.

The coupling with the atmosphere is respresented in a rather crude way and by wind stress only. Consistent with the scenario developed above, we assume that the atmosphere has no internal dynamics and responds passively to the oceanic changes. Since our ocean model carries no sea surface temperature, we parameterize the SST variations in terms of the strength of the western boundary current which is taken to be proportional to the streamfunction difference across the two gyres near the eastern boundary. The actual wind stress curl is given by:

$$\nabla \times \boldsymbol{\tau} = \alpha \cdot g(q) \cdot \nabla \times \overline{\boldsymbol{\tau}}, \quad q = [(\partial\psi/\partial y)_W/(\partial\overline{\psi}/\partial y)_W] \qquad (3)$$

Bars denote in (3) quantities that were derived from the steady state solution of the uncoupled system. The subscript "W" indicates that the pressure gradients are computed near the western boundary. The coupling strength is given by the parameter α. In order to describe the non-linear equilibration of the growth, we assume a function g(q) proportinal to $[q/(1+\gamma\cdot|q|)]$, with γ a parameter controling the degree of the non-linearity.

The equations are solved numerically on a $2^o \times 2^o$ grid. Preliminary results indicate that such a simple coupled model oscillates at decadal time scales at sufficiently high values of the coupling strength α. An example is shown in Fig. 8 which shows the normalized streamfunction anomaly in the center of the subtropical gyre as function of longitude and time for a rectangular basin that extends 50° in the meridional and 100° in zonal direction. The other parameters are given in the caption of Fig. 8. Although the oscillation is weakly damped for the set of parameters chosen, a clear periodicity of the order of 30 years can be readily seen. Thus, the results of our simple coupled model support our hypothesis that decadal variability in mid-latitudes can originate from unstable air-sea interactions in mid-latitudes themselves. A complete investigation of the sensitivity of our simple coupled model is underway and will be described by Münnich et al. (in preparation).

5   The decadal variability in the North Atlantic

The type of decadal variability described above can exist in principal also in the North Atlantic. Indeed, observations show some evidence of a decadal cycle in the North Atlantic. This is shown in Fig. 9 which displays observations of subsurface temperature anomalies in the North Atlantic at 125 m depth (Levitus et al. (1994)). The subsurface temperature measurements show some remarkable oscillatory behaviour during the last few decades,

a)

r=0.91

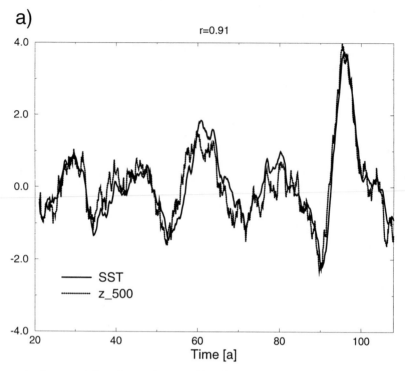

Figure 10: *Canonical Correlation Analysis (CCA) of low-pass filtered (retaining variability with time scales longer than 5 years) North Atlantic SST and 500 hPa height anomalies simulated by the CGCM. The SST and height anomalies were projected onto the first five EOFs prior to the analysis. a) canonical time series.*

with a period of the order of about 15 years. However, the observational record is certainly much too short, in order to prove the existence of a decadal climate cycle in the North Atlantic.

We investigated therefore additionally the variability simulated by our coupled ocean-atmosphere general circulation model "ECHO" in the multi-decadal integration described above. In order to derive the major modes of ocean-atmosphere co-variability, we performed a Canonical Correlation Analysis (CCA, see Barnett and Preisendorfer (1987)) between low-pass filtered simulated North Atlantic SST and 500 hPa height anomalies. The coupled run shows a rather stable decadal cycle in the North Atlantic, as can be inferred from the time series of the leading CCA mode (Fig. 10a). The canonical correlation amounts to about 0.9, and the period of the decadal variations is of the order of 17 years, which is consistent with the observations of Levitus et al. (1994) shown in Fig. 9.

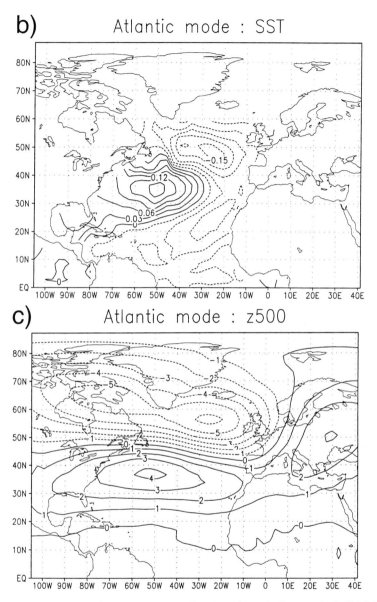

Figure 10: *continued: b) canonical SST (predictor), and c) canonical 500 hPa height (predictand) pattern. The units in (b) and (c) are C and gpm, respectively, and the values are representative of a one-standard deviation change.*

The corresponding SST anomaly pattern (Fig. 10b) shows a positive anomaly along the path of the (model) Gulf stream in the western and central part of the basin centered near 35° N which is surrounded by neg-

NORTH AMERICAN DECADAL MODE
TEMPERATURE/PRECIPITATION INDEX

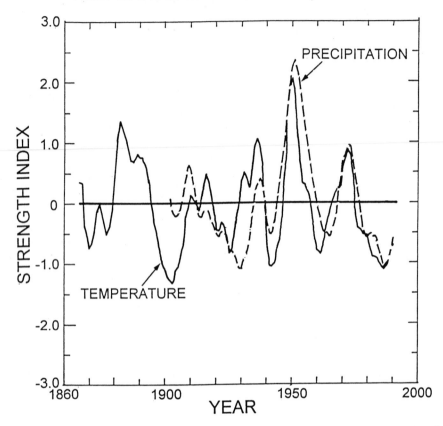

Figure 11: *Normalized near surface temperature and rainfall indices of the North Pacific decadal mode, as derived from observations over North America (details can be found in Latif and Barnett (1996)).*

ative anomalies. Fortunately, the center of the negative SST anomaly in the northeast coincides approximately with the location of weather ship "C", so that the observations shown by Levitus et al. (1994) can be regarded as a good index of the decadal mode. Overall, the characteristics of the decadal mode simulated in the North Atlantic are very similar to that in the Pacific (Figs. 1 and 2), with anomalously high pressure over anomalously warm SST (Fig. 10c).

The anomalous height pattern bears some resemblence to the North Atlantic Oscillation (NAO, see e. g. van Loon and Rodgers (1978)) which is (like the PNA) an eigenmode of the atmospheric circulation and characterized by a dipole pattern, with opposite changes in the Iceland low and Azores high. The centers of action, however, are displaced equatorward by about 10° in the coupled model simulation. All major features in the canonical SST and 500 hPa height fields are statistically highly significant, with explained variances up to 60% relative to the low-pass filtered values.

The mechanism behind the decadal variability in the North Atlantic is similar to that for the North Pacific (see section 3). The variations in heat content or subsurface temperature are qualitatively similar to those found in the Pacific (Fig. 3) and show basically the same propagation characteristics (not shown). Thus, we belive that the decadal variability in the North Atlantic is generated by the same mechanism as its counterpart in the North Pacific. This would imply that the period of the North Atlantic mode is about half the period of the North Pacific mode, since the basin size of the North Atlantic is about half the basin size of the North Pacific. Our results indicate, however, that the time scales of the two oscillations are very close to each other. The reasons for this are still under investigation.

6 Summary and discussion

A hierarchy of numerical models was used to understand the decadal variability observed in the Northern Hemisphere, and we propose a mechanism for its generation. Our results suggest that a considerable part of the decadal variability in both the North Pacific and North Atlantic can be attributed to cycles with periods of approximately 15 to 20 years. The decadal modes originate from unstable ocean-atmosphere interactions and must therefore be regarded as inherently coupled phenomena. The existence of such cycles implies the potential for long-range climate forecasting at decadal time scales over North America and Europe.

The scenario for the generation of the decadal mode is similar to that developed by Bjerknes (1964) for the Atlantic Ocean. According to this picture, the memory of the coupled system resides in the ocean. The ocean adjusts slowly to past variations in the surface wind stress field, and these slow variations in the wind-driven ocean circulation are crucial in setting the time scale of the decadal mode. Wave and advective processes are both found to be important in the ocean adjustment. The wave adjustment, however, appears to be the dominant process, as discussed by Latif and Barnett (1996). The atmosphere responds passively to the changes in the lower boundary conditions. However, since the decadal mode is an inherently coupled phenomenon, the feedback of the atmosphere onto the ocean is a crucial part of the dynamics of the decadal mode.

Our results are in conflict with several previous studies, since we found the tropics played a minor role for the generation of the decadal mode. The studies of Trenberth and Hurrell (1994) and Graham (1994) argue basically that low-frequency changes in tropical Pacific SST introduce the signal into the North Pacific through a changed atmospheric circulation. Jacobs et al. (1994) argues also that decadal variability in the North Pacific is forced by the tropics, but through the ocean by the propagation of planetary waves in the aftermath of strong ENSO extremes, such as the 1982/1983 warm event. We did not find much evidence for an active role of the tropics in the generation of the decadal mode in the North Pacific. Our view of an independent mid-latitudinal mode is supported by the findings of Robertson (1996). He investigated a multi-century integration with another CGCM and found a similar mode in the North Pacific to that described here. However, the integration Robertson (1996) analyzed shows virtually no ENSO-type variability in the tropical Pacific so that some of the proposed tropical forcing mechanisms cannot operate in that integration.

Another point of controversy which might originate from our study is the question of whether the atmosphere is sensitive to mid-latitudinal SST anomalies. Our results suggest indeed that mid-latitudinal SST anomalies force a significant atmospheric response, as implied by the early study of Palmer and Sun (1985). Further, we were able to reproduce the dominant atmospheric response pattern found in the coupled integration and observations by forcing our atmosphere model in a stand-alone mode by the characteristic North Pacific SST pattern of the decadal mode. We speculate that changes in the surface baroclinicity and resultant changes in the transient activity are crucial in establishing the time-mean response. If correct, this would imply that atmosphere models need to simulate the eddy activity rather well when they are used in a coupled model to study decadal variability. This would require quite high resolution for climate integrations, at least of the order of the T-42 resolution we used in our coupled integration.

Finally, we would like to discuss critically the predictability of the decadal modes. Our results suggest that their predictability is probably less than that of ENSO, for example, if one measures the predictability

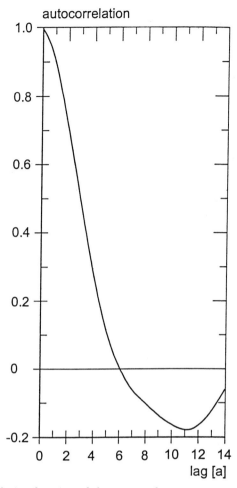

Figure 12: *Autocorrelation function of the near surface temperature index of the North Pacific decadal mode (shown in Fig. 11).*

limits in terms of fractions of the cycle period. This can be inferred from Fig. 12 showing the autocorrelation function of an index time series of the North Pacific decadal mode (Fig. 11), as derived from observations (see Latif and Barnett (1996) for details). Although the structure of the autocorrelation function shows some weak evidence for a periodicity of approximately 20 years, the minimum correlation at lead times of about 10 years are insignificant and of the order of -0.2 only. This suggests that the predictability of the decadal mode is considerably less than one-half cycle. In contrast, the autocorrelation functions of typical ENSO indices show significant correlations at lead times of the order of about two years, which corresponds to approximately half an ENSO cycle (e.g. Wright, 1985). Thus, the decadal mode is strongly damped and the internal "noise" in the coupled ocean-atmosphere system will limit its predictability considerably.

Acknowledgement

We would like to thank Drs. D. Cayan, C. Frankignoul, P. Müller, W. White, and E. Zorita for many fruitful discussions. We thank Drs. T. Stockdale, J. Wolff, and G. Burgers for helping in the development of the coupled general circulation model. We thank the Theoretical Applications Division of the Los Alamos National Laboratory for making most of the computer time available for the CGCM run, Dr. D. Poling for tending the run and managing the extensive output files, and Dr. C. Keller for facilitating the cooperation on the CGCM project. In addition, we thank M. Junge, J. Ritchie, T. Tubbs, and M. Tyree for providing computational assistance. This work was supported in part by the National Science Foundation under grant NSFATM 9314495, the U.S. Department of Energy's CHAMP program under grant DE-FG03-91ER61215 (T.P.B.), NOAA via the Lamont/SIO Consortium program (M. M.), the German Climate Computer Centre (DKRZ), the European Community under grants EV5V-CT92-0121 and EV5V-CT94-0538, and the Bundesminister für Forschung und Technologie under grant 07VKV01/1 (M.L.).

References

ANDERSON, D.L.T. AND A.E. GILL, 1975: Spin-up of a stratified ocean with application to upwelling. *Deep-Sea Research*, **22**, 583-596.

ANDERSON, D.L.T., K. BRYAN, A.E. GILL, AND R.C. PACANOWSKI, 1979: The transient response of the North Atlantic - some model studies. *J. Geophys. Res.*, **84**, 4795-4815.

BARNETT, T. P., 1983: Interaction of the Monsoon and Pacific trade wind system at interannual time scales. Part I: The equatorial zone. *Mon. Wea. Rev.*, **111**, 756-773.

BARNETT, T. P. AND R. PREISENDORFER, 1987: Origins and levels of monthly forecast skill for United States surface air temperatures determined by canonical correlation analysis. *Mon. Wea. Rev.*, **115**, 1825-1850.

BJERKNES, J., 1964: Atlantic Air-Sea Interaction. *Adv. in Geophys.*, **10**, 1-82.

BRANKOVIC, C., T.N. PALMER, AND L. FERRANTI, 1994: Predictability of seasonal atmospheric variations. *J. Climate*, **7**, 217-237.

DELWORTH, T., S. MANABE, AND R.J. STOUFFER, 1993: Interdecadal variations of the thermohaline circulation in a coupled ocean-atmosphere model. *J. Climate*, **6**, 1993-2011.

DKRZ, 1992: The ECHAM-3 atmospheric general circulation model. Edited by Deutsches Klimarechenzentrum Modellbetreuungsgruppe. *Technical report no. 6. Available from: DKRZ, Bundesstr. 55, D-20146 Hamburg, Germany.*

GILL, A,. 1982: Atmosphere-Ocean Dynamics. *Academic Press, New York.*

GRAHAM, N.E., 1994: Decadal-scale climate variability in the 1970's and 1980's: Observations and model results. *Clim. Dyn.,* **10,** 135-162.

HOREL, H.D. AND J.M. WALLACE, 1981: Planetary-scale atmospheric phenomena associated with the Southern Oscillation. *Mon. Wea. Rev.,* **109,** 813-829.

HULME, M. AND P.D. JONES, 1993: A historical monthly precipitation dataset for global land areas: Applications for climate monitoring and climate model evaluation. pp. A/14-A/17, *In: Analysis methods of precipitation on a global scale, Report of a GEWEX Workshop, 14-17 September, 1992, Koblenz, Germany, WMO-TD-No. 558, Geneva.*

JACOBS, G.A., H.E. HURLBERT, J.C. KINDLE, E.J. METZGER, J.L. MITCHELL, W.J. TEAGUE, AND A.J. WALLCRAFT, 1994: Decade-scale trans-Pacific propagation and warming effects of an El Niño anomaly. *Nature,* **370,** 360-363.

JIANG, S., F.-F. JIN, AND M. GHIL, 1995: Multiple equilibria, periodic and aperiodic solutions in a wind-driven, double-gyre, shallow-water model. *J. Phys. Oceanogr., submitted.*

JONES, P.D., R.S. BRADLEY, H.F. DIAZ, P.M. KELLY, T.M.L. WIGLEY, 1986: Northern Hemisphere surface air temperature variations: 1851-1984. *J. Clim. Appl. Met.,* **25,** 161-179.

KUSHNIR, Y., 1994: Interdecadal variations in North Atlantic sea surface temperature and associated atmospheric conditions. *J. Climate,* **7,** 141-157.

LATIF, M. AND T.P. BARNETT, 1994: Causes of decadal climate variability over the North Pacific and North America. *Science,* **266,** 634-637.

LATIF, M. AND T. P. BARNETT, 1996: Decadal variability over the North Pacific and North America: Dynamics and predictability. *J. Climate, accepted.*

LATIF, M., T. STOCKDALE, J.O. WOLFF, B. BURGERS, E. MAIER-REIMER, M.M. JUNGE, K. ARPE, AND L. BENGTSSON, 1994: Climatology and variability in the ECHO coupled GCM. *Tellus,* **46A,** 351-366.

LEVITUS, S., T. P. BOYER, AND J. ANTONOV, 1994: World ocean atlas 1994. Volume 5: Interannual variability of upper ocean thermal structure. *U. S. Department of Commerce, NOAA, Washington D. C.*

MILLER, A.J., D.C. CAYAN, T.P. BARNETT, N.E. GRAHAM, AND J.M. OBERHUBER, 1994: Interdecadal variability of the Pacific Ocean: model response to observed heat flux and wind stress anomalies. *Clim. Dyn.,* **9,** 287-302.

NAMIAS, J., 1959: Recent seasonal interactions between North Pacific waters and the overlying atmospheric circulation. *J. Geophys. Res.,* **64,** 631-646.

NAMIAS, J., 1969: Seasonal interactions between the North Pacific and the atmosphere during the 1960s. *Mon. Wea. Rev.,* **97,** 173-192.

NEELIN, J. D., M. LATIF, AND F.-F. JIN, 1994: Dynamics of coupled ocean-atmosphere models: The tropical problem. *Annu. Rev. Fluid. Mech.*, **26**, 617-659.

NITTA, T. AND S. YAMADA, 1989: Recent warming of tropical sea surface temperature and its relationship to the northern hemisphere circulation. *J.Meteor. Soc. Japan*, **67**, 375-383.

PALMER, T.N. AND Z. SUN, 1985: A modelling and observational study of the relationship between sea surface temperature in the northwest Atlantic and the atmospheric general circulation. *Quart. J.R. Meteorol. Soc.*, **111**, 947-975.

PHILANDER, S. G. H., 1990: El Niño, La Niña, and the Southern Oscillation. *Academic Press, San Diego, Ca. 92101.*

ROBERTSON, A.W., 1996: Interdecadal variability over the North Pacific in a multi-century climate simulation. *Climate Dynamics, in press.*

ROECKNER, E., K. ARPE, L. BENGTSSON, S. BRINKOP, L. DTMENIL, M. ESCH, E. KIRK, F. LUNKEIT, M. PONATER, B. ROCKEL, R. SAUSEN, U. SCHLESE, S. SCHUBERT, AND M. WINDELBAND, 1992: Simulation of the present-day climate with the ECHAM model: Impact of model physics and resolution. Report-No. 93, *October, 1992, 171 pp. Available from Max-Planck-Institut für Meteorologie, Bundesstr. 55, D-20146 Hamburg, Germany.*

SCHOPF, P. S. AND M. J. SUAREZ, 1988: Vacillations in a coupled ocean-atmosphere model. *J. Atmos. Sci.*, **45**, 549-566.

TRENBERTH, K.E. AND J.W. HURRELL, 1994: Decadal atmosphere-ocean variations in the Pacific. *Clim. Dyn.*, **9**, 303-319.

VAN LOON, H. AND J. G. RODGERS, 1978: The seasaw in winter temperatures between Greenland and Northern Europe. Part I: General description. *Mon. Wea. Rev.*, **106**, 296-310.

WEISSE, R. AND U. MIKOLAJEWICZ, 1994: Decadal variability of the North Atlantic in ocean general circulation model. *J. Geophys. Res.*, **99**, 12.411-12.421.

WHITE, W.B. AND T.P. BARNETT, 1972: A servomechanism in the ocean/atmosphere system of the mid-latitude North Pacific. *J. Phys. Oceanogr.*, **2(4)**, 372-381.

WRIGHT, P.B., 1984: Relationships between indices of the Southern Oscillation. *Mon. Wea. Rev.*, **112**, 1913-1919.

YANG, J. AND J. D. NEELIN, 1993: Sea-ice interactions with the thermohaline circulation. *Geophys. Res. Letters*, **20**, 217-220.

THE CLIMATE RESPONSE TO THE CHANGING GREEN-HOUSE GAS CONCENTRATION IN THE ATMOSPHERE

Lennart Bengtsson
Max Planck Institute for Meteorology
Hamburg, FRG

Contents

1 Introduction

The question whether climate changes, on timescales shorter than say several thousand years, at all can take place has not until comparatively recently been carefully considered. The question is related to the accelerating increase in the so-called greenhouse gases, carbon dioxide, methane, nitrous oxide and CFCs and to what extent and how fast this increase is likely to lead to climate changes. It is indeed of considerable interest that a few gases, preferably water vapour and carbon dioxide which exist in quite small concentrations, carbon dioxide only little more than 0.03%, are

NATO ASI Series, Vol. I 44
Decadal Climate Variability
Dynamics and Predictability
Edited by David L. T. Anderson and Jürgen Willebrand
© Springer-Verlag Berlin Heidelberg 1996

absolutely crucial to life on our planet since only due to their ability to efficiently absorb terrestrial radiation the average surface temperature of the earth is increased by 33^oC or from -18^oC to $+15^oC$. The most common gases of our atmosphere, oxygen and nitrogen, which take up more than 99% of the total volume, are almost completely transparent to both the radiation from the sun and the terrestrial radiation. The first attempt to estimate the effect of atmospheric greenhouse gases on climate was made by the Swedish physicist Arrhenius (1896). Arrhenius undertook a careful examination of available measurements concerning the absorption of terrestrial radiation in the atmosphere and was able, on the basis of these measurements, to correctly quantify the role of carbon dioxide and water vapour. By making use of a simple energy balance model he also calculated the change in the surface temperature which would result in an increase respectively a decrease in the atmospheric concentration of CO_2 including the feedback from water vapour. Based on available data for surface albedo and cloud distribution, Arrhenius calculated that a doubling of CO_2 from 300 ppm to 600 ppm would result in a warming of about 5^oC. A similar reduction in the concentration to 150 ppm would give a corresponding cooling. Arrhenius himself did not believe that man-made release of carbon dioxide into the atmosphere could have any effect, due to the very small amounts of fossil coal burning at the time (500 Mtons annually), on the concentration of carbon dioxide in the atmosphere because the amount was such a tiny part compared to the natural exchange between the atmosphere, the oceans and the land surface. Today the situation is different with a 20-fold increase in the burning of fossil fuels. With the exception of some interesting papers by Callendar (1938, 1949), who indeed was the first to raise the issue whether the anthropogenic production could influence climate, and pioneering work thereafter e.g. by Bolin and Eriksson (1959), Möller (1963), Manabe (1971), Manabe and Wetherald (1975), it was not until the 1980ies that the issue started to attract world-wide scientific as well as public interest.

2 Basic energy processes of the climate system

The concept of climate, as will be considered here, will be defined as the totality of weather conditions which are experienced during a given time interval. These weather conditions are quantified by different meteorological variables such as surface pressure, wind, temperature, humidity,

The global heat balance
(observed)

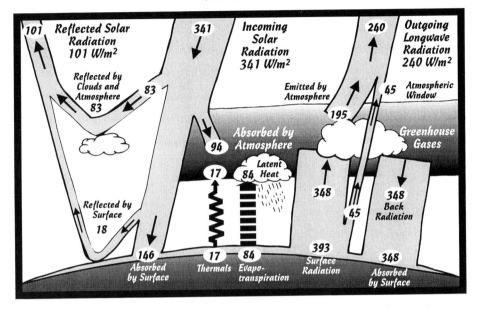

Figure 1: *The global heat balance of the atmosphere annually averaged. The actual values are compiled from different sources. For further information see text. Units are in Wm^{-2}*

cloudiness, precipitation, etc. The climate is then described by suitably selected statistical quantities including mean values, variances and covariances and extreme values with respect to the meteorological variables over conveniently selected time-periods. There is no stringent definition of the time interval. Short changes, however, over weeks or months are seldom characterized as climate, while year-to-year occasionally, decade-to-decade frequently and century-to-century usually are considered as climate. Traditionally, in classical climatology, periods of 30 years (1960-1990 etc.) are used to describe the present climate. However, it may be more convenient to use a wider definition of climate and generally consider phenomena on timescales beyond which only predictions in a statistical sense are possible. Weather and climate are the manifestation of complex, mainly chaotic processes on a rotating planet driven by temperature differences resulting from differential heating by the sun. Fig. 1 describes the energetics of the atmosphere averaged over the year and for the whole earth. However in spite of major improvements in measurement technology and observational systems in recent years there are still substantial inaccuracies in our knowledge of the 3-dimensional radiational forcing and the hydrological cycle. The global average incoming radiation from the sun is 341 W/m^2 (1/4 of the solar constant), of which about 30% is reflected back to space due to reflections from clouds, from the surface of the earth and from back-scattering by the air and dust particles in the air (planetary albedo). Of the remaining 240 W/m^2 some 146 W/m^2 reach the surface while the remaining part is absorbed in the atmosphere. The same amount of heat, 240 W/m^2, leaves the planet through terrestrial radiation. However, that takes place in a complex way, since the surface is cooled (and the atmosphere correspondingly heated) by both surface radiative emission and fluxes of sensible and latent heat. The atmosphere in return radiates back to the surface (due to water vapour and greenhouse gases) and the outgoing net surface long wave radiation amounts only to some 45 W/m^2. The sensible heat flux from the surface is 14-20 W/m^2 and the latent heat flux is estimated to be as high as 80-88 W/m^2. It is important to note that the moisture flux (evaporation) is cooling the surface of the earth some 80% more than the net radiative cooling. Data for absorption of short wave radiation in the atmosphere have recently been found to be higher than previously estimated. The same is true for the long wave radiation. The figures here are based on recent estimates, e. g. from Hartmann (1993), Ohmura and Gilgen (1993), Hahn et al.(1994), Giorgetta and Wild (1995). In general the accuracies of most energy fluxes are of the order of some 5 W/m^2. The ERBE data for example has a residual error of 6 W/m^2 in the globally averaged energy balance. The different heating between pole and equator generates kinetic energy through the work of the pressure and the Coriolis force. However, only some 3 W/m^2 is being converted into kinetic energy, so the earth is a very inefficient engine having an efficiency factor of less than 1%! The overall direct effect of the greenhouse gases can be calculated under simplified although not particularly realistic conditions. Assuming radiation balance under clear sky it is found that the average temperature of the earth's surface now at $+15^oC$ would fall to -18^oC. The relative contributions from the optically active gaseous components are: water vapour (H_2O) -21 K, carbon dioxide (CO_2) -7 K and the remaining part from the other greenhouse gases including methane (CH_4), nitrous oxide (N_2O) and ozone (O_3), (Kondratyev and Moskalenko, 1984).

Our closest planets in the solar system are also very much influenced by the greenhouse gases, particularly Venus (Fig. 2). However, it is not

	Surface pressure	Greenhouse gas	Surface temperature without greenhouse effect	Observed temperature	Grennhouse effect
Venus	90.000	90.00% CO_2	−46 ˚C	477 ˚C	523 ˚C
Earth	1.000	0.04% CO_2 1.00% H_2O	−18 ˚C	15 ˚C	33 ˚C
Mars	0.007	80.00% CO_2	−57 ˚C	−47 ˚C	10 ˚C

Figure 2: *The greenhouse effect for the three planets Venus, Earth and Mars.*

possible from this calculation to draw the conclusion that the temperature of the earth would drop by 33 K if we were to remove the greenhouse gases or that the temperature would drop by 7 K if we would remove all CO_2 from the atmosphere. Such a change would in fact set into motion complex feedback processes changing the surface fluxes of heat and moisture and modifying the albedo and cloudiness. It is not unlikely that the removal of all CO_2 from the atmosphere through a series of processes including interactions with the oceans, the land surfaces, the biosphere and the cryosphere could finally put the earth into an eternal glaciation! Realistic climate investigations must therefore pay attention to the totality of the climate system and the interaction between its different subsystems. We will discuss this further under section 4.

3   Climate change mechanisms

Over any period for which there exist observations or indirect evidence of the climate there are indications of climate changes. Measurements of surface temperature for substantial parts of the globe exist for somewhat more than 100 years, for certain limited observation records for some 250 years or longer. For the last 120 years these records show indication of a temperature increase of some 0.5^oC but with distinct variations. Of particular interest is a period of cooling for the northern hemisphere between 1940 and 1970. A marked warming has taken place since the end of the 1970ies. (Fig. 3). For longer periods back, the climate has been reconstructed by means of indirect measurements. Such reconstructions make use of a multitude of information concerning ice distribution, vegetation,

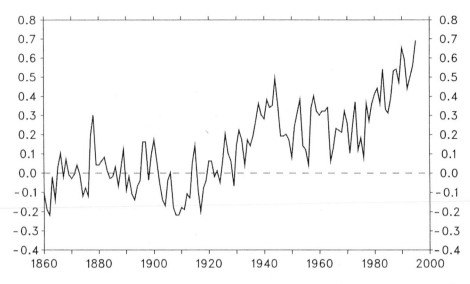

Figure 3: *Annual global mean surface temperature 1860-1995 (Parker, pers. comm.). Reference period 1860-1890.*

sea-level height etc. Of particular importance have been measurements of the relation of two isotopes of oxygen; ^{18}O and ^{16}O in sea-level deposits and in ice-core measurements at Antarctica and Greenland. This relation provides information at which temperature condensation is taking place. There exists today among the paleoclimatologists a broad consensus in the reconstruction of the Earth's temperature for say the last few million years (Berger 1980). Below we will shortly discuss possible physical explanations to the climate changes in the past.

3.1   Changes in the solar radiation

The radiation from the sun is the primary source of energy for the earth's climate system and variations in the amount of solar radiation are consequently the prime cause of climate change. There are two distinct sources of this variability. The first is related to variations in the orbital parameters which affect the earth's climate on timescales of 10,000 to 100,000 years. These changes have to do with the precession of the orbit of the earth, changes in the tilt of the earth axis against the ecliptic plane, and the variation in the ellipticity of the orbit of the earth. The precession movements and the tilt of the earth axis affects the geographical distribution of radiation, while variation in the ellipticity of the orbit affects the annual variability of the incoming radiation. Milankovic (1930) and later Berger (1980) have carefully calculated this effect and have been able to demonstrate that it is at least qualitatively able to explain the glaciation cycle. The second source of variability is due to changes in the total solar irradiance. There are some changes in the near ultraviolet part of the solar spectrum which are taking place over the 11-year sunspot cycle. These changes are generally insignificant and induce variations of no more than 0.01% in the total irradiance. Of greater importance, in terms of the direct effect on climate, are changes affecting the total irradiance. Spaceborn measurements available for the last decade indicate changes of the order of 1 W/m^2 in the value of the solar constant, Foukal and Lean (1990). However, it is important to stress that only ca. 17% of this amount will enter the climate system. The question of larger changes over longer periods has been discussed considerably, see for example Marshall Institute (1989), but no conclusive evidence has yet been provided. Such evidence must in any case be based on real observations of the irradiance or based upon verifiable theories of the energetics of the solar atmosphere. To use proxy climate data, say from the last several hundred years, to postulate changes in the solar constant is a fallacy, since climate changes over the last few hundred years can be explained by natural, inherent fluctuations in the climate system.

3.2 Changes in the greenhouse gases

Concentration of carbon dioxide and methane during the last 160,000 years has been determined from the core measurements from Vostok, Antarctica (Chappelaz et al., 1990). These measurements indicate considerable long term variations. They are also broadly correlated by the estimated temperature changes, in turn based on the measured ^{18}O oxygen-isotope concentration. At the end of the last glaciation the concentration of CO_2 and CH_4 did grow to their preindustrial values 280 ppm and 0.65 ppm, respectively. These values then hardly varied, and certainly not during the millennium prior to the beginning of the industrial age around 1750. The CO_2 concentration since then has increased virtually exponentially and amounts presently to 365 ppm (1996).

Table 1 gives some key data for the most important greenhouse gases, their present concentration, their present annual increase, their residence

Trace gases	Chemical symbol	Present concentration (1992)	Atmopheric residence time (year)	Present annual trends	Increased terrestrial radiation (1750 - 1990)	Most important anthropogenic sources
Carbon dioxide	CO_2	355 ppm	50 – 200	0.4%	1.56 W m-2	Fossil fuels, deforestation land use and land erosion
Methane	CH_4	1.71 ppm	12 – 17	0.8%	0.47 W m-2	Rice paddies, fossil fuels, biomass burning, gas leaks
Nitrous oxide	N_2O	0.31 ppm	120	0.25%	0.14 W m-2	Fertilizer, fossil fuels, biomass burning
Chlorofluorcarbons (CFCs)	CCl_3F CCl_2F_2 etc.	~800 ppt	10 – 50000	2% – 7%	0.28 W m-2	Refrigerants solvents (aerosol propellants)
Other greenhouse gases					0.25 W m-2	
Total					2.75 W m-2	

Table 1: *Greenhouse gases in the atmosphere compiled after IPCC 1994.*

time in the atmosphere and the direct enhancement of the terrestrial radiation due to this increase. The direct radiation effect can be calculated quite accurately through a detailed integration of the equation for radiative transfer. It is interesting to note that for the period 1750-1990 the relative contribution of CO_2 is only about 60%. Another aspect which needs to be stressed, is that the relative absorptive efficiency of the different gases varies; for CO_2 it is proportional to the logarithm of the concentration, for CH_4 and N_2O to the square root and for the CFCs to the concentration itself. This is the reason why the CFC's with their very small concentration (8×10^{-7}) yet have such a relatively high impact. It is interesting to note that already Arrhenius was able to deduct the absorptive properties of CO_2 in this respect. In previous model studies with the greenhouse gases the other greenhouse gases have been represented by CO_2 equivalents. This is a crude simplification since the effect, in particular on the vertical temperature distribution, is quite different. In the experiment described in section 5 this has been remedied.

Anthropogenic sulfate burden in mg/m^2 , annual mean

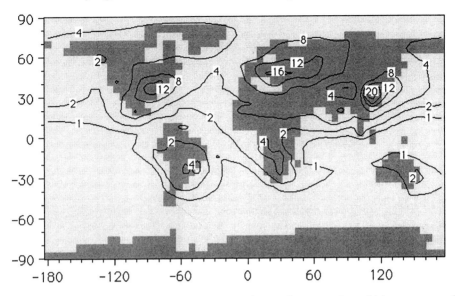

Figure 4: *The total amount of anthropogenic sulfate in the atmosphere. Values are annual averages expressed as mg sulfate per square meter. The maxima are associated with the main industrialised regions (Feichter, pers. comm.).*

3.3 Changes in atmospheric aerosols

Aerosol particles play an important role in the climate system because of their interaction through absorption and scattering with solar and terrestrial radiation, as well as through their influence on cloud processes and thereby, indirectly, on radiative fluxes. The total amount of suspended particles in the air varies from less than $1\mu g/m^3$ in the polar regions and at mid-ocean, to $1mg/m^3$ in desert dust outbreaks or in dense fumes from for example forest fires. However, the residence time of aerosol particles in the atmosphere and of their precursor gases is only of the order of days and weeks. The residence time for some of the greenhouse gases for comparison is of the order of 100 years (table 1). The short lifetime of tropospheric aerosols means that the effect is geographically restricted over and downstream to the source. Considerable differences exist therefore from area to area and between the two hemispheres.

It has been established from analysis of Greenland ice core that amounts of sulfate, nitrate and trace metals, derived mainly from atmospheric aerosols, have been increasing since the beginning of industrialisation (Neftel et al. 1985). The most important contribution comes from sulfate where today the anthropogenic contribution dominates. The annual release of SO_2, mainly from fossil fuel combustion, amounts to some 80 M tons and is now larger than the natural emission of sulfate components. Fig. 4 shows the annual mean in the total amount of anthropogenic sulfate in mg/m^2 (Feichter pers. inf., 1996). Over the most polluted regions of Europe and North America the sulfate levels have gone up by more than a factor of 10. Experiments have been undertaken to estimate the effect of the increase in the sulfate on atmospheric radiation. Since the effect of the aerosols mainly is to modify the reflectivity of solar radiation directly or indirectly via increased cloudiness, sulphate aerosols cool the atmosphere. The present effect is localized to certain areas of the Northern Hemisphere over and downstream the source region. In these areas it is likely that it practically can offset the present warming by the greenhouse gases. The calculations, however, are difficult to do and the result is very preliminary (Charlston et al. 1991).

Stratospheric aerosols have a more global effect due to their much longer residence time (several years). They mainly enter the stratosphere in relation to major volcanic eruptions of the explosive type whereby large amounts of sulphur particles can be emitted high up in the atmosphere. In recent years there have been two major eruptions of this kind, El Chicon in 1982 and Pinatubo in 1991. Fig. 5 shows the radiative effects from these eruptions. It follows that a series of major eruptions occurring over a longer period of time could create an overall cooling effect. Again, however, the effect from a single event is limited to at most a few years.

3.4   Internal, natural variations

Meteorological processes are typically chaotic, and infinitesimal errors in the initial data or in the governing equations are rapidly growing. This is the main reason why weather forecasts cannot be made very much longer than a few weeks ahead. Since the errors due to non-linear interaction are rapidly spreading to the whole spectrum of atmospheric motions it follows that even the large scale features, which dominate the circulation over months and seasons, also change. Differences in the initial state, by say $10^{-2}K$, can lead to different types of circulation particularly at high latitudes which can result in a strong westerly flow over Western Europe or alternatively in a blocking pattern. Even in such a basic quantity as the global averaged surface temperature such chaotic variations are clearly vis-

Temperature changes of the lower stratosphere

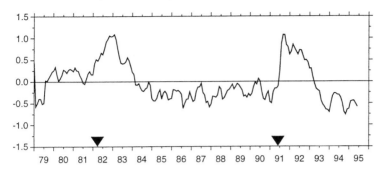

Temperature changes of the lower troposphere

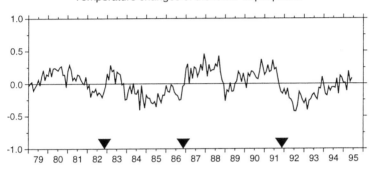

Figure 5: *The two curves show the global change in temperature in the lower stratosphere and the lower troposphere for the period 1979- 1995. Data have been obtained by passive microwave radiometry from satellites (NOAA-6 and NOAA-7) by Spencer and Christy (1990). The two marked warmings in the stratosphere are caused by the volcanic eruptions in 1982 (El Chicon) and 1991 (Mount Pinatubo). The response in the troposphere is not clearly detectable because of influence from El Niño events. The volcanic eruptions and the peaks of the El Niño events are marked with arrows.*

ible on annual and longer timescales. However, there exists a principal difference between tropics and extra-tropics. Tropical circulation is strongly constrained by the temperature of the surface of the oceans, and repetitive calculations using prescribed ocean surface temperatures (Bengtsson et al., 1996) reproduce quite accurately the averaged atmospheric circulation over longer periods than months and seasons and are thus essential for determining the climate. This is not the case at middle and high latitudes where the forcing from the atmospheric boundary conditions is not as clearly seen. Studying the complete climate system, we see that natural variations in the oceans, such as the El Niño phenomenon, significantly affect the tropical climate on timescales of several years and fluctuations in the thermohaline circulation which are common in the North Atlantic on even longer timescales. Bryan (1986) has shown, by using an idealized model of the Atlantic Ocean and a given forcing of the circulation with fresh water fluxes, that the ocean can develop at least two completely different stable sets of circulation depending on the initial state. One of these states is characterized by the present vigorous North-Atlantic circulation with a strong northward transport of heat, the other state has practically no circulation at all and hardly any heat transport. Whether major changes in the thermo-haline circulation will take place on timescales of centuries and shorter is hardly likely, although a major melting of Arctic ice or increase in freshwater due to atmospheric processes such as increased precipitation cannot be ruled out. Other examples of long term fluctuations are the draughts in the Sahel region. When such draughts are established, they have a tendency of becoming very long-lasting due to the strong feedback between local evapotranspiration and convective precipitation, convective precipitation being the dominating precipitation mechanism in this part of the world. At higher latitudes such feedbacks can enhance and prolong a dry summer, but are generally broken during the winter when the large scale synoptic circulation can transport moisture over long distances. In conclusion, the variation in climate during the last century or so can by and large be explained by natural internal variations.

4   Modelling climate change

The complexity of climate processes and the many interactions and feedbacks in the climate system require a consistent quantitative approach. Series of different models have been developed over the years from simple conceptual ones to full scale 3-dimensional models. For an in-depth evaluation of climate models the reader is referred to available textbooks such as Trenberth (1992). The models which best hold the promise of being able to deal with the complexity of the climate system are the comprehensive 3-dimensional global coupled models, incorporating the interaction between the atmosphere, oceans, cryosphere and the land surfaces. These are the only models which can reproduce realistically the characteristics of climate variability in time and space and the many feedback processes of the climate system. Dynamical processes in the atmosphere and ocean must be properly resolved, and physical processes and exchange mecha-

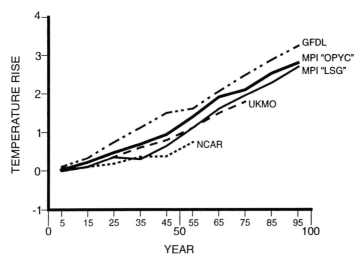

Figure 6: *Decadal mean changes in globally averaged surfaced temperature (^{o}C), in various coupled ocean-atmosphere experiments used in transient climate change integration in accordance with the IPCC Scenario A, "Business as usual". This corresponds approximately to an annual increase of the atmospheric CO2 concentrations of 1%. Note that the scenarios employed differ from model to model, and that the effect of temperature drift in the control simulation has been removed. GFDL (Manabe et al., 1991), MPI OPYC (Lunkeit, 1993), MPI LSG (Cubasch et al., 1992), UKMO (Murphy and Mitchell, 1995), NCAR (Washington and Meehl, 1991). Modified figure and caption from IPCC, 1992.*

nisms in and between atmosphere, ocean, land and the cryosphere must be consistently parameterized with particular emphasis on minimizing systematic errors. During the last years the modelling emphasis has been on transient time-dependent climate change experiments in which the evolution of the climate response to a gradual change in the greenhouse gases is examined. The transient-response experiments provide a more physically realistic framework for evaluating climate change than so- called equilibrium model simulations common in previous climate change studies. The transient-response experiments require long integrations with comprehensive 3-dimensional fully coupled global climate models and hence are computationally very demanding.

Transient simulations with coupled climate models (IPCC, 1992) in which, however, neither aerosols nor ozone changes have been included, suggest a warming rate of $0.3^{o}C$ per decade (Fig. 6). These values are approximately 60% of the model's equilibrium warming (where known) with double CO_2 when run with simple mixed-layer oceans. These models have been integrated from present to the end of the last century assuming

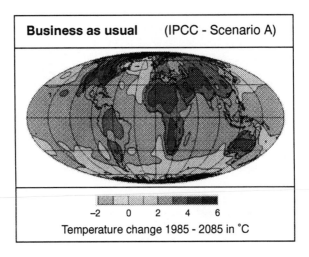

Figure 7: *Surface air temperature change for the last decade of a 100-year run of a climate change experiment, Business as usual, equivalent to a compound increase of atmospheric CO_2 of about 1.3% annually (from Cubasch et al., 1992).*

roughly an increase in the greenhouse gas concentration of around 1%/year. The lower values of warming compared to the equilibrium experiments take into account the thermal inertia of the deep ocean, and the models are therefore not at equilibrium at the time of effective CO_2 doubling. The warming does not proceed steadily, but all models exhibit variability on interannual, decadal and even longer time-scales. The intra-decadal variability of the models is found to be comparable with the observed natural variability of 0.3 to 0.4^oC. There are considerable spacial variations in the warming, with generally higher values over land than over sea, Fig. 7.

Minimum values are found in ocean areas with typical strong deep vertical mixing such as in the North Atlantic and around Antarctica. A particular problem is the initialization of coupled models. This is primarily because the state of the ocean is insufficiently known due to lack of observations. Moreover, the ocean due to its high thermal inertia, is not in balance with the present atmospheric forcing. Therefore, these deficiencies make it difficult to initialize the ocean correctly. In order to take this into consideration, recent greenhouse simulation experiments Cubasch et al. (1994) have incorporated this aspect by undertaking an ensemble of calculations starting from different ocean states, 50 years apart (Fig. 8).

The broadness of the temperature band is a measure of the accuracy in the calculation depending on the state of the ocean. It is likely that the in-

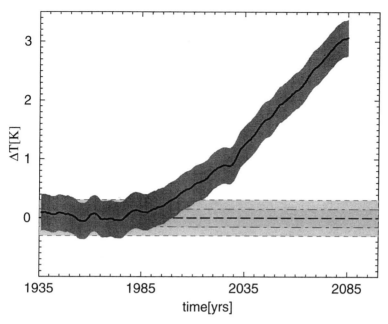

Figure 8: *Global mean annually averaged surface temperature changes for a climate change experiment (Business as usual alternative, equivalent to a compound increase of atmospheric CO_2 of about 1.3% annually) relative to the interannual variability of a control integration. Temperature changes are defined relative to the initial decade of the experiment (1935-46). The shaded area denotes the 95% confidence limits in the climate change experiment as well as in the control experiment (Cubasch et al., 1995).*

accuracy estimated in this experiment is too high due to an overly large low frequency variability of the ocean model used in this experiment. Coupled models used in climate change studies have undergone a rapid development in recent years and have in several respects obtained a considerable degree of realism. Of particular importance has been the capability of the models to reproduce coupled modes such as ENSO (Roeckner et al. 1995). Below we will describe some recent climate experiments carried out the at Max Planck Institute for Meteorology in Hamburg with a new coupled general circulation model. (CGCM)

4.1   Atmosphere and land surface

The atmospheric component of the CGCM is the 4th generation MPI model (ECHAM4), which is the most recent in a series evolving from the operational spectral transform model from ECMWF (Roeckner et al. 1996). ECHAM4 is based on the primitive equations and has vorticity, divergence, surface pressure, temperature, water vapour and cloud water as prognostic variables. Except for the water components, the prognostic variables are represented by spherical harmonics at triangular truncation at wavenumber 42 (T42). In order to reduce truncation errors for the advection of water vapour and cloud water a semi-Lagrangian scheme is applied (Williamson and Rasch, 1994). A semi-implicit time stepping scheme is used with a weak time filter. The time step is 24 minutes for dynamics and physics, except for radiation which is only calculated every 2nd hour. A hybrid sigma-pressure coordinate system is used (Simmons and Strüfing, 1983) with 19 irregularly spaced levels up to a pressure level of 10 hPa. The pa-rameterization of radiation has been applied from Morcrette (1991) with a number of modifications, such as the consideration of additional green-house gases (methane, nitrous oxide and 16 different CFCs) as well as various types of aerosols. Moreover, the water vapour continuum has been revised to include temperature weighted band averages of e-type absorp-tion and a band dependent ratio of (p-e)-type to e-type continuum ab-sorption (Giorgetta and Wild,1995). The single scattering properties of cloud droplets and ice-crystal are derived from Mie theory with suitable adaption to the broad band model (Rockel et al.,1991). The effective ra-dius of droplets and ice crystal is parameterized in terms of the liquid and ice water content, respectively. The main objective in the development of the radiation code was to reproduce the satellite and ground based ob-servations of the Earth's radiation within observational constraints while satisfying a physically stringent theory. Particularly important were the changes to the continuum absorption, which increased the downward long wave radiation at the surface by some 10-15 W/m^2 in agreement with observations. The turbulent transfer of momentum, heat, moisture and cloud water within and above than atmospheric boundary layer is com-puted with a higher-order closure scheme. The eddy diffusion coefficients are calculated as function of the turbulent kinetic energy (Brinkop and Roeckner, 1995). The effect of orographically excited gravity waves on the momentum budget is parameterized on the basis of linear theory and dimensional considerations (Palmer et al., 1986, Miller et al, 1989). The vertical structure of the momentum flux induced by the gravity waves is calculated from a local Richardson number, which describes the onset of turbulence due to convective instability and the breakdown approaching a critical level. The parameterization of cumulus convection is based on the concept of mass flux and comprises the effect of deep, shallow and mid-level convection on the budget of heat, water vapour and momentum (Tiedtke, 1989). Cumulus clouds are represented by a bulk model including the effect of entrainment and detrainment on the updraft and downdraft convective mass fluxes. Mixing due to shallow stratocumulus convection is considered as a vertical diffusion process with the eddy diffusion coefficients depending on the cloud water content, cloud fraction and the gradient of relative humidity at the top of the cloud. The closure for deep convection and organized entrainment in the original scheme has been modified and is now based on buoyancy instead of the moisture budget, and organized detrainment is computed for a spectrum of clouds detraining at different heights (Nordeng, 1994). Stratiform clouds are predicted per se in accordance with a cloud water equation including sources and sinks due to condensation/evaporation and precipitation formation both by coalescence of cloud droplets and sedimentation of ice crystals (Sundqvist, 1978; Roeckner et al., 1991). Sub-grid scale condensation and cloud formation is taken into account by specifying appropriate thresholds of relative humidity depending on height and static stability. Convective cloud water detrained in cumulus anvils as well as in shallow non-precipitating cumulus clouds is used as a source term in the stratiform cloud water equation. The land surface model considers the budget of heat and water in the soil, snow over land and the heat budget of permanent land and sea ice (Dümenil and Todini, 1992). The heat transfer equation is solved in a five-layer model assuming vanishing heat flux at the bottom. Vegetation effects such as the interception of rain and snow in the canopy and the stomatal control of evapotranspiration are grossly simplified. The local run-off scheme is based on catchment considerations and takes into account sub-grid scale variations of field capacity over inhomogeneous terrain. In the coupled model the hydrological cycle is closed by a river routing scheme (Sausen et al., 1994) which directs the local runoff into the oceans.

4.2   Ocean and sea ice

The oceanic model is based on the OPYC model (Oberhuber, 1993a, 1993b). It consists of three sub-models; the interior ocean, the surface mixed layer and the sea-ice respectively. The governing equations are solved on an Arakawa B-grid with no-slip horizontal boundary condition. An implicit time stepping scheme is used and an alternating direction implicit solution technique. Poleward of $36°$, the horizontal resolution is identical to the Gaussian grid of the atmospheric model (2.8° for the T42 model). At low latitudes the meridional spacing is gradually decreased down to 0.5° at the equator. The reason to this is the requirement to resolve equatorial waves, required to reproduce El Niño phenomena realistically. Vertically, 11 interior layers and special mixed surface layer are used. The model for the interior ocean uses the primitive equation in the flux form of the conservation laws for momentum, mass, heat and salt at isopycnal layers. These quantities are the prognostic variables together with sea level height. Horizontal mixing of momentum is a function of the local Rossby deformation radius, while horizontal diffusion of temperature and salinity depends on the deformation of the flow. Vertical mixing follows the concept of entrainment and detrainment for which budgets of turbulence and mean potential energy are being solved. A standard convective adjustment scheme is employed which instantaneously removes vertical instabilities. The model for the interior ocean is coupled to a mixed layer model, since the isopycnal coordinates break down near the surface when strong turbulence is present. A special mixed layer model calculates fluxes in and out of the uppermost isopycnal layer according to the budget for turbulent kinetic and mean potential energy. Wind stirring, surface buoyancy due to heat and fresh water fluxes, sub-surface stability and flow shear affect these calculations. The sea-ice model calculates the thickness and concentration of ice and its momentum. The amount of snow on ice is also calculated. A viscous-plastic rheology is used to parameterize the stress tensor, while the thicknesses of ice and snow and the concentration of ice are computed from respective continuity equations. Further parameterization relates the heat fluxes to changes in ice and snow thickness as well as to lead size and to changes in salinity due to brine rejection. The conversion of snow to ice and the surpression of snow is included in a parametric form. The thermodynamic part consists of a prognostic computation of the temperature profile, taking into account the heat capacity and conductivity of the slab and the net surface flux.

4.3   The coupling

Prior to the coupling the oceanic model has been integrated for about 1000 years by prescribing a combination of observed variables and simulated fluxes (with the atmospheric model). The dynamical components such as wind stress and friction velocity have been derived from the atmospheric model (in turn forced with climatological sea surface temperatures, SSTs). The fluxes of heat and freshwater on the other hand are based on a combination of (i) observed climatology (Oberhuber, 1988), (ii) a bulk flux parameterization and (iii) additional relaxation towards observed SSTs (averaged for the years 1979-1988) and surface salinity (Levitus, 1982). When the restoring boundary conditions are replaced by the fluxes computed in the coupled model, a drift towards a new model state is taking place.This drift is partly due to inconsistencies between the dynamical fluxes of the coupled system and the mixture of different forcing mechanisms employed in the spin up phase. The drift is also occurring due to model errors such as deficiencies in the simulation of clouds leading to errors in the amount of heat reaching the ocean surface. Other uncertainties are related to the hydrological cycle over the oceans and difficulties to reproduce the salinity distribution due to different problems with sharp ocean currents and transport of sea ice. Major reductions have taken place in reducing the systematic errors in recent models and remaining problems are mainly confined to higher latitudes. (Latif et al. 1994). The technique to circumvent the problem with climate drift has been to employ a so called flux adjustment (e.g. Sausen et al., 1988), whereby the calculated fluxes are corrected with the fluxes occurring in the uncoupled state as derived from climatology. In this study an approach has been applied whereby only the mean annual drift has been adjusted. The importance of leaving the annual cycle free is particularly valid where for example SST undergoes major changes independent of the annual heat flux, such as during El Niño (Neelin et al., 1992).

In the current model an alternative approach (Bacher and Oberhuber, 1995) is applied. It differs from the traditionally one in two major aspects, namely (i) the flux adjustment is computed by gradual updating during a 100 year spin up run of the coupled model and (ii) only the annual mean of heat and fresh water is corrected while other quantities including wind stress are generated by the coupled model without any corrections altogether. After the 100 years spin up with variable adjustment, the model has been integrated for another 100 years with the flux adjustment fixed in time (Fig. 9). A relatively small secular drift was found with a cooling trend (all layers) of about 0.1^oC, a reduction of the upper ocean salinity by ca. 0.2 psu and a salinity increase with about the same rate in the deep ocean.

Fig. 10 shows the evolution of the globally averaged surface air temper-

Coupling strategy (ECHAM4 / OPYC3)

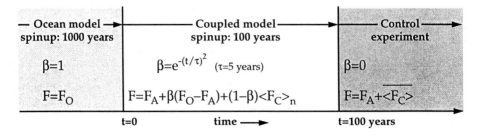

$$F = F_A \quad : \quad \text{AGCM}$$

$$F = F_{O_n} \quad : \quad \text{OGCM (incl. relax. towards } SST_{OBS})$$

Fluxes of heat and freshwater

$F = F_A$:   AGCM

$F = F_{O_n}$:   OGCM (incl. relax. towards SST_{OBS})

$<F_C>_n = \frac{1}{n}\int_0^n (F_O - F_A)\, dn$: $0 < n < 100$ years

$\overline{<F_C>}$ = estimated from history of $<F_C>_n$

Figure 9: *Coupling strategy between the atmospheric model ECHAM4 and the ocean model OPYC3. FA is atmospheric (AGCM) fluxes of heat and freshwater and FO is the same for the ocean (OGCM), b is an adjustment factor. For further explanation see text. After Bacher and Oberhuber (1996).*

ature of the control experiment. Superimposed upon a weak cooling trend, similar to that of the ocean, are interannual and interdecadal fluctuations with approximately the same amplitude as those seen in the observational records during the last 100 years (Fig. 1).

Due to the flux adjustment, the annual mean SST distribution simulated in the coupled model is very close to that obtained with the original spin up of the ocean model. As compared to observations, the annual SST deviations are generally below $1^{\circ}C$. Larger errors both in the ocean as well as in the coupled model are found in the regions of the western boundary currents which are not sufficiently resolved by the model. However, since these errors are confined to small areas, the long term atmospheric impact is insignificant, Table 2. This can also be seen, for example, from Fig. 11, which shows a comparison of the annual mean precipitation as simulated with the atmospheric and the coupled model, respectively. Although pre-

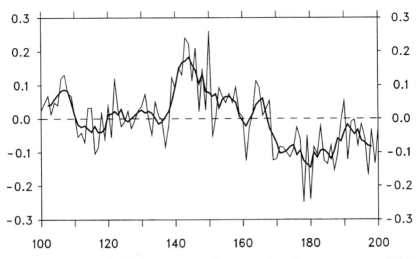

Figure 10: *Derivation of global mean annually averaged surface temperature (K) from the 100 year years mean (14.7°C) between years 100 and 200 of the control experiment of the coupled model. The smoothed curve (bold) represents 5-year measuring means. From Roeckner et al., 1995.*

	Sfc. air temp. (˚C)	Precip. (mm/d)	Column water vapor (kg/m²)	Column cloud water (g/m²)	Total cloud cover (%)	Net SW radiation [SWCF] (W/m²)	Net LW radiation [LWCF] (W/m²)
CGCM	14.7	2.82	24.5	78.8	59.8	237.9 [−49.9]	−235.7 [28.9]
AGCM	14.4	2.81	24.4	78.9	59.8	236.9 [−49.3]	−235.3 [28.8]

Table 2: *Global annual mean climate variables as simulated in the CGCM and ECHAM4 respectively. The averaging period in 100 years for the CGCM and 30 years for the atmospheric model. The last two columns refer to the top-of-atmosphere longwave and shortwave radiation fluxes. Cloud forcing is given in brackets. From Roeckner et al., 1995.*

cipitation is one of the climate variables which is known to respond notably to SST changes, its large scale distribution in the coupled model is rather similar to that of the atmospheric model. This applies to other climate variables as well.

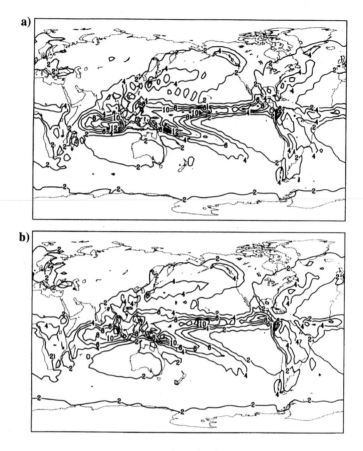

Figure 11: *Annual mean precipitation (mm/day) as simulated in the coupled control experiment (50-year mean) (top) and the stand alone atmospheric model (ECHAM4) forced with observed SSTs 1979-94 (mean of two realizations). From Roeckner et al. 1995.*

5 Climate change scenarios

The following climate change experiment has been carried out with the coupled model described in section 4. The first part consist of a control experiment whereby the solar constant and the concentration of the greenhouse gases are prescribed according to current conditions (IPCC, 1994). Moreover, since both the spin up of the ocean model and the calculation of flux adjustments are based on the currently observed SST-

Source (time period)	CO_2	CH_4	N_2O	CFCs	$\Sigma^{(1)}$	Total: including band overlap
IPCC'94 (1750-1990)	1.56	0.47	0.14	0.28	2.45	2.45
ECHAM4 (1750-1990)	1.55	0.57	0.15	0.30	2.57	2.56
[2]ECHAM4 (1860-1990) (1750-1860)	1.37 0.18	0.47 0.10	0.13 0.02	0.30 0.00	2.27 0.30	2.26 0.30
IPCC/IS92a (1990-2100)	4.50	0.76	0.37	0.15	5.78	5.78
ECHAM4/IS92a (1990-2100)	4.72	0.69	0.40	0.10 [3]	5.91	5.86

[1] In IPCC, overlap is included already in the individual components

[2] year 1860: Start of the scenario experiment

[3] updated CFC-scenario according to IPCC'95

Table 3: *Greenhouse gas radiative forcing (W/m^2) at the tropopause level. The years 1860-1990 are based on observational data. 1990-2100 according to IPCC scenario IS92a.*

distribution (averaged between 1979 and 1988), the coupled control experiment can be considered as a simulation of the present climate. In addition to the coupled control experiments, two Atmospheric Model Intercomparison Project (AMIP) -type experiments (Gates, 1992) have been performed with the atmospheric component (ECHAM4). In these experiments, observed monthly mean SSTs and sea ice extents are prescribed according to an extended AMIP SST data set and sea ice data set (Jan. 79 through May 94). Operational ECMWF analyses (1980-1994) of sea level pressure, geopotential height, wind and temperature are used as an observational reference.

The climate change experiment started in Jan. 1860. However since the model has been initialized with present SSTs, and present greenhouse gas concentrations, we have to assume that these data were also valid initially. The concentrations of greenhouse gases are therefore systematically higher than in the present climate. This is not likely to be of any significance since we have used the actual changes in the direct forcing that have taken place up to 1990 and estimated increases after that date (Table 3). An actual initialization of the initial conditions of the coupled system at 1860 is presently not feasible due to lack of relevant observations. It could of

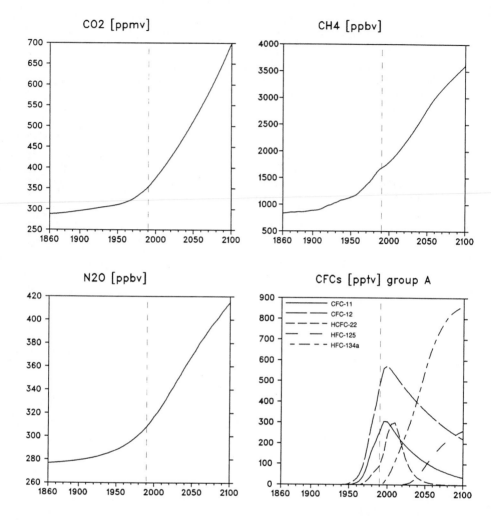

Figure 12: *Greenhouse gas concentrations as prescribed in the climate change experiment. The years 1860-1990 according to observations, the years 1990-2100 according to IPCC scenario IS92a. From IPCC 1994.*

course be calculated from the known radiative conditions and greenhouse gas concentration at the time, but this is excessively computationally demanding and has not yet been undertaken.

Table 3 gives the radiative forcing (W/m^2) for the different greenhouse gases at the tropopause level. For the period 1750 (beginning of the industrialization) to 1990 the concentrations are based om measurements.

The global heat balance
(model simulated)

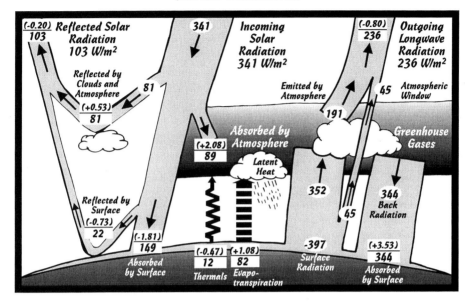

Figure 13: *The same as for Fig. 1 but for the coupled model experiment. Lower figures show simulated data for the reference period 1860-1890. Upper figures show predicted changes for the period 2010-2020 relative to the reference period. Units Wm^{-2}.*

For the period beyond 1990 the concentrations are obtained from IPCC scenario IS92a (Fig. 12), approximatively equivalent to an increase of ca. 1%/year. Note however, the fall in the CFC's during the next century due to the international agreement to reduce CFC's in the atmosphere (Montreal Protocol). There is an insignificant difference between the IPCC data and those used in the experiment due to the form of the radiation code being used. The sum of the different contributions includes overlapping of the different absorption bands. In this paper we will not evaluate the whole experiment (it has in fact not been completed at the time of writing) but concentrate on the period until 2020. We are confident that the greenhouse gas emission estimates are reasonably realistic up to that time and probably also the calculations of greenhouse gas concentrations in the atmosphere.

Fig. 13 shows the global radiation balance showing the mean radiation balance for the period 1860-1890 and the changes in radiation conditions compared to the period 2010-2020. As can be seen by comparing with Fig. 1, the calculated radiation balance agrees quite well with the observational estimates as obtained from ERBE and from surface observations (e. g. Hartmann (1993), Ohmura and Gilgen (1993), Hahn et al.(1994)). The accuracy of the observational data is of the order of 5 W/m^2 in global annual averages for most quantities. Absorption of both short and long wave radiation in the atmosphere are still open to discussion and more measurements are required to obtain a better understanding of the 3-dimensional distribution of radiation forcing. Table 4 shows the changes of the radiation at the top of the atmosphere, in the atmosphere and at the surface of the Earth, respectively. Due to the ongoing increase in the greenhouse gas forcing, the model is not in radiative equilibrium. The mean net heating at the ground during the 2010-2020 is 1.08 W/m^2 as the result of different forcing processes as we will discuss below.

5.1   Short wave radiation

There is an increase in the short wave cloud forcing at the top of the atmosphere by 0.65 W/m^2 due to enhanced cloud albedo. There is also an increase in the short wave radiative forcing in the atmosphere due to a higher absorption by water vapour in the atmosphere by 2.15 W/m^2. The surface albedo is slightly reduced, 0.73 W/m^2, due to a general reduction in snow- and ice-cover. The effect is a small reduction in planetary albedo, an increased warming of the atmosphere and a reduced warming at the surface. The reduced warming at the surface, 1.81 W/m^2, is both caused by increased absorption in the atmosphere and to increased reflection by clouds (short wave cloud forcing). The overall changes in short wave radiative processes are so small that it is probably not possible to detect them by any known observational system.

5.2   Long wave radiation

The enhanced high-cloud emissivity is leading to a slight increase in the long wave cloud forcing by 0.25 W/m^2 at the top of the atmosphere. This is less than the increase cooling due to the increased cloud albedo effect, so it follows that the overall effect of clouds is to counteract the greenhouse warming, that is a negative feedback. At the surface the direct and indirect long wave radiative forcing due to greenhouse gases and water vapour is equal to 4.57 W/m^2, in turn reduced by the increased long wave cloud

Top of atmosphere (TOA)	Change (W/m²)	Main causes
SW (clear sky)	+ 0.921	Enhanced absorption by water vapor
LW (clear sky)	+ 0.578	Enhanced water vapor greenhouse
SW cloud forcing	− 0.653	Enhanced cloud albedo
LW cloud forcing	+ 0.253	Enhanced high-cloud emissivity
Net TOA radiation	**+ 1.099**	
GHG forcing (tropopause)	*3.358*	
Atmosphere (ATM)		
SW (clear sky)	+ 2.147	Enhanced water vapor absorption
LW (clear sky)	− 3.990	Enhanced water vapor emission
SW cloud forcing	− 0.072	Competition by water vapor
LW cloud forcing	+ 1.291	Enhanced high-cloud emissivity
Net ATM radiation	**− 0.624**	Balanced by sensible and latent heat
GHG forcing (troposphere)	*2.810*	
Surface (SFC)		
Sensible heat flux	+ 0.466	Surface layer more stable?
Latent heat flux	− 1.080	Warmer SST
SW (clear sky)	− 1.226	Enhanced water vapor absorption within the atmosphere
LW (clear sky)	+ 4.568	Direct and indirect radiative forcing through GHG and water vapor
SW cloud forcing	− 0.581	Larger cloud albedo
LW cloud forcing	− 1.038	Competition by enhanced water vapor absorption in the atmosphere
Net SFC radiation	+ 1.723	
Net SFC heat	**+ 1.109**	Almost identical to net TOA radiation
GHG forcing (SFC)	*0.548*	

Table 4: *Change of heat budget in the climate change experiment (decadal mean 2010-2020 changes from pre-industrial mean 1860- 1890).*

forcing at the surface calculated to 1.04 W/m^2. The increase in the net long wave radiative forcing at the surface is thus 3.53 W/m^2, representing a clear positive feedback mechanism.

5.3 Heat fluxes

The sensible heat flux is reduced by 0.47 W/m^2. The reason appear to be due to the stabilization of the boundary layer. The latent heat flux from the ground is increasing by 1.08 W/m^2 related to increased SSTs and to

a)

Total Surface Flux

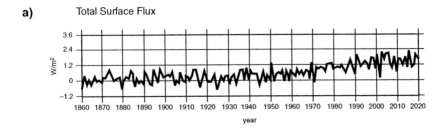

b)

Integrated Ocean Heat Input

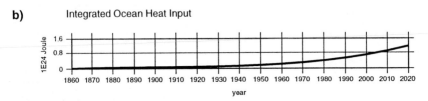

Figure 14: *(a). Total annually net surface heat flux in Wm-2 into the world's oceans for the period 1860-2020. (b) integral ocean heat input in unit 1024 Joule. Note the approximative linear increase in the total heat from 1990 consistent with the linear increase in the greenhouse gas forcing.*

an intensification of the hydrological cycle in particular over land.

We have calculated the total warming of the climate system in the form of accumulated warming of the ocean for the whole period from 1860 to 2020. Figure 14 shows the net surface heat flux in W/m^2 and the total accumulated heat in units of 1024 Joule as a function of time. The total warming at 2020 amounts to around 1.2×1024 Joule. If the oceans would be uniformly heated this would correspond to an average heating of the whole ocean by 0.21^oC. This is not the case since the warming generally stabilize the oceans giving an increase of 1.4^oC at the surface.

5.4 The hydrological cycle

Fig. 15 shows the estimated hydrological cycle as estimated from Baumgartner and Reichel (1975), Chahine (1992) and Bromwich (1990). The hydrological cycle over land is obtained from routine meteorological observations and the river run off from available hydrological records. Net water vapour transport from oceans towards land is obtained as a residual. The hydrological cycle over oceans is essentially obtained from energy balance calculations since actual representative observations over oceans hardly exist. Accumulation on glaciers is a gross accumulation since calving of icebergs has not been incorporated. The unit is $1000 \ km^3/year$.

Global water cycle

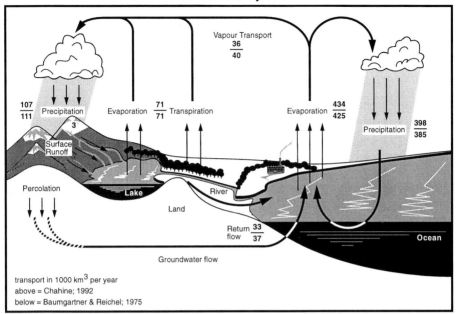

Figure 15: *Global annual hydrological cycle for the marine and continental hemisphere, respectively. Upper figures show empirical estimate from Chahine (1992) and lower figures from Baumgartner and Reichel (1975). Snow accumulation according to Bromwick (1990). Units are given in $10^3 km^3$/year or 10^{15} kg/year.*

The atmospheric model reproduces the observations quite well over land but is some 5% higher over sea (both precipitation and evaporation). It is presently not possible to conclude whether this reflects a systematic model error or whether the observational estimates are incorrect. There are small differences (2-4 units) between the atmospheric model and the coupled model (not shown). These differences are due to minor model differences such as the specification of slightly different coastlines in the coupled model, and are insignificant in this context.

Fig. 16 shows the hydrological cycle for the coupled model for the reference period 1860-1890 as well as the changes between this time and the decade 2010-2020. The evaporation over sea increases by some 4 units, while precipitation over ocean is practically the same. Over land, on the other hand, there is a marked activation of the hydrological cycle with an increase in precipitation by 6.2 units and in evapotranspiration of 3.2

Global water cycle

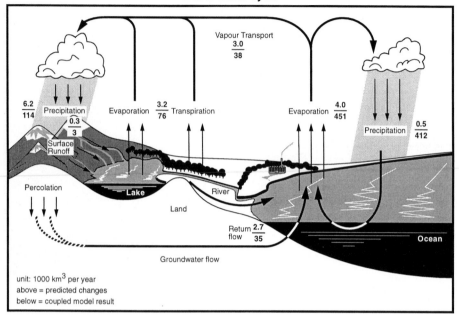

Figure 16: *Global annual mean hydrological cycle for the marine and continental hemisphere, respectively. Lower figures show simulated data for the reference period 1860-1890. Upper figures show the predicted changes for the period 2010-2020 relative to the reference period. Units are given in 103 km³/year or 1015 kg/year.*

units. Accumulation of snow on glaciers is also increasing by some 10% or 0.3 units. The river run off goes up by 2.7 units. The increased activation of the hydrological cycle over land and the associated increase in the net transport of water vapour into land are associated with the relatively stronger sensible heat forcing over land. This enhances a general monsoonal flow component in the lower troposphere and an associated net outflow aloft. The net subsidence caused by this return flow is probably the likely reason for the unchanged precipitation over the oceans in spite of higher SSTs. Another consequence of the warming is a marked increase in atmospheric water vapour. Over the 30-year period 1990-2020 the globally vertically integrated increase amounts to more than 7%. There is also a minor increase in cloud water, while the total cloud cover is practically unchanged.

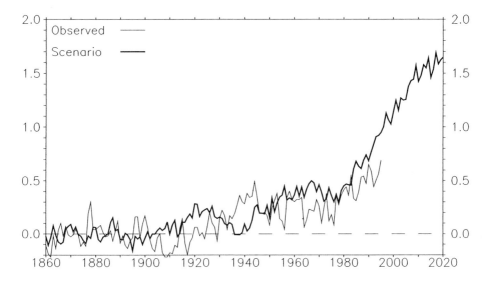

Figure 17: *Predicted and observed annual global surface temperature changes. The observed values are the same as in Fig. 3.*

5.5 Temperature changes

The surface temperature increase for the decade 2010-2020 compared to the reference period 1860-1890 is 1.56^oC with an average increase over land by 2.35^oC and over sea by 1.24^oC. As can be seen from Fig. 17, the simulated temperature up to present time agrees well with observations, suggesting that the typical variations in temperatures up to around 1980 are dominated by natural temperature fluctuations, while thereafter the effect of the greenhouse warming is becoming more distinct. Of particular interest is the rapid warming between the 1980ies and the second decade of the next century, amounting to a global average increase of 1^oC!

Fig. 18 shows the temperature increase over this period in a zonally averaged vertical cross section. A maximum warming of around 2^oC occurs in the upper tropical troposphere. This maximum warming occurs at the upper part of the deep tropical convective regions and is caused by the reduced slope of the moist adiabats caused by the higher surface temperatures. Another area of maximum warming occurs over the high latitudes of the Northern Hemisphere. This warming is presumably related to a general reduction in the Arctic sea-ice cover. In this period the Arctic sea ice cover is calculated to decrease by $0.9 \times 106\,km^2$ or by some 10% compared to the

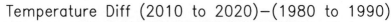

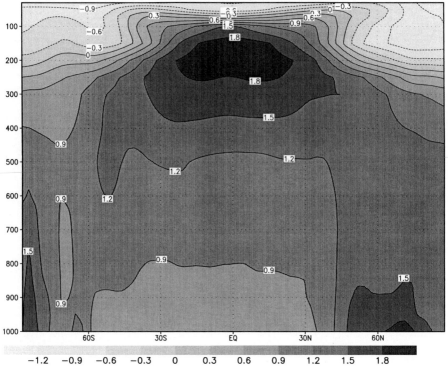

Figure 18: *Zonally averaged cross section of the predicted decadal mean temperature change between the decades 2010-2020 and 1980- 1990.*

reference period 1860-1890. In the stratosphere there is a cooling, increasing upward. This is due to the increased outgoing radiation at these levels from enhanced greenhouse gases and water vapour. The maximum cooling takes place at around 40 km. Unfortunately, this part is not represented by the model since the top level is at 10 hPa or 30 km. Fig. 19 shows the temperature change at 30 hPa for the whole integration period. The fact that the maximum cooling take place at the Northern Hemisphere here is probably a sampling problem due to the very large low frequency internal variability in the stratospheric circulation at the Northern Hemisphere.

A typical characteristic of the general circulation is the considerable variations which occur over a wide range of time scales from days towards several decades. In a recent study by Manabe and Stouffer (1996) where several 1000 year long integrations have been carried out by a hierarchy of

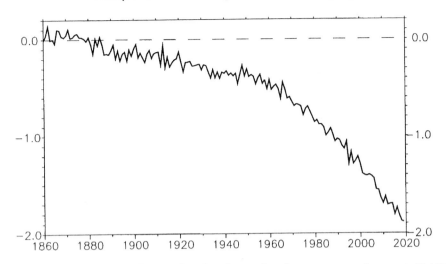

Figure 19: *The same as Fig. 17 but for the predicted temperature change at 30 hPa. Reference temperature* −58.1°C *(1860-1890).*

climate models it was found that even atmospheric models with climatological SSTs have considerable variability on decadal timescales. At high latitudes the size of the variability is practically the same as the variability of a fully coupled model. A 100 year integration with an atmospheric model very similar to the one used in this experiment gave a very similar result, demonstrating that atmospheric processes on its own can generate substantial variability on decadal and longer time scales. For this reason is it difficult to relate a particular climate aspect, such as an extreme in the middle and high latitude general circulation to the surface boundary conditions or to any other specific factor.

This aspect can be illustrated by a result from the coupled model. Fig. 20 shows the temperature difference between the 30 year mean 1960- 1990 and the 30 year mean 1930-1960. While the global temperature in this period increased by 0.27°C in the experiment the temperature decreased over the whole area from Scandinavia to central Canada between 50−70°N. Around Greenland the fall in temperature is as high as 1°C! It is interesting that a decrease in temperature over this length of time actually took place in this area in reality. This is of course just coincidental, but illustrate nevertheless the realism of the present simulation.

Annual surface air temperature difference
(1960 to 1990) − (1930 to 1960)

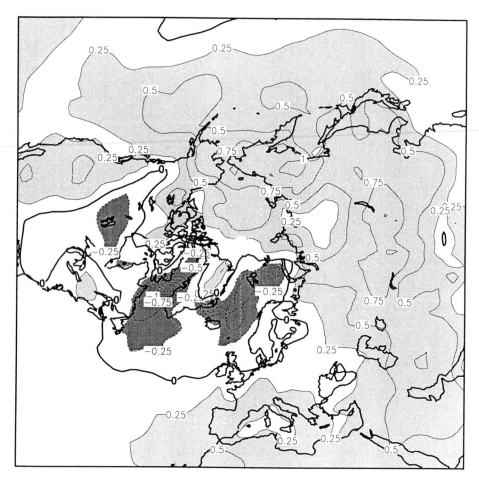

Figure 20: *Predicted surface temperature changes between the 30 years mean 1960-1990 and 1930-1960. Note the relatively large area of cooling over the North Atlantic-Canadian sector reflecting natural variations in the climate system.*

6   Concluding remarks

In this paper we have presented some results from a recent high resolution transient climate change experiment at the Max Planck Institute for Meteorology in Hamburg. The experiment was started in 1860 and carried through until 2020. For the period 1860-1990 actual greenhouse gas concentrations were used, thereafter a continued increase based upon the most likely emission scenario. The overall findings are in broad agreement with observed global surface temperature suggesting that the calculated warming until present is fully consistent with the observed warming of about 0.6^oC since the middle of the last century. Major regional variations occur at high latitudes including cooling trends lasting several decades. Experiments with atmospheric models only indicate that atmospheric processes may play an important role in generating the low frequency variability which is typical for high latitude climate patterns on decadal time scales. A preliminary analysis of radiative forcing patterns suggest a strong enhancement of the greenhouse warming from atmospheric water vapour, while the cloud forcing indicate a negative feedback through an increased short wave cloud forcing. The hydrological cycle is becoming more active in particular over land, where both precipitation and evapotranspiration are increasing. Evaporation is also increasing over ocean areas while precipitation is more or less unchanged. A possible explanation is an enhanced monsoonal forcing due to the more rapid increase of surface energy fluxes over land. The experiment indicate that a more rapid warming has commenced in recent years. Although the global warming has increased during the last years in consistency with these results, it cannot be taken as a proof, since recent warming could also be due to natural fluctuations. However, systematic global monitoring over the next decades is of particular importance in that respect. Upper tropical tropospheric temperature, integrated amount of atmospheric water vapour, Arctic ice cover and ice volumes, ocean temperatures are examples of parameters where a climate warming is likely to be indicated. In this study we have calculated the climate change over the period 1860-2020. An alternative approach is to compare the climate change experiment with a control experiment using unchanged greenhouse gas concentrations. By doing so we may also incorporate any possible effect due to long term climate drift of the coupled model not eliminated by the flux adjustment. Such a minor drift appears to occur in the control experiment, indicating a cooling of around $0.1^oC/100 years$. The cooling is mainly concentrated to the Southern Hemisphere and suggest that the spin-up time of the coupled model probably was too short. If the trend is the same in the climate change experiment the global warming at 2020 is underestimated by some 0.15^oC. It is furthermore suggested that the hydrological cycle over oceans is slightly enhanced (by about 2 units for both precipitation and evaporation). No changes occur over land. The climate effect of sulfate aerosols has been investigated in recent studies by Taylor and Penner (1994), Bengtsson et al. 1995, and Mitchell et al. 1995. In these studies the direct effect of sulfate aerosols has been considered. The experiments have been done in two steps, first by calculating the atmospheric load of sulfate from a stand-by chemical model using known sources of sulfate dioxide emission. Secondly, a simple empirical relation between the overall sulfate concentration and the clear air albedo have been employed, implying higher albedo at high sulfate concentrations. The overall effect on climate is a minor cooling. It has a strong regional variation due to the short residence time of sulfate aerosols in the troposphere (5-7 days) and is therefore mainly found near and downstream the areas of major emission (mainly industrial and densely populated areas of Europe, United States, China). Due to lack of accurate observational data these experiments are less reliable than the greenhouse experiments. Furthermore, other aerosols like soot, not yet considered in the experiments, may partly counteract the effect of sulphate aerosols. For the time being we must consider the results obtained so far in these experiments as indicative only.

Acknowledgement

The author acknowledges the assistance of staff at the Max Planck Institute for Meteorology and DKRZ, in particular Erich Roeckner and Josef Oberhuber. Kornelia Müller and Norbert Noreiks are specially thanked for their technical assistance.

References

ARRHENIUS, S., 1896: On the Influence of Carbonic Acid in the Air upon the Temperature of the Ground. *Philosophical Magazine and Journal of Science, 236-276.*

BACHER, A. AND J. M. OBERHUBER, 1996: Global coupling in the ECHAM4/OPYC3 atmosphere-sea ise-ocean GCM with annual mean flux correction restricted to heat and freshwater. *In preparation.*

BAUMGARTNER, A. AND E. REICHEL, 1975: The World Water Balance. *Elsevier, New York.*

BENGTSSON, L., P. CRUTZEN, M. KANAKIDOU, H. KELDER, J. LELIEVELD, F. RAES, H. RODHE AND E. ROECKNER, 1995: Study of the indirect and direct climate influences of anthropogenic trace gas emissions (SINDICATE). *Final Report of a Project of the EC Environmental Research Programme, Max-Planck-Institut für Meteorologie, Hamburg.*

BENGTSSON, L., K. ARPE, E. ROECKNER AND U. SCHULZWEIDA, 1996: Climate predictability experiments with a general circulation model. *Climate Dynamics.* **12,** 261-278.

BERGER, A., 1980: The Milankovic astronomical theory of paleoclimates: a modern review. *Vistas in Astronomy*, **24**, 103-122.

BOLIN, B. AND E. ERIKSSON, 1959: Changes in the Carbon Dioxide Content of the Atmosphere and Sea due to Fossil Fuel Combustion. *The Atmosphere and Sea in motion. Rockefeller Institute Press, 130- 146.*

BRINKOP, S. AND E. ROECKNER, 1995: Sensitivity of a general circulation model to parameterizations of cloud-turbulence interactions in the atmospheric boundary layer. *Tellus*, **47A**, 197-220.

BROMWICH, D. H., 1990: Estimates of Antarctic precipitation. *Nature*, **343**, 627-629.

BRYAN, F., 1986: High latitude salinity effects and interhemispheric thermohaline circulations. *Nature*, **305**, 301-304.

CALLENDAR, G.S., 1938: The artificial production of carbon dioxide and its influence on temperature. *Quart. J. R. Met. Soc.*, **64**, 223-240.

CALLENDAR, G.S., 1949: Can carbon dioxide influence climate? *Weather*, **4**, 310-314.

CHAHINE, M., 1992: The hydrological cycle and its influence on climate. *Nature*, **359**, 373-380

CHAPPELLAZ, J., J.M. BARNOLA, D. RAYNAUD, Y.S. KOROTKEVICH AND C. LORIUS, 1990: Ice-core record of atmospheric methane over the past 160.000 years. *Nature*, **345**, 127-131.

CHARLSON, R. J., J. LAGNER, H. RODHE, C. B. LEOVY AND S. G. WARREN, 1991: Perturbation of the Northern Hemisphere radiative balance by backscattering from anthropogenic sulphate aerosols. *Tellus*, **43A-B(4)**, 152-163.

CHEN, C.-T., E. ROECKNER AND B. J. SODEN, 1995: A comparison of satellite observations and model simulations of column integrated moisture and upper tropospheric humidities. *Max-Planck-Institut für Meteorologie, Report No. 155, Hamburg.*

CUBASCH, U., K. HASSELMANN, H. HÖCK, E. MAIER-REIMER, U. MIKOLAJEWICZ, B. D. SANTER AND R. SAUSEN, 1991: Time-dependent greenhouse warming computations with a coupled ocean-atmosphere model. *Climate Dynamics*, **8**, 55-69.

CUBASCH, U., B. D. SANTER, A. HELLBACH, G. HEGERL, H. HÖCK, E. MAIER- REIMER, U. MIKOLAJEWICZ, A. STÖSSEL AND R. VOSS, 1994: Monte Carlo climate change forecasts with a global coupled ocean-atmosphere model. *Climate Dynamics*, **10**, 1-19.

CUBASCH, U., G. HEGERL, A. HELLBACH, H. HÖCK, U. MIKOLAJEWICZ, B. D. SANTER AND R. VOSS, 1995: A climate change simulation from 1935. *Climate Dynamics*, **11**, 71-84.

DÜMENIL, L. AND E. TODINI, 1992: A rainfall-runoff scheme for use in the Hamburg GCM. In: Advances in theoretical hydrology. A tribute to James Dooge, Ed. J. P. O'Kane. *European Geophysical Society Series on Hydrological Sciences, 1, Elsevier 1992, 129-157.*

FOUKAL, P. AND J. LEAN, 1990: An empirical model of total star irradiance variations between 1874 and 1988. *Science*, **247**, 556-558.

GATES, W. L., 1992: AMIP: The atmospheric model intercomparison project. *Bull. Am. Met. Soc.*, **73 (12)**, 1962-1970.

GIORGETTA, M. AND M. WILD, 1995: The water vapor continuum and its representation in ECHAM4. *Max-Planck-Institut für Meteorologie, Report No. 162, Hamburg.*

HAHN, C. J., WARREN, S. G. AND J. LONDON, 1994: Climatology data for clouds over the globe from surface observation, 1982-1991: The Total Cloud Edition. *NDP-026A, The Carbon Dioxide Information Analysis Center, Oak Ridge National Laboratory, Oak Ridge.*

HARTMANN, D. L., 1993: Radiation effects on clouds on Earth climate. In Aerosol, Clouds, Climate interaction. *Ed. P. Hobbs.*

IPCC, 1990: Climate Change, The IPCC Scientific Assessment. Ed. Houghton, J., G. J. Jenkins and J. J. Ephraums, *Cambridge University Press,; 364.*

IPCC, 1992: Climate Change, The Supplementary Report to the IPCC Scientific Assessment. Ed. Houghton, J., B. A. Callendar and S. K. Varnay, Cambridge University Press, *198.*

IPCC, 1994: Climate Change 1994. Ed. Houghton, J., L.K. Meira Filho, J. Bruce, H. Lee, B.A. Callender, E. Haites, N. Harris and K. Maskell, *Cambridge University Press.*

KONDRATYEV, K.Y. AND N. I. MOSKALENKO, 1984: The role of carbon dioxide and other minor gaseous components and aerosols in the radiation budget. *The Global Climate, Ed. J. T. Houghton. Cambridge University Press, 225-233.*

KUSHNIR, Y. 1993: Interdecadal variations in North Atlantic Sea Surface Temperature and Associated Atmospheric Conditions. *J. of Climate*, **7**, 141-157.

LEVITUS, S., 1982: Climatological atlas of the world ocean. NOAA Professional Paper No. 13, *US Govt Printing Office, Washington DC.*

LUNKEIT, F., 1993: Simulation der interannualen Variabilität mit einem globalen gekoppelten Ozean-Atmosphärenmodell. *Berichte aus dem Zentrum für Meeres- und Klimaforschung, Reihe A, Nr. 8, Meteorologisches Institut, Hamburg.*

MANABE, S., 1971: Estimate of future changes in climate due to increase of carbon dioxide concentration in the air. *In Man's Impact on the Climate, eds, W. H. Mathews, W. W. Kellogg and G.D.Robinson, MIT Press, Cambridge, MA, 249-264.*

MANABE, S. AND R. T. WETHERALD, 1975: The effects on doubling the CO_2 concentration on the climate of a general circulation model. *J. Atmos. Sci.* **32**, 3-15.

MANABE, S., R. J. STOUFFER, M. J. SPELMAN AND K. BRYAN, 1991: Transient responses of a coupled ocean-atmosphere model to graduate changes of atmospheric CO_2. Part I: Annual mean response. *J. Climate*, **4**, 785-818.

MANABE, S. AND R. J. STOUFFER, 1996: Low frequency variability of surface air temperature in a 1000 year integration of a coupled ocean- atmosphere model. *Accepted for publication in J. Climate.*

MARSHALL INSTITUTE, 1989: Scientific Perspectives on the Greenhouse Problem. *Ed. E. Seitz, Marshall Institute, Washington D.C..*

MILANKOVIC, M., 1930: Mathematische Klimalehre und astronomische Theorie der Klimaschwankungen, *Handbuch der Klimatologie, BdI, Teil A, Hrg. Köppen/Geiger, Berlin, 1930.*

MILLER, M. J., T. N. PALMER AND R. SWINBANK, 1989: Parameterization and influence of sub-grid scale orography in general circulation and numerical weather prediction models. *Met. Atm. Phys.*, **40**, 84-109.

MITCHELL, J. F. B., T.C. JOHNS, J. M. GREGORY AND S. TETT, 1995: Transient climate response to increasing sulfate aerosols and greenhouse gases. *Nature*, **376**, 501-504.

MÖLLER, F., 1963: On the influence of changes in CO_2 concentration in air on the radiative balance of the earth's surface and on the climate. *J. Geophys. Res.*, **68**, 3877-3886.

MURPHY, J. M., 1990: Prediction of the transient response of climate to a gradual increase in CO_2, using a coupled ocean/atmospheric model with flux correction. *In: Research Activities in Atmospheric and Ocean Modelling. CAS/JSC Working Group on Numerical Experimentation, Report No. 14, WMO/TD No 396, 9.7-9.8.*

MURPHY, J. M. AND J. F. B. MITCHELL, 1995: Transient response of the Hadley Centre coupled model to increasing carbon dioxide. Part II - Temporal and spatial evolution of patterns. *J. Climate*, **8**, 57-80.

NEELIN, J.D., M. LATIF, M.A.F. ALLAART, M. CANE, U. CUBASCH, W.L. GATES, P.R. GENT, M. GHIL, C. GORDON, N.C. LAU, N.C. MECHOSO, G.A. MEEHL, J.M. OBERHUBER, S.G.H. PHILANDER, P.S. SCHOPF, K.R. SPERBER, A. STERL, T. TOKIOKA, J. TRIBBIA AND S.E. ZEBIAK, 1992: Tropical air-sea interaction in general circulation models. *Clim. Dyn.*, **7**, 73-104.

NEFTEL, A., J. BEER, H. OESCHGER, F. ZURCHER AND R. C. FINKEL, 1985: Sulphate and nitrate concentrations in snow from South Greenland 1895-1978. *Nature*, **314**, 611-613.

NORDENG, T., 1994: Extended versions of the convective parameterization scheme at ECMWF and their impact on the mean and transient activity of the model in the tropics.ECMWF, Research Department, Technical Memorandum No.206, October 1994, *41pp., submitted to Q. J. R. Meteorol. Soc.*

OBERHUBER, J.M., 1988: An atlas based on the 'COADS' data set: The budgets of heat, buoyancy and turbulent kinetic energy at the surface of the global ocean. *Max-Planck-Institut für Meteorologie, Report No. 15, Hamburg.*

OBERHUBER, J.M., 1993A: Simulation of the Atlantic circulation with a coupled sea ice-mixed layer-isopycnal general circulation model. Part I: Model description. *J. Phys. Oceanogr.*, **22**, 808-829.

OBERHUBER, J.M., 1993B: The OPYC Ocean General Circulation Model. Deutsches Klimarechenzentrum GmbH, *Technical Report No. 7, Hamburg.*

OHMURA, A. AND H. GILGEN, 1993: Re-evaluation of the global energy balance. *Geophysical Monograph 75, IUGG Volume* **5**, 93-110.

PALMER, T.N., G. J. SHUTTS AND R. SWINBANK, 1986: Alleviation of a systematic westerly bias in general circulation and numerical weather prediction models through an orographic gravity wave drag parameterization. *Quart. J. Roy. Meteor. Soc.*, **112**, 1001-1031.

ROCKEL, B., E. RASCHKE AND B. WEYNES, 1991: A parameterization of broad band radiative transfer properties of water, ice and mixed clouds. *Beitr. Phys. Atmosph.*, **64**, 1-12.

ROECKNER, E., M. RIELAND AND E. KEUP, 1991: Modelling of cloud and radiation in the ECHAM model. ECMWF/WCRP. "Workshop on clouds, radiative transfer and the hydrological cycle", 12-15 Nov. 1990, *199-222, ECMWF, Reading, U.K.*

ROECKNER, E., J. M. OBERHUBER, A. BACHER, M. CHRISTOPH, I. KIRCHNER, 1995: ENSO variability and atmospheric response in a global coupled atmosphere-ocean GCM. *Max-Planck-Institut für Meteorologie, Report No. 178, Hamburg.*

ROECKNER, E., K. ARPE, L. BENGTSSON, M. CHRISTOPH, M. CLAUSSEN, L. DÜMENIL, M. ESCH, M. GIORGETTA, U. SCHLESE, AND U. SCHULZWEIDA, 1996: The Max Planck Institute for Meteorology fourth generation atmospheric general circulation model (ECHAM4): Model description and climatology. *(In preparation).*

SAUSEN, R., R. K. BARTHELS AND K. HASSELMANN, 1988: Coupled ocean- atmosphere models with flux corrections. *Climate Dynamics,* **2**, 154- 163.

SAUSEN, R., S. SCHUBERT AND L. DÜMENIL, 1991: A model of the river runoff for use in coupled atmosphere-ocean models. *Meteorologisches Institut, University of Hamburg, Report No 9, 15-33.*

SIMMONS, A. J. AND R. STRÜFING, 1983: Numerical forecast of stratospheric warming events usinf a model with a hybrid vertical coordinate. *Quart. J. Roy. Met. Soc.,* **109**, 81-111.

SPENCER, R. W. AND J. R. CHRISTY, 1990: Precise Monitoring of Global Temperature Trends from Satellites. *Science,* **247**, 1558-1562.

SUNDQVIST, H., 1978: A parameterization scheme for non-convective condensation including prediction of cloud water content. *Quart. J. Roy. Meteor. Soc.,* **104**, 677-690.

TAYLOR, K. E. AND J. E. PENNER, 1994: Response of the climate system to atmospheric aerosols and greenhouse gases. *Nature,* **369**, 734-737.

TIEDTKE, M., 1989: A comprehensive mass flux scheme for cumulus parameterization in large-scale models. *Mon. Wea. Rev.,* **117**, 1779- 1800.

TRENBERTH, K. (ED.), 1992: Climate system modeling. *Cambridge University Press.*

WASHINGTON, W. M. AND G. A. MEEHL, 1991: Climate sensitivity due to increased CO2: Experiments with a coupled atmosphere and ocean general circulation model. *Climate Dyn.,* **4**, 1-38.

WILLIAMSON, D.L. AND P. J. RASCH 1994: Water vapour transport in the NCAR CCM2 *Tellus,* **46A**, 34-51.

ANALYSIS OF THERMOHALINE FEEDBACKS

JOCHEM MAROTZKE
Massachusetts Institute of Technology
Cambridge, USA

Contents

Abstract

Feedbacks between atmospheric meridional transports and the thermohaline circulation (THC) are analysed, using a four-box ocean-atmosphere model in one hemisphere. The ocean model is Stommel's; the atmospheric model is similar to the one developed earlier by Marotzke and Stone and gives the surface heat and freshwater fluxes as residuals of the atmospheric energy and moisture budgets, assumed in balance. Radiation at the top of the atmosphere depends linearly on surface temperature; atmospheric meridional heat and moisture transports are proportional to an arbitrary positive power of the meridional temperature gradient, which is taken to be identical to the oceanic temperature gradient. The coefficients in the power laws for atmospheric meridional transports are chosen such that all coupled models have one steady state in common. Upon linearization, a Newtonian cooling law is derived for anomalous differential surface heat flux. The timescale on which the ocean temperature gradient is restored decreases with increasing power, n, in the atmospheric heat

NATO ASI Series, Vol. I 44
Decadal Climate Variability
Dynamics and Predictability
Edited by David L. T. Anderson and Jürgen Willebrand
© Springer-Verlag Berlin Heidelberg 1996

transport parameterization. The limits, $n = 0, \infty$, correspond approximately to the uncoupled cases of fixed surface heat flux and fixed surface temperature, respectively. A power, m, of zero in the atmospheric moisture transport is equivalent to fixed surface freshwater flux. Stronger dependence of moisture transport on temperature gradient (increasing m) destabilises the THC. When moisture transport depends weakly on temperature gradients ($m \leq 1$), the THC is more stable to perturbations as n decreases (weaker restoring). For $m \geq 2$, however, the THC is more unstable with smaller n, because the destabilising effect of anomalous moisture transport outweighs the stabilising effect of anomalous thermal forcing of the THC. If zonal mixing in the atmosphere is incomplete, the zonal mean temperature deviates from the ocean temperature. Meridional atmospheric transports are less sensitive to variations in the ocean temperature gradient if the ocean area is small, and Newtonian cooling is weaker. Simultaneously, a smaller ocean can compensate a given atmospheric energy budget imbalance only through a greater change in surface heat flux, which translates into stronger Newtonian cooling. When flux adjustments are applied to obtain the correct model climate despite incorrect atmospheric transports, the effect is that of choosing incorrect n or m, so transient behavior and model sensitivity are wrong although the mean state is correct. It is speculated that climate drifts seen even in flux-adjusted coupled models are amplifications of residual drifts in an incomplete, uncoupled spinup. The linearised model is non-normal, which leads to amplifications of small perturbation through interference of non-orthogonal eigenfunctions. Considerable excursions in temperature and salinity gradients can occur, which are density compensated and hence are not effectively counteracted by changes in the circulation unless the atmospheric transports respond to the changes in ocean temperature.

1 Introduction

This paper provides a conceptual framework to understand some feedbacks involving meridional transports in the atmosphere and the thermohaline circulation. Along the way, a strategy is presented that could be used to assess the strength of various feedbacks in a wide range of models. The analysis is motivated by the quest for a dynamical understanding of variations in North Atlantic Deep Water (NADW) formation and Atlantic meridional overturning, which are both crucial for the northward heat transport in the Atlantic (Hall and Bryden, 1982; Roemmich and Wunsch, 1985). Paleoclimatic observations indicate that the strength of NADW formation may have varied significantly in the geologic past (see, for example, Boyle, 1990, or Sarnthein *et al.*, 1994). Besides natural fluctuations in overturning strength, anthropogenic ones may occur in the future: The coupled climate model of the Geophysical Fluid Dynamics Laboratory (GFDL; Manabe and Stouffer, 1994) shows a temporary reduction by one half of the Atlantic thermohaline circulation, in response to a gradual doubling of atmospheric

CO_2 concentration. If the CO_2 content is gradually quadrupled, the Atlantic thermohaline circulation ceases and never recovers.

In the Manabe and Stouffer (1994) experiments, the weakening of the thermohaline circulation is caused by the increase in atmospheric temperature and hence water vapor content, leading to enhanced poleward moisture transport and lower high-latitude salinities. But the resilience of NADW formation to the increased freshwater input in the sinking region depends on various feedback processes, which are possibly incorrectly represented in the coupled general circulation model (GCM), as demonstrated by the large flux adjustments applied to model current climate. There is a need to identify the important processes and assess how well any comprehensive model represents them.

Here, I use the simplest possible model of interactions between atmospheric transports and the thermohaline circulation to analyze some feedback processes that I believe are important. Different model versions employ different assumptions about atmospheric transports, which are shown to correspond to different stages of ocean-atmosphere coupling. I then investigate the transient behavior of the coupled models near the steady state corresponding to vigorous NADW formation.

The model builds upon the ones developed earlier by Nakamura, Stone, and Marotzke (1994, NSM hereinafter) and by Marotzke and Stone (1995, MS hereinafter), but differs in that a greater variety of atmospheric transport formulations is analyzed in detail. The paper is organised as follows. Section 2 presents the model, closely following MS. Section 3 discusses the stability of the equilibrium corresponding to vigorous NADW formation, by performing numerical integrations of the various model versions. Section 4 discusses the effects that land can have on ocean-atmosphere coupling, and section 5 addresses climate drifts as seen in coupled GCMs despite the application of flux adjustments. Section 6 analyzes a particular type of variability, called 'neutrally buoyant mode' by Saravanan and McWilliams (1995), in terms of the superposition of non-orthogonal eigenfunctions. A few concluding remarks follow in section 7. Section 3 revisits topics discussed earlier in NSM and MS, but the discussion of competing feedbacks is more definitive. The material in sections 4, 5, and 6 is novel.

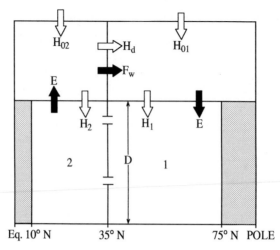

Figure 1: *Vertical cross-section of the model. See text for definitions.*

2   Model formulation

2.1   Basic equations

The model consists of two ocean boxes and two atmospheric boxes. The ocean boxes are well mixed and have depth D; box 1 represents the high-latitude ocean and box 2 the low-latitude ocean (Fig. 1). H_1 and H_2 are ocean heat gain through the surface, and H_{01} and H_{02} are atmospheric energy gain at the top. H_d is the meridional energy transport in the atmosphere. E is net evaporation at low latitudes, and net precipitation at high latitudes, and F_W is the meridional atmospheric moisture transport.

The conservation equations for the ocean are (*cf.*, Stommel, 1961; Marotzke, 1990)

$$
\begin{aligned}
\dot{T}_1 &= & H_1 + |q|(T_2 - T_1), & \qquad (1)\\
\dot{T}_2 &= & H_2 - |q|(T_2 - T_1), & \qquad (2)\\
\dot{S}_1 &= - & H_s + |q|(S_2 - S_1), & \qquad (3)\\
\dot{S}_2 &= & H_s - |q|(S_2 - S_1), & \qquad (4)
\end{aligned}
$$

The flow strength, q, is related to the meridional density gradient by a linear law,

$$q = k[\alpha(T_2 - T_1) - \beta(S_2 - S_1)], \tag{5}$$

where a linear equation of state has been assumed, and α and β are, respectively, the thermal and haline expansion coefficients. Some support for the simple relationship (5) derives from the recent results of Hughes and Weaver (1994). H_S, the virtual surface salinity flux, is related to the surface freshwater flux E through

$$H_s = S_0 \frac{E}{D}, \tag{6}$$

where S_0 is a constant reference salinity and E is water loss (gain) of the low (high) latitude ocean, measured in meters per second. Notice that in order to convert H_1 and H_2 (which have units $°Cs^{-1}$) into physical heat fluxes, denoted $\tilde{H}_1$ and $\tilde{H}_2$, they are multiplied by the heat capacity of a unit water column, which is

$$c\rho_0 D \cong 4 \times 10^6 Jm^{-3}K^{-1} \times 5 \times 10^3 m \cong 2 \times 10^{10} Jm^{-2}K^{-1} \tag{7}$$

We will use the term surface heat flux interchangeably for the heat fluxes proper and for the induced temperature tendencies.

We assume that *heat and moisture capacities of the atmosphere are negligible*. This is justified if one only considers timescales longer than the atmosphere's equilibration time, which is given by radiative cooling for temperature (order 1 month) and the typical lifetime of a water particle in the atmosphere (order 1 week, Gill, 1982, p. 31). We assume that we can parameterize the energy fluxes at the top of the atmosphere and the meridional energy and water transports in the atmosphere; as a result, the air-sea exchanges can be determined as the residuals of the steady-state atmospheric heat and moisture budgets. For the moment, we assume perfect longitudinal mixing in the atmosphere and zero temperature difference between sea and atmospheric surface temperatures, so that the zonally averaged atmospheric surface temperatures are equal to the oceanic temperatures. Furthermore, we assume a constant atmospheric lapse rate. The planetary-scale meridional temperature gradients in the atmosphere are then completely determined by the temperatures in the ocean boxes. In section 4, the condition of perfect zonal homogenization will be relaxed, and air temperatures over land computed separately.

The radiation at the top of the atmosphere is parameterized as a linear function of surface temperature (Wang and Stone, 1980)

$$H_{01} = A_1 - BT_1, \tag{8a}$$

$$H_{02} = A_2 - BT_2, \tag{8b}$$

where A_1 and A_2 are net incoming radiation at high and low latitudes, respectively, for a surface temperature of 0°C. A_1 is negative, and A_2 is positive. BT_1 and BT_2 mark longwave fluxes at high and low latitudes, respectively, caused by deviations of surface temperature from zero.

The parameterization of meridional atmospheric transports is based on the same concepts as in NSM and MS, namely that baroclinic eddies are the main transport mechanism, and that eddy activity depends mainly on the large-scale, zonal mean meridional temperature gradient (*e.g.*, Stone and Miller, 1980). We deviate from NSM and MS, however, in assuming a general power law for both meridional heat and moisture transports.

$$H_d = \tilde{\chi}_n (T_2 - T_1)^n \qquad n \geq 0 \tag{9}$$
$$F_W = \tilde{\gamma}_m (T_2 - T_1)^m \qquad m \geq 0 \tag{10}$$

Meridional transports increase with temperature gradient, but at an unspecified rate, which is expressed by the powers and the constant coefficients, $\tilde{\chi}_n$ and $\tilde{\gamma}_m$ (the tilde is used for later convenience). The substantial approximation has been made here that latent heat and moisture transports are independent of temperature (as opposed to temperature gradient) and hence, due to the Clausius-Clapeyron equation, of specific humidity. Ideally, both coefficients in (9) and (10) should depend on temperature also, but for simplicity this is not done here. Also, the powers of the heat and moisture transport laws should be identical, because both transports are accomplished by the same physical process (baroclinic eddies bring warm, moist air northward and cold, dry air southward). Varying m and n independently allows us to isolate the feedbacks more clearly, and also to identify several approximations that have been employed. NSM use parameterizations for atmospheric eddy transports in which $n \approx m \approx 3.5$, and they do include the dependence of saturation water vapor pressure on temperature. MS use (9) and (10) with $n = m = 1$ (linear atmospheric model); we will, for the moment, proceed with this special case and return to the general case later on. For brevity, we define $\tilde{\chi} \equiv \tilde{\chi}_1, \tilde{\gamma} \equiv \tilde{\gamma}_1$.

2.2 Surface heat fluxes

The heat budget for the high-latitude atmospheric box reads

$$\int_{F_{01}} H_{01} da - \int_{F_1} \tilde{H}_1 da + H_d = 0. \tag{11}$$

The first integral is the (negative) energy gain at the top of the atmosphere, integrated over the entire area, F_{01}, north of the latitude circle dividing the boxes. The second integral is the (likewise negative) heat gain of the ocean, integrated over the ocean portion, F_1, of the high-latitude box. The third term is the integrated meridional energy flux and given by eq. (9). It is now assumed that H_{01} and H_1 are spatially constant (or, rather, that only the area averaged values matter), and we obtain from (11)

$$F_{01} H_{01} - F_1 \tilde{H}_1 + H_d = 0, \tag{12}$$

and hence

$$\tilde{H}_1 = \frac{1}{\varepsilon} H_{01} + \frac{1}{\varepsilon F_{01}} H_d, \tag{13a}$$

where $\varepsilon \equiv F_1/F_{01}$, the relative ocean coverage of the high-latitude area. Analogous considerations hold for the low-latitude boxes, and we assume, for simplicity, that $F_2 = F_1$ and $F_{02} = F_{01}$. Hence,

$$\tilde{H}_2 = \frac{1}{\varepsilon} H_{02} - \frac{1}{\varepsilon F_{01}} H_d. \tag{13b}$$

Using the parameterizations (9) for H_d and (8) for H_{01} and H_{02}, one finally arrives at expressions for the surface heat fluxes that are functions of the oceanic temperatures only,

$$\tilde{H}_1 = \frac{1}{\varepsilon}(A_1 - BT_1) + \frac{\tilde{\chi}}{\varepsilon F_{01}}(T_2 - T_1), \tag{14a}$$

$$\tilde{H}_2 = \frac{1}{\varepsilon}(A_2 - BT_2) - \frac{\tilde{\chi}}{\varepsilon F_{01}}(T_2 - T_1). \tag{14b}$$

The focus will be on the meridional temperature difference,

$$T \equiv T_2 - T_1, \tag{15}$$

for which we obtain, from oceanic heat conservation, eqs. (1) and (2), and the parameterizations for the surface fluxes, eq. (14),

$$\dot{T} = H_2 - H_1 - 2|q|T = \frac{1}{\varepsilon}\left(\frac{A_2 - A_1}{c\rho_0 D}\right) - \frac{1}{\varepsilon}\left(\frac{2\chi + B}{c\rho_0 D}\right)T - 2|q|T, \quad (16)$$

where $\chi \equiv \tilde{\chi}F_{01}^{-1}$. Assume, first, that neither atmosphere nor ocean transport heat, i.e., that both χ and q are zero. The steady-state meridional temperature gradient would then be determined by the radiation balance alone, and given as

$$T_R \equiv T|_{\chi=0,q=0} = \frac{A_2 - A_1}{B}. \quad (17)$$

For a sensible choice of parameters (see Table 1), T_R is about 75°C. Next, assume that only the atmosphere transports heat horizontally, but not the ocean. The steady-state temperature, thus defined, would be the equilibrium temperature, T_E, introduced by Bretherton (1982). It is derived from a balance purely between dynamical and radiative transports in the atmosphere; the ocean's role in heat transport is neglected. From (16), T_E is readily found as

$$T_E \equiv T|_{q=0} = \frac{A_2 - A_1}{2\chi + B}. \quad (18)$$

With the parameters of Table 1, T_E is about 30°C; atmospheric transports reduce the purely radiative temperature contrast by more than one half.

From eq. (16), the surface heat fluxes driving the meridional temperature gradient can be rewritten,

$$H_2 - H_1 = \lambda(T_E - T), \quad (19)$$

with

$$\lambda \equiv -\left(\frac{\partial(H_2 - H_1)}{\partial T}\right) = \frac{1}{\varepsilon}\left(\frac{2\chi + B}{c\rho_0 D}\right), \quad (20)$$

and T_E given by (18), and we obtain

$$\dot{T} = \lambda(T_E - T) - 2|q|T. \quad (21)$$

Equation (19) is a Newtonian cooling law for ocean temperature, which has been widely used as a boundary condition on SST in numerical ocean models (*e.g.*, Haney, 1971; Marotzke and Willebrand, 1991). Notice, however, that eq. (19) connects spatial differences in surface heat flux to temperature differences. It is readily shown (MS) that the spatial mean heat flux can likewise be written as a Newtonian law for the mean temperature, but spatial mean and difference are restored with different time constants. This 'scale dependence' of the restoring coefficient is equivalent to the heat fluxes, H_1 and H_2, not being computable by a purely local relationship, but involving horizontal atmospheric transports [see eq. (14); see Marotzke, 1994, for a detailed discussion].

2.3   Surface freshwater fluxes

In analogy to the high-latitude atmospheric heat budget, eq. (11), we write

$$\int_{F_{01}} E\,da + F_W = 0, \tag{22}$$

where E is the evaporation minus precipitation. The integral in (22) also comprises precipitation over land, which through river runoff influences the oceanic freshwater budget. To compute E over the ocean, one has to consider the ratio, ε_W, of the ocean area to the catchment area of the ocean basin. Taking E to be constant along a latitude circle, the range of ε_W is

$$\varepsilon \le \varepsilon_W \le 1, \tag{23}$$

meaning that in one extreme ($\varepsilon_W = 1$) the ocean basin receives only the moisture transported in the atmosphere right above it, whereas in the other extreme ($\varepsilon_W = \varepsilon$, meaning that the catchment area is the entire latitude circle) it catches all the river runoff as well. With the Atlantic in mind, ε_W in the range 0.3 to 0.5 is a reasonable value since the Atlantic (including the Arctic) receives runoff from almost the entire Americas and a large portion of Asia (*e.g.*, Baumgartner and Reichel, 1975; Broecker *et al.*, 1990). Here, $\varepsilon_W = \varepsilon = 0.5$ is chosen, which corresponds to a single ocean basin of the size of the Pacific that receives all river runoff, from a continent the same size as the ocean, through zonal river flow. This choice of ocean area is made for convenience in the perturbation experiments described in Section 3.

Thus, we obtain for E over the ocean

$$E = \frac{1}{\varepsilon_W}\frac{F_W}{F_{01}}, \tag{24}$$

Quantity	Symbol	Value
High-latitude radiative forcing	A_1	-39 Wm^{-2}
Low-latitude radiative forcing	A_2	91 Wm^{-2}
Longwave radiation coefficient	B	1.7 Wm^{-2}K^{-1}
Atmospheric heat transport efficiency	χ	1.3 Wm^{-2}K^{-1}
Atmospheric moisture transport efficiency	γ	2.8×10^{-10} ms^{-1}K^{-1}
Total high-latitude area	F_{01}	1.25×10^{14} m^2
Fractional ocean area	ε	0.5
Fractional catchment area	ε_W	0.5
Thermal expansion coefficient	α	1.8×10^{-4} K^{-1}
Haline expansion coefficient	β	0.8×10^{-3} psu^{-1}
Hydraulic constant	k	$2 \times 10^{-8}s^{-1}$
Heat capacity per unit volume of water	$\rho_0 c$	4×10^6 Jm^{-3}K^{-1}

Table 1: *List of model parameters*

and combining (24) with the conversion formula from freshwater fluxes into equivalent surface salinity fluxes, eq. (6), we obtain

$$H_s = \frac{1}{\varepsilon_W} \frac{S_0}{D} \frac{F_W}{F_{01}}. \tag{25}$$

It has been tacitly assumed that whatever water reaches the high-latitude ocean through atmospheric transports has originated from the low latitudes of the same ocean, so the same ε_W applies at high and low latitudes. In particular, this assumption eliminates cross-basin atmospheric water vapour transports as proposed, for example, by Broecker *et al.* (1990) to be crucial for the maintenance of the global thermohaline circulation. The present model is too simple to address this issue, which is therefore sidestepped.

Finally, the parameterization (10) with $m = 1$ for F_W is inserted into (25), which, together with the abbreviation (15) for $T_2 - T_1$, and the definition $\gamma \equiv \tilde{\gamma}/F_{01}$, yields

$$H_s = \frac{1}{\varepsilon_W} \frac{S_0}{D} \gamma T. \tag{26}$$

Total salt content of the model ocean is constant; introducing

$$S \equiv S_2 - S_1, \tag{27}$$

the difference between the salt conservation equations, (3) and (4), gives

$$\dot{S} = \frac{2}{\varepsilon_W} \frac{S_0}{D} \gamma T - 2|q|S. \tag{28}$$

The coupled model has been reduced to two prognostic equations, (21) and (28), for the meridional temperature and salinity differences, respectively. Inserting the hydraulic flow law, eq. (5), gives

$$\dot{T} = \lambda(T_E - T) \quad - 2k|\alpha T - \beta S|T, \tag{29}$$
$$\dot{S} = \frac{2}{\varepsilon_W} \frac{S_0}{D} \gamma T \quad - 2k|\alpha T - \beta S|S. \tag{30}$$

Note that eq. (30) for the salinity contrast is homogeneous; the meridional salinity gradient is forced by the temperature gradient, which itself is maintained by solar radiation and eroded by longwave radiation and meridional atmospheric transports [implicit in the definitions for T_E and λ, eqs. (18) and (20), respectively], and by oceanic transports. The heat and salt budgets are coupled not only through the oceanic flow, q, but also through the temperature-driven freshwater flux.

2.4 Equilibrium solutions

The structure of the equilibrium solutions is best displayed graphically. Fig. 2 shows the T-S phase space between the origin and a little beyond $\beta S = \alpha T_E$, $\alpha T = \alpha T_E$, for the choice of parameters of Table 1. The main diagonal (dashed line) marks $\beta S = \alpha T$ where the influences of temperature and salinity gradients on density gradients exactly cancel, and there is no flow ($q = 0$). To the left of the diagonal, the flow is temperature dominated (thermally direct), with near-surface flow toward high latitudes. The heavy and thin solid curves represent, respectively, the equations $\dot{S} = 0$ and $\dot{T} = 0$. The curves intersect at the solution points; solutions A and C are stable to infinitesimal perturbations, while the other is unstable. The equilibrium curves are now discussed for $q > 0$.

The shape of the temperature equilibrium curve, which is not a straight line, is readily understood as follows. It meets the main diagonal at $(\alpha T_E, \alpha T_E)$ where the flow vanishes; hence, $T = T_E$. Generally, a smaller salinity gradient causes stronger flow, which in turn leads to a stronger reduction of the temperature gradient. Hence, T deviates more from T_E along the $\dot{T} = 0$ curve as one moves to the left in Fig. 2. For the parameters used here, the ocean circulation can further reduce the meridional temperature gradient by up to 10°C.

The shape of the $\dot{S} = 0$ curve can be deduced qualitatively by rearranging the right side of (30), which for $q > 0$ can be written as

$$\dot{S} = 2k\alpha(S_{crit} - S)T + 2k\beta S^2, \tag{31}$$

where

$$S_{crit} \equiv \frac{1}{k\alpha D}\frac{\gamma}{\varepsilon_W}S_0 \tag{32}$$

is a critical salinity gradient above which the steady-state salinity gradient must lie. A physical interpretation is readily given. There are two influences of the temperature gradient on the salinity budget, salinity advection by the thermally driven part of the flow, and the surface freshwater flux. As S gets smaller and approaches S_{crit}, the two influences of the temperature gradient on the salinity budget almost cancel, and a very large temperature contrast is required to balance salinity advection by the salinity driven part of the flow. $S < S_{crit}$ would mean that $T < 0$ (polar box warmer than equatorial box) in equilibrium, which is nonsensical. A *large* salinity difference, on the other hand, requires a large temperature difference to keep the flow q small (and positive), so salinity advection can be balanced by surface freshwater flux despite the large salinity gradients. (A large temperature gradient also causes strong freshwater flux, but its effect on the salinity budget is smaller than the effect of larger temperature-driven salinity advection.) It follows that on the $\dot{S} = 0$ curve, T goes to infinity for S approaching S_{crit} and for S going to infinity, so the $\dot{S} = 0$ curve must have a minimum at some intermediate point, which is readily shown to be at the line making a 22.5° angle with the αT axis (dash-dotted in Fig. 2).

For a more detailed discussion of the equilibrium solutions of eqs. (29) and (30) and their parameter sensitivities, see MS. Here, it suffices to note that, like the uncoupled box models (Stommel, 1961; Marotzke, 1990; Huang et al., 1992), eqs. (29) and (30) also admit a stable solution with $q < 0$, characterised by $\alpha T < \beta S$ and equatorward surface flow. A transition between the two stable solutions is possible through large enough perturbations. Our focus, however, will be on the stable solution with high-latitude sinking and the feedbacks affecting it.

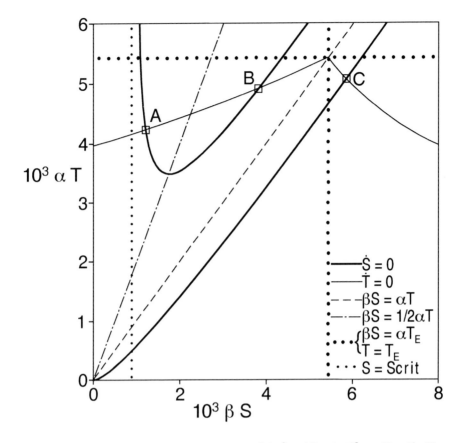

Figure 2: *T-S phase space between the origin and* $(10^3 \times \beta S = 8, 10^3 \times \alpha T = 6)$. *Heavy solid curves:* $\dot{S} = 0$. *Thin solid curves:* $\dot{T} = 0$. *Thin dotted line:* $S = S_{crit}$. *Heavy dotted lines:* $\beta S = \alpha T_E$ *and* $T = T_E$. *Dashed:* $\beta S = \alpha T$ *(no flow,* $q = 0$); *to the left of this line, the flow is thermally dominated ('high-latitude sinking'), to the right, it is salinity dominated ('low-latitude sinking'). Dash-dotted:* $\beta S = 1/2\alpha T$, *which intersects the* $\dot{S} = 0$ *curve at its minimum,* $(2S_{crit}, 4S_{crit})$. *Steady-states lie at the intersections of the* $\dot{S} = 0$ *and* $\dot{T} = 0$ *curves, and are marked A (thermally dominated, stable), B (thermally dominated, unstable), and C (salinity dominated, stable).*

3 Stability of the high-latitude sinking equilibrium

3.1 Atmospheric transport anomalies

The MS model with linear atmospheric transports is a convenient tool to define a reasonable stable equilibrium of the coupled box model. We now return to the general power laws for meridional atmospheric transports, eqs. (9) and (10). Following the strategy already used by NSM, we choose the coefficients, $\tilde{\chi}_n$ and $\tilde{\gamma}_m$, such that models with arbitrary m and n have the high-latitude sinking equilibrium identical to the model with $m = n = 1$:

$$H_d(\bar{T}) = \tilde{\chi}_1 \bar{T}^1 = \tilde{\chi}_n \bar{T}^n \quad \text{if} \quad \tilde{\chi}_n \equiv \tilde{\chi}_1 \bar{T}^{1-n} \tag{33}$$
$$F_W(\bar{T}) = \tilde{\gamma}_1 \bar{T}^1 = \tilde{\gamma}_m \bar{T}^m \quad \text{if} \quad \tilde{\gamma}_n \equiv \tilde{\gamma}_1 \bar{T}^{1-m}, \tag{34}$$

meaning that for a given steady-state temperature gradient $\bar{T}$, the meridional transports are identical for arbitrary powers n and m. All other coefficients are unchanged. In particular, the longwave loss does not change between different models. By construction, then, all models have the same steady-state temperature and the same steady-state surface fluxes, hence also the salinities and the flow strength are identical. This equilibrium common to all models is an ideal starting point for investigating the effect of the atmospheric transports on the transient behavior.

Notice that a model with arbitrary n and m could have been chosen to define the reference steady state. In general, it is required that the relationship

$$\tilde{\chi}_{n_1} \equiv \tilde{\chi}_{n_2} \bar{T}^{n_2 - n_1} \tag{35}$$

hold (similarly for moisture transports); eq. (33) is the special case $n_2 = 1$. Notice also that while all models with coefficients defined by eqs. (33) and (34) share the high-latitude sinking equilibrium, there is no guarantee that the steady state is stable to infinitesimal perturbations for all n or m.

Numerical solutions for various nonlinear atmospheric transports will be presented below. For analytical considerations it is convenient to linearise the atmospheric transport anomalies about the equilibrium temperature gradient $\bar{T}$,

$$H'_{d,n} \equiv H_{d,n}(\bar{T} + T') - H_{d,n}(\bar{T}) = \tilde{\chi}_n(\bar{T} + T')^n - \tilde{\chi}_n\bar{T}^n = n\tilde{\chi}_n\bar{T}^{n-1}T' + \dots \tag{36}$$

Use of eq. (33) for $\tilde{\chi}_n$ gives

$$H'_{d,n}(T) \approx n\tilde{\chi}_1 T'. \tag{37}$$

Similarly, one obtains for the moisture transport anomalies

$$F'_{W,m}(T) \approx m\tilde{\gamma}_1 T'. \tag{38}$$

Equations (37) and (38) state that atmospheric heat and water transport anomalies are, respectively, n or m times the anomalies of the linear case, for a given anomaly in temperature gradient. All other results of the linear model remain valid; in particular, the (anomalous) differential surface heat flux is still governed by a Newtonian cooling law. But now the restoring coefficient, λ, is given by

$$\lambda = \frac{1}{\varepsilon}\left(\frac{2n\chi_1 + B}{c\rho_0 D}\right), \tag{39}$$

meaning that a higher power in the transport law (9) reasults in a stronger restoring.

Some special cases are readily identified. The choice $n = 0$ corresponds to fixed atmospheric heat transport, $H'_{d,0} = 0$. As a consequence, all temperatures are restored locally on the radiative timescale, as in the model of Zhang et al. (1993). If additionally $B = 0$, surface heat fluxes are completely prescribed. The case $m = 0$ corresponds to fixed atmospheric moisture transport, $F'_{W,m} = 0$, and hence fixed surface freshwater flux. The limit $n \to \infty$ requires $T' \to 0$, which is infinitely strong restoring, to keep atmospheric heat transport anomalies finite. The temperature *gradient* is fixed (the temperatures themselves are fixed if B is very large, too); as a consequence, the surface freshwater flux is constant irrespective of m, and we recover the mixed boundary conditions. Thus, the two physically meaningful uncoupled models, with either temperature or surface heat flux prescribed, correspond to the extreme cases $n = \infty$ or $n = 0$, respectively. For either choice, the surface freshwater flux is also prescribed.

Model #	Definition	Comments
1	$\chi_0, \gamma_0, B = 0$	Fixed surfacefluxes
2	$\chi_\infty, B \to \infty$	Fixed temperatures
3	χ_0, γ_0	Radiative restoring
4	χ_1, γ_0	Diffusive + fixed E-P
5	χ_1, γ_1	MS
6	χ_3, γ_1	
7	χ_1, γ_3	
8	χ_3, γ_3	$\cong$ NSM

Table 2: *Definition of the models. The indices on χ and γ indicate the powers of the heat and moisture transport laws, respectively.*

3.2 Modelling strategy

In the following, 8 combinations of n and m are used in the power laws for atmospheric transports, (9) and (10). The models are listed in Table 2, along with some comments. Models 1-3 have been characterised above; model 4 is the simplification to two horizontal boxes of linear diffusive atmospheric transport, as recently used with oceanic GCMs by Rahmstorf and Willebrand (1995) and Pierce *et al.* (1996). Model 5 is the reference one (MS); Models 6 and 7 use combinations of linear and cubic laws, and Model 8 is closest to the one used by NSM.

The goal is to characterise the stability of the high-latitude sinking equilibrium of each model, and to assess how the feedbacks reinforce or counteract each other. To this end, steady state A of Fig. 2 is subjected to an initial anomaly in salinity gradient, such that the amount ΔS is taken out of the polar box and added to the equatorial one. There are at least three measures of stability that are readily used for this type of experiment,

i) The critical perturbation in salinity, ΔS_c, defined as the minimum necessary to induce a transition to the low-latitude sinking equilibrium. The smaller ΔS_c, the less stable is the model.

ii) The time it takes a model to make a transition, for a perturbation greater than ΔS_c. The shorter the transition time, the less stable is the model.

iii) The time constant of the exponential decay of small (subcritical) perturbations. The slower the decay, the less stable is the model.

There is no guarantee that the three criteria always give the same ordering, but a combination of the three allows a comprehensive stability analysis.

3.3 Thermohaline feedbacks

A negative (positive) feedback is present when a perturbation weakens (enhances) itself through the changes it causes. To discuss the feedbacks of the present model, it is useful to linearise the oceanic equations as well as the atmospheric transports. One obtains for a thermally direct mean state ($\bar{q} > 0$)

$$
\begin{align}
\dot{T}' &= H_T'(T') - 2q'\bar{T} - 2\bar{q}T' - ..., \tag{40} \\
\dot{S}' &= H_S'(T') - 2q'\bar{S} - 2\bar{q}S' - ..., \tag{41} \\
q' &= k(\alpha T' - \beta S'), \tag{42}
\end{align}
$$

with the surface flux anomalies given by

$$
H_T'(T') \approx -\lambda T' = -\frac{2n\chi_1 + B}{\varepsilon c\rho_0 D}T', \tag{43}
$$

$$
H_S'(T') \approx \frac{2m\gamma_1 S_0}{D\varepsilon_W}T'. \tag{44}
$$

Feedback #0: Mean Flow The mean circulation tends to eliminate all anomalous temperature and salinity gradients. This feedback is almost but not entirely trivial. It is always present, always negative (for example, positive T' leads to negative $\dot{T}'$), and represented by the respective terms #3 on the right hand sides of the linearised heat and salt equations, (40) and (41).

The other feedbacks all involve anomalous circulation.

Feedback #1: Ocean Heat Transport *Weaker overturning circulation, $q' < 0 \Rightarrow \dot{T}' > 0 \Rightarrow$ Larger meridional temperature gradient $\Rightarrow$ Stronger overturning.*

This is a negative feedback, purely oceanic, and represented by term 2 in the heat conservation equation (40). It is suppressed by prescribing sea surface temperature and conceptually assumes fixed surface heat fluxes.

Feedback #2: Salinity Transport *Weaker overturning circulation, $q' < 0 \Rightarrow \dot{S}' > 0 \Rightarrow$ Larger salinity gradient $\Rightarrow$ Even weaker overturning.*

This is a positive feedback and purely oceanic; it is represented by term 2 in the salt conservation equation (41) and is suppressed if sea surface salinity is prescribed as effectively done in GCMs with restoring surface salinity boundary condition. It is the fundamental feedback responsible for the existence of multiple equilibria of the thermohaline circulation under mixed boundary conditions. Stommel (1961) did not, however, discuss this process explicitly, which was apparently first described by Walin (1985). Marotzke (1990) analyzed how the interaction between feedbacks #0 and #2 leads to Walin's (1985) analytical stability criterion for the 2-box model with prescribed temperature and freshwater flux. Notice that without feedback #0, there would be no stabilising process at all under mixed boundary conditions. Notice, also, that feedback #2 conceptually assumes fixed P-E.

We now turn to the coupled feedbacks, which involve changes in the meridional transports in the atmosphere and hence changes in the surface heat and freshwater fluxes. All these changes are driven by SST changes; therefore their connection with anomalies in the circulation comes through the first three stages of feedback #1.

Feedback #3: Atmospheric Heat Transport *Weaker overturning circulation, $q' < 0 \Rightarrow \dot{T}' > 0 \Rightarrow$ Larger meridional temperature contrast $\Rightarrow$ Increased atmospheric heat transport $\Rightarrow$ Smaller meridional temperature contrast $\Rightarrow$ Even weaker overturning circulation.*

This is a *positive*, coupled atmosphere-ocean feedback. It involves the fundamental *negative* feedback in the atmosphere between the meridional temperature gradient and eddy activity and hence heat transport, which tends to wipe out anomalous temperature gradients and in its linearised form acts as a Newtonian damping. It is counterintuitive that this feedback, which in its purely atmospheric part is a paradigm of a stabilising effect, destabilises the thermohaline circulation. The apparent paradox is resolved if one considers feedbacks #1 and #3 in conjunction: The change in atmospheric heat transport is induced by the SST change, which is instrumental in setting up the oceanic negative feedback. As the anomalous atmospheric heat transports limits the change in SST, it weakens the negative oceanic feedback - and hence constitutes a positive feedback.

Notice that changes in longwave radiation act in the same sense as changes in dynamical atmospheric transports: A greater atmospheric temperature gradient means lower (higher) temperature at high (low) latitudes, hence smaller (greater) longwave loss at high (low) latitudes, which reduces

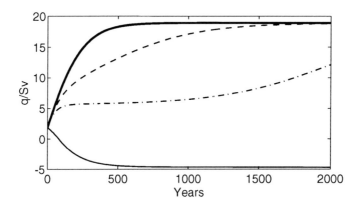

Figure 3: *Time series of overturning strength after perturbing the high-latitude sinking equilibrium by -1.7 psu in the polar box and +1.7 psu in the equatorial box, for 4 models with fixed surface freshwater flux. Heavy solid: model #1, fixed heat flux; thin solid: #2, fixed temperatures; dashed: #3, radiative restoring; dash-dotted: #4, linear atmospheric heat transport.*

the anomaly in temperature gradient.

The relative strengths of feedbacks #1 and #3 depend on the strength of the Newtonian damping - if Newtonian damping is infinitely fast ($n = \infty$), the stabilising feedback #1 is inactive; if the damping is weak (radiative, $n = 0$) or absent (surface heat fluxes fixed), the destabilising feedback #3 plays a minor role or none at all. Since the change in atmospheric transports is a response to anomalous meridional temperature contrasts, which at most can eliminate its cause completely but can never overshoot, feedback #3 cannot be stronger than feedback #1.

These considerations are illustrated by Fig. 3, showing overturning time series of four models with fixed surface freshwater fluxes, after an initial salinity anomaly of -1.7 psu has been added to the polar box at equilibrium, and +1.7 psu to the equatorial box. Model #2 (fixed temperatures) makes a rapid transition to the low-latitude sinking state, model #4 (linear atmospheric heat transport) hovers near the unstable equilibrium for almost 2000 years before returning to the original steady state, while model #3 (radiative restoring) and, even sooner, model #1 (fixed heat flux) quickly return to the high-latitude sinking case. Notice that model #1 has only one equilibrium and therefore must return.

Table 3 shows the same order in stability for models 1, 2, 3, and 4, for two choices of moisture transport efficiency γ_1 in the spinup. The stronger the Newtonian damping, the less stable are the models, consistent with the findings of, for example, Zhang et al. (1993), Power and Kleeman (1994), and Mikolajewicz and Maier-Reimer (1994), who all compare strong vs. weak Newtonian damping of SST in an ocean GCM, and those of Rahmstorf and Willebrand (1995) and Pierce et al. (1996), who couple diffusive, linear energy balance models with fixed moisture transports to ocean GCMs.

Analytically, the same dependence of model stability on the choice of n can be shown for infinitesimal perturbations if one considers the steady limit of eq. (40) for temperature anomalies, which is a good approximation since typically $\lambda \gg q$. It follows that for a given perturbation S',

$$T' = \frac{2k\beta\bar{T}}{\lambda + 2k\alpha\bar{T} + 2\bar{q}}S' \approx \frac{2k\beta\bar{T}}{\lambda}S'. \tag{45}$$

For fixed surface freshwater fluxes, the tendency equation (41) for salinity anomalies gives

$$\dot{S}' = -2k\alpha\bar{S}T' + 2(k\beta\bar{S} - \bar{q})S'. \tag{46}$$

The stabilising first term on the right-hand side of (46) is proportional to the temperature anomaly, and in the diagnostic limit it is approximately inversely proportional to the restoring coefficient λ and hence to n. All three stability criteria thus give the same result.

The last feedback this model can exhibit involves the response of the atmospheric moisture transport to changes in circulation and hence temperatures:

Feedback #4: Atmospheric Moisture Transport *Weaker overturning circulation, $q' < 0 \Rightarrow \dot{T}' > 0 \Rightarrow$ Larger meridional temperature contrast $\Rightarrow$ Increased atmospheric moisture transport $\Rightarrow$ Greater surface salinity contrast $\Rightarrow$ Even weaker overturning circulation.*

This also is a positive, coupled ocean-atmosphere feedback, termed EMT (for eddy moisture transport) feedback by NSM. It is suppressed by prescribing either SST or SSS. Its effect is demonstrated by Fig. 4, showing the flow strengths in response to an initial salinity perturbation of -1.6 psu (polar box) and +1.6 psu (equatorial box) of three models with the same (linear) atmospheric heat transport, but with moisture transports that are constant (model #4), linear (#5), or cubic (#7). Model #7 undergoes a

Model #	Definition	ΔS_c Case 1:	ΔS_c Case 2:
1	Fixed surface flux	∞	∞
3	χ_0, γ_0, radiative restoring	2.4	1.9
4	χ_1, γ_0, diffusive + fixed E-P	1.8	1.4
5	χ_1, γ_1, MS	1.6	1.2
6	χ_3, γ_1	1.4	1.0
8	$\chi_3, \gamma_3, \cong$ NSM	1.150	0.73
7	χ_1, γ_3	1.149	0.68
2	Fixed temperatures	1.13	0.77

Table 3: *Perturbations in initial salinity, ΔS_{crit}, necessary to induce a transition to the low-latitude sinking equilibrium. Cases 1 and 2 are two different equilibria; Case 1 uses the standard set of parameters (Table 1), which gives $(\bar{E} - \bar{P}) = 0.42$ m/yr. Case 2 uses $\gamma = 3.2 \times 10^{-10}$ $ms^{-1}K^{-1}$, which gives $(\bar{E} - \bar{P}) = 0.49$ m/yr. The models are listed with increasing sensitivity in Case 1; notice that models #8, #7, and #2 have a different order in Case 2.*

transition to the low-latitude sinking state within 500 years, #5 does this in 1000 years, while #4 slowly returns to the original steady state. The same order in stability is evident from Table 3. For small perturbations, we note from the diagnostic temperature perturbation equation (45) that positive anomalies in salinity contrast lead to positive anomalies in temperature contrast, which also leads to increased surface freshwater flux, through eq. (44), by an amount that is proportional to m.

So far, everything seems unambiguous and clear-cut. We have identified two stabilising feedbacks, both purely oceanic (#0, mean flow feedback; #1, ocean heat transport feedback), and three destabilising ones (#2, salinity transport feedback; #3, atmospheric heat transport feedback; #4, atmospheric moisture transport feedback). But we have not yet completely related the strengths of the various feedbacks to the choice of model parameters. Consider the last three rows of Table 3. For Case 1, model #2 (mixed boundary conditions) is the most unstable, followed by model #7 (linear heat transport, cubic moisture transport) and by model #8 (cubic heat and moisture transports). Case 2, however, with a slightly larger moisture transport per unit temperature gradient, shows model #2 more stable than the other two. Going from mixed boundary conditions to the 'fully coupled' model thus has effects on the stability that sensitively depend on model parameters, since one negative feedback (#1) and two positive feedbacks (#3 and #4) are added. Saravanan and McWilliams's

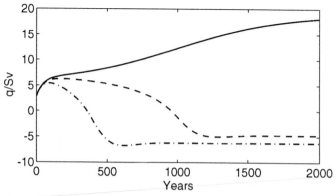

Figure 4: *Time series of overturning strength after perturbing the high-latitude sinking equilibrium by -1.6 psu in the polar box and +1.6 psu in the equatorial box, for 3 models with linear atmospheric heat transport. Solid: model #4, fixed moisture transport; dashed: #5, linear moisture transport; dash-dotted: #7, cubic moisture transport.*

(1995) model acts like Case 1 in this respect, while the NSM model acts like Case 2.

The most surprising result from Table 3 is probably that model #8 is more stable than model #7, that is, the model with *stronger* Newtonian damping of the temperature gradient is *more* stable. Comparing models #5 and #6 on one hand, and models #7 and #8 on the other, shows that the effect on the stability of going from a linear to a cubic heat transport law depends on the moisture transport law: If the latter is linear (or constant, models #1-#4, Fig. 3), a stronger Newtonian damping *destabilises*; if it is cubic, stronger Newtonian damping *stabilises*. The effect is very small but noticeable for Case 1 and clearer for Case 2.

Fig. 5 illustrates this point. It shows the time histories of flow rates q for models #2, #7, and #8, after a salinity anomaly of -1.149 psu has been added to the polar box of the Case 1 equilibrium and an anomaly of +1.149 to the equatorial box. Model #2 makes a transition within 1500 years, model #7 hovers near 10 Sv (the unstable equilibrium) for over 3000 years before collapsing, while model #8 returns.

What is going on? The clearest answer comes from the linearised perturbation analysis. The combination of the linearised tendency equation for salinity gradient, (41), perturbation flow law, (42), anomalous moisture transport, (44), and the abbreviation (32) for S_{crit} gives

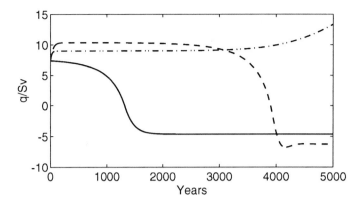

Figure 5: *Time series of overturning strength after perturbing the high-latitude sinking equilibrium by -1.149 psu in the polar box and +1.149 psu in the equatorial box, for the 3 most unstable models. Solid: model #2, fixed temperatures; dashed: #7, linear atmospheric heat and cubic moisture transports; dash-dotted: #8, cubic atmospheric heat and moisture transports.*

$$\dot{S}' = 2k\alpha(mS_{crit} - \bar{S})T' + 2(k\beta\bar{S} - \bar{q})S', \qquad (47)$$

where in the diagnostic limit T' is given by eq. (45) and approximately inversely proportional to the strength of the Newtonian damping. Analogous to eq. (31) for the total salinity gradient in model #5 (linear atmospheric transports), the terms in the bracket of eq. (47) describe the two competing effects that a temperature gradient perturbation has on the salinity gradient: It induces oceanic perturbation flow, which advects mean salinity gradient and reduces the perturbation salinity gradient. But the anomaly in temperature gradient also creates anomalous atmospheric moisture transport, which increases the salinity gradient perturbation. In steady state and for $m = 1, S_{crit} < \bar{S} < 2S_{crit}$ (see section 2.4 and Fig. 2), that is, temperature-driven salinity advection outweighs the surface moisture input. This relationship, however, need not be true for perturbations. In fact, for $m \geq 2, mS_{crit} > \bar{S}$, so the coefficient of T' in eq. (47) is positive. With the help of the diagnostic temperature perturbation equation (45), eq. (47) can be rewritten

$$\dot{S}' = \frac{4k^2\alpha\bar{T}\beta(mS_{crit} - \bar{S})}{\lambda + 2k\alpha\bar{T} + 2\bar{q}}S' + 2(k\beta\bar{S} - \bar{q})S'. \qquad (48)$$

The coefficient of the first term on the right-hand side is negative and hence stabilising for $m \leq 1$, and positive (destabilising) for $m \geq 2$. Either way, the effect is largest for small λ; weak restoring is stabilising if $m \leq 1$ and destabilising if $m \geq 2$. As a consequence, models with either fixed surface freshwater flux (Zhang *et al.*, 1993; Power and Kleeman, 1994; Mikolajewicz and Maier-Reimer, 1994; Rahmstorf and Willebrand, 1995; Pierce *et al.*, 1996) or linear dependence of atmospheric moisture transport on temperature gradient (MS, Lohmann *et al.*, 1996b) are in danger of falsely concluding that a weaker control of SST by the atmosphere means a stabilization of the thermohaline circulation.

The two competing effects of temperature on salinity have been noted previously by NSM and MS, without however giving a satisfactory explanation of when either effect dominates. Krasovskij and Stone (1995, pers. comm.) have recently found a physically equivalent effect of nonlinearity in the atmospheric heat *and* moisture transports, on whether a larger atmospheric equilibrium temperature gradient stabilises or destabilises the high-latitude sinking.

Notice that all stabilising feedbacks turn destabilising and vice versa if the low-latitude sinking equilibrium is considered (apart from the mean flow feedback, #0, which always stabilises). Qualitatively, this can be understood as follows. The high-latitude sinking state is dominated by temperature effects, so ocean heat transport must be stabilising (#1) if the steady state be stable. Temperature and salinity are antagonists, so the effect of ocean salinity transport is destabilising (#2). Atmospheric heat transport changes in response to a temperature change and must counteract its cause, reducing the negative ocean heat transport feedback (#3). Atmospheric moisture transport, finally, transports buoyancy in the same direction as atmospheric heat transport, influences ocean density in the same sense, and hence destabilises (#4). The low-latitude sinking state, in contrast, is dominated by salinity, so the ocean salinity transport must now be *stabilising*. Ocean heat transport opposes and *destabilises*, and is in turn counteracted by both atmospheric transports, which *stabilise*.

4 Land effects

4.1 Basic equations

So far, it has been assumed that the atmosphere completely homogenises temperature in the zonal direction, so that temperatures over land are identical to the oceanic temperatures at the same latitude. This assumption is questionable even in the case of only one ocean basin, and this section analyzes the consequences of admitting zonal temperature contrasts. The ratio of ocean area to total area, ε, figures most prominently in the strength of the Newtonian cooling coefficient, λ [eq. (20)], and we expect modifications in this dependence when the zonal transport efficiency is finite.

Consistent with the spirit of this paper, we formulate the simplest possible model with separate atmospheric temperatures over land, $T_{L,1}$ and $T_{L,2}$ (Fig. 6). Atmospheric transports depend linearly on temperatures. There are heat transports across the ocean-land boundaries, $H_{R,i}$ and $H_{L,i}$, $i = 1, 2$. Only their zonal differences are relevant and assumed to be

$$H_{R,i} - H_{L,i} = \tilde{\mu}(T_i - T_{L,i}), \qquad i = 1, 2 \tag{49}$$

The zonal mean temperature $[T_i]$ is

$$[T_i] \equiv \varepsilon T_i + (1 - \varepsilon)T_{L,i}, \qquad i = 1, 2. \tag{50}$$

The meridional heat transport is now given by

$$H_d = \tilde{\chi}\left([T_2] - [T_1]\right) = \tilde{\chi}\varepsilon(T_2 - T_1) + \tilde{\chi}(1 - \varepsilon)(T_{L,2} - T_{L,1}) \tag{51}$$

and is assumed evenly distributed over a latitude circle, so that the fraction ε occurs over the ocean and the fraction $1 - \varepsilon$ over land. The atmospheric heat budgets over land then read

$$A_1 - BT_{L,1} + \chi\varepsilon(T_2 - T_1) + \chi(1 - \varepsilon)(T_{L,2} - T_{L,1}) + \frac{\mu}{1 - \varepsilon}(T_1 - T_{L,1}) = 0, \tag{52}$$

$$A_2 - BT_{L,2} - \chi\varepsilon(T_2 - T_1) - \chi(1 - \varepsilon)(T_{L,2} - T_{L,1}) + \frac{\mu}{1 - \varepsilon}(T_2 - T_{L,2}) = 0. \tag{53}$$

From the atmospheric heat budgets over the ocean, the surface heat fluxes are

$$\tilde{H}_1 = A_1 - BT_1 + \chi\varepsilon(T_2 - T_1) + \chi(1 - \varepsilon)(T_{L,2} - T_{L,1}) - \frac{\mu}{\varepsilon}(T_1 - T_{L,1}), \tag{54a}$$

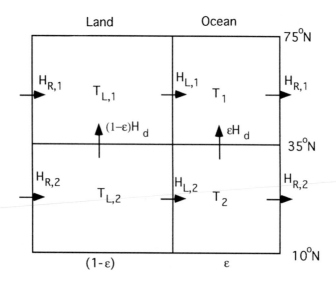

Figure 6: *Plan view of the land-ocean model of section 4. Displayed are only temperatures and heat transports. Notice the cyclic continuity.*

$$\tilde{H}_2 = A_2 - BT_2 - \chi\varepsilon(T_2 - T_1) - \chi(1-\varepsilon)(T_{L,2} - T_{L,1}) - \frac{\mu}{\varepsilon}(T_2 - T_{L,2}). \tag{54b}$$

4.2 Spatially averaged quantities

Solving for land temperatures is straightforward but somewhat tedious. Again, it is most instructive to consider spatial means and meridional differences separately. The sum of (52) and (53) gives the (meridional and hence global) mean land temperature, $T_L^* \equiv (T_{L,1} + T_{L,2})/2$, as a weighted average of global mean ocean temperature, $T^* \equiv (T_1 + T_2)/2$, and global mean atmospheric equilibrium temperature, $T_E^* \equiv (A_1 + A_2)/2B$,

$$T_L^* = \frac{BT_E^* + \{\mu/(1-\varepsilon)\}T^*}{B + \mu/(1-\varepsilon)}. \tag{55}$$

The global mean land-ocean temperature difference is

$$T_L^* - T^* = \frac{B}{B + \mu/(1-\varepsilon)}(T_E^* - T^*), \tag{56}$$

so the global mean ocean heat uptake,

$$\tilde{H}^* \equiv \frac{1}{2}(\tilde{H}_1 + \tilde{H}_2) = B(T_E^* - T^*) + \frac{\mu}{\varepsilon}(T_L^* - T^*), \qquad (57)$$

can be written as

$$\tilde{H}^* = B\left\{1 + \frac{\mu/\varepsilon}{B + \mu/(1-\varepsilon)}\right\}(T_E^* - T^*) \equiv \lambda^*(T_E^* - T^*), \qquad (58)$$

which is a Newtonian cooling law with a rather complex coefficient, λ^*. The following limits can now be considered:

1. **Very Efficient Zonal Mixing, $\mu \gg B$:**

   One obtains $T_L^* \cong T^*, \lambda^* \cong B/\varepsilon$. This is the case analyzed by MS and in sections 2 and 3. It says that 'global' scale SST anomalies are damped faster than purely on the radiative timescale, a point perhaps not appreciated before in the discussions of scale-dependent SST damping (Bretherton, 1982; Willebrand, 1993; Marotzke, 1994; Rahmstorf and Willebrand, 1995; MS).

2. **No Zonal Mixing, $\mu = 0$:**

   It follows that $T_L^* = T_E^*, \lambda^* = B$; land and ocean are uncoupled

3. **Ocean-Covered Planet, $\varepsilon \to 1$:**

   Now, $T_L^* \to T^*, \lambda^* \to B$, and land plays no role.

4. **Very Small Ocean, $\varepsilon \ll 1$:**

   This yields $T^* \to T_E^*, T_L^* \to T_E^*, \lambda^* \gg B, \mu$. The entire heat budget is controlled by radiation, and SST anomalies are restored very fast.

4.3 Meridional gradients

The meridional temperature gradient over land, denoted T_L, is obtained by taking the difference between the heat budgets over land, eqs. (53) and (52), which gives a weighted mean between the oceanic temperature gradient T and the atmospheric equilibrium temperature gradient T_E [defined in eq. (18)],

$$T_L \equiv T_{L,2} - T_{L,1} = \frac{T_E(2\chi + B) + T\{\mu/(1-\varepsilon) - 2\varepsilon\chi\}}{2\chi + B + \mu/(1-\varepsilon) - 2\varepsilon\chi}. \qquad (59)$$

From eq. (59), one obtains

$$T_L - T = \frac{(2\chi + B)}{B + 2\chi(1 - \varepsilon) + \mu/(1 - \varepsilon)}(T_E - T). \tag{60}$$

From the diagnosed surface heat fluxes, eqs. (54a) and (54b), an equation for differential surface heat flux is obtained,

$$\tilde{H}_T \equiv \tilde{H}_2 - \tilde{H}_2 = (2\chi + B)(T_E - T) + \{\mu/\varepsilon - 2\chi(1 - \varepsilon)\}(T_L - T). \tag{61}$$

which, using eq. (60), can be rewritten as a Newtonian cooling law,

$$\tilde{H}_T = (2\chi + B)\frac{B + \mu/\{\varepsilon(1 - \varepsilon)\}}{B + 2\chi(1 - \varepsilon) + \mu/(1 - \varepsilon)}(T_E - T). \tag{62}$$

Comparison with eq. (20) for the case of perfect zonal mixing suggests the introduction of an effective ocean area ratio, ε_L, defined as

$$\varepsilon_L = \frac{B + 2\chi(1 - \varepsilon) + \mu/(1 - \varepsilon)}{B + \mu/\{\varepsilon(1 - \varepsilon)\}}, \tag{63}$$

so the restoring law is now

$$\tilde{H}_T = \frac{2\chi + B}{\varepsilon_L}(T_E - T). \tag{64}$$

An illustration of eq. (60) is given in Fig. 7a, showing the difference between land and ocean temperature gradients, for T=25°C, as a function of ε and the logarithm of the ratio between μ and χ. Fig. 7b gives the resulting zonal mean temperature gradient and Fig. 7c the effective ocean area ratio, eq. (63), both in the same representation as Fig. 7a. As before, we now discuss some limiting cases, using Fig. 7 for illustration.

1. **Very Efficient Zonal Mixing, $\mu \gg \chi$:**

   One readily obtains $T_L \cong T, \varepsilon_L \cong \varepsilon$ the results from section 2. Fig. 7a shows that if zonal mixing is at least an order of magnitude stronger than meridional mixing, land and ocean temperature gradients are less than 1°C apart. The zonal mean temperature gradient is then equal to the ocean temperature gradient (Fig. 7b), and the effective ocean area ratio is equal to the actual ratio (Fig. 7c).

2. **No Zonal Mixing, $\mu = 0$:**

   The difference between land and ocean temperature gradients is

$$T_L - T = \frac{(2\chi + B)}{B + 2\chi(1 - \varepsilon)}(T_E - T) \geq (T_E - T). \tag{65}$$

The temperature gradient over land is now greater than T_E, an unexpected result, which is understood as follows. With no ocean at all, the zonal mean temperature gradient must be T_E so that locally the meridional atmospheric transport matches the radiative budget. Assume now that T_L *were* equal to T_E, despite an ocean with temperature gradient $T < T_E$. The zonal mean temperature gradient would then be less than T_E, and hence meridional heat transport over land too low to achieve energy balance. Consequently, the land temperature gradient must be larger than T_E, without, however, raising the zonal mean temperature gradient to T_E (Fig. 7b). Fig. 7a illustrates that with no zonal transport, greater ε leads to more extreme gradients over land. For finite μ, this effect is counteracted by the moderating influence the ocean has on land temperatures, which is stronger for greater ε.

Equation (63) for the effective area ratio gives in this limit

$$\varepsilon_L = \{B + 2\chi(1 - \varepsilon)\}/B, \tag{66}$$

reflecting that land temperatures and hence the meridional heat transport respond only weakly to a change in the temperature gradient in a small ocean basin. Anomalies are damped away only slightly more rapidly than on the radiative timescale.

3. **Ocean-Covered Planet, $\varepsilon \to 1$:**

Now we have $T_L \to T, \varepsilon_L \to \varepsilon$ and land plays no role. Fig. 7a indicates that if μ goes to zero also, some care must be taken. Since the physical picture is clear, we will ignore the question of properly defining the simultaneous limit.

4. **Very Small Ocean, $\varepsilon \ll 1$:**

We obtain for the effective ocean area ratio from eq. (63)

$$\varepsilon_L \cong \frac{2\chi + B + \mu}{B + \mu/\varepsilon}. \tag{67}$$

The value of ε_L is now very sensitive to μ (see Fig. 7c); if the zonal heat transport efficiency is roughly equal to or greater than the meridional and radiative ones, the second term in the denominator of eq. (67) dominates, and ε_L is proportional to ε and very small as well. Consequently, the Newtonian damping is very strong. If, however, μ/ε is much smaller than B, ε_L is equal to $(2\chi + B)/B$, and the Newtonian damping coefficient in eq. (62) is just B, which means very weak damping. Physically, these two limiting cases can be understood as follows. If zonal mixing is very efficient, even a small ocean basin dominates the zonal mean temperatures, and a change in ocean temperature gradient leads to a proportional change in meridional heat transport and longwave radiation. These will not balance over land, and the residual must go into or come out of the ocean. But land has a much greater area, and so the change in surface heat flux is amplified by ε. If, on the other hand, zonal mixing is extremely weak and the ocean basin small, a change in ocean temperature has no effect on either meridional heat transport or land temperature, so the only response is changed local longwave radiation.

The ocean area ratio causes two competing effects on the efficiency of the Newtonian damping. A smaller ocean experiences stronger surface heat flux variations due to imbalances in the heat budget over land than a larger ocean, tantamount to stronger Newtonian damping. Simultaneously, a smaller ocean has less influence on the zonal mean temperature gradient and hence on the atmospheric heat budget, meaning weaker Newtonian damping. Which effect dominates depends on the efficiency of zonal heat transport; Fig. 7c shows that for roughly isotropic atmospheric transport efficiency, the effective area ratio is quite insensitive to the actual area ratio over a wide parameter range.

The effect of finite zonal homogenization on the moisture transport and the associated feedback #4 is more difficult to answer with this simple model because the ocean-to-catchment area ratio ε_W enters as another free parameter (see Section 2.3). A small ocean basin influences the zonal mean temperature gradient and hence the meridional moisture transport only weakly. If the ocean receives no river runoff ($\varepsilon_W = 1$), this means that feedback #4 is weak as well. If, however, river runoff is substantial ($\varepsilon_W = \varepsilon$, in the extreme case), the same intensification of surface freshwater flux occurs as for surface heat flux, and even a small change in atmospheric moisture transport might cause a significant variation in ocean freshwater forcing. Clearly, the simple model has reached the limit of its usefulness here, in particular since we consider a single basin only and do not include water exchange between the catchment areas of different oceans. The need for a careful examination with idealised 3-dimensional oceanic *and* atmospheric models is evident.

5 Flux adjustments and climate drifts

NSM and MS have demonstrated that erroneous transports in either subsystem lead to incorrect climate sensitivity and stability in a coupled model as used here, even when flux adjustments are applied to simulate the current equilibrium climate. Complementary discussions of the dynamical consequences of flux adjustments in coupled models have been given in Neelin and Dijkstra (1995) and Schneider (1996). Here, the focus is on the relationship between flux adjustments and - applied flux adjustments notwithstanding - drifts of the simulated 'current' climate, which have occurred in the models of the United Kingdom Meteorological Office (UKMO, Murphy, 1995) and the Max Planck Institute for Meteorology (MPI) in Hamburg (Santer *et al.*, 1994).

In both the UKMO and the MPI models the ocean part was spun up separately before coupling. The UKMO ocean model was run for 150 years using restoring boundary conditions on both T and S, with a relaxation timescale of 15 days. The MPI ocean model was run for 5000 years using restoring boundary conditions on both T and S; then the implied surface freshwater fluxes were diagnosed and used as boundary conditions for another 2000 years (Cubasch *et al.*, 1992). 75 years after coupling, the UKMO model showed pronounced warming in the upper 500 m of the Antarctic Circumpolar Current (ACC) region, of typically 1° to 2°C at the sea surface; below 500 m cooling occurred. Simultaneously, salinity increased in the upper 500 m and decreases below, with a resulting increase in convection and meridional overturning (from 5 Sv to 15 Sv). The MPI model SST showed, 100 years after coupling, cooling in the Arctic of 5°C and warming in the Ross Sea of 6°C. The drift had added to it an oscillatory component in the Northern Hemisphere (Santer *et al.*, 1994).

Only speculations concerning the origin of the drift were given in the cited papers. However, Santer *et al.* (1994) note that the drift was largest where the flux adjustment fields were largest as well. Murphy (1995) con-

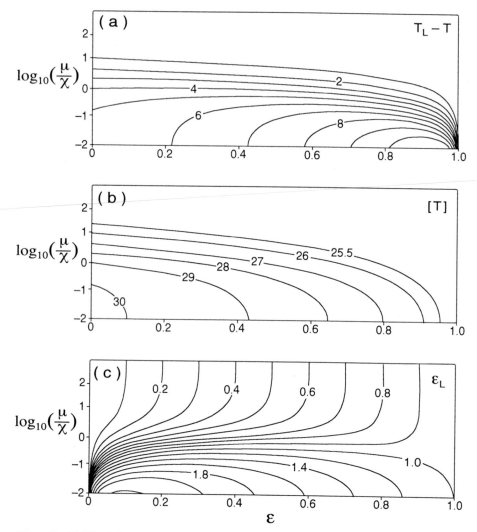

Figure 7: (a) The difference between land and ocean temperature gradients, for $T = 25°C$, as a function of ocean area ratio ε and the logarithm of the ratio between the zonal and meridional heat transport efficiencies, $log_{10}(\mu/\chi)$. (b) Zonal mean temperature gradient, for $T = 25°C$, against ε and $log_{10}(\mu/\chi)$. (c) Effective ocean area ratio ε_L against ε and $log_{10}(\mu/\chi)$.

sidered it unlikely that the relatively short spin-up duration was responsible for the drift, citing as evidence a later coupled run, which showed very similar drifts although it followed a 475-year spinup. He used as corroboration that the drifts during the coupled stage were of opposite sign to the drifts before the coupling.

Rahmstorf (1995a) has analyzed the drifts that occur when a global ocean GCM, spun up under mixed boundary conditions, is coupled to a diffusive energy balance model, the parameters of which are chosen such that the resulting surface heat fluxes match those of the spinup exactly. In the classification of the models of section 3, this corresponds to the shift from model #2 to model #4, that is, to a different atmospheric transport model. Normally, the different atmospheric model would lead to a different model climate; insisting, as we do in section 3 and as Rahmstorf (1995a) does, that the same ocean climate including surface fluxes be obtained, is conceptually equivalent to flux adjustment (this will be made explicit below). Rahmstorf (1995a) then shows that the drifts in his model occur because of a shift in convection patterns. Implied is an unconditional instability of the convection patterns of the spinup; if the spinup is in mathematically exact steady-state, a change in the formulation of the boundary conditions leaves the model in equilibrium, provided the new conditions are exactly consistent. But even roundoff error can destroy this exact match.

This section investigates an alternative hypothesis for why coupled GCMs might drift despite flux adjustments. The coupled box model is used to show that a drift is likely when the switch from one atmospheric model to another is made before the spinup is fully equilibrated. Strong motivation comes from the paper that presented the first systematic introduction and analysis of flux adjustments or 'flux correction' (Sausen et al., 1988), which had an apparent predecessor in form of an MPI internal report (Sausen *et al.*, 1987). In this, two different spinup runs of an ocean GCM were described, before the GCM was coupled to a simple diagnostic atmospheric model (but more complex than Rahmstorf's). If the spinup was run for 3000 years only, even the flux-adjusted coupled model showed drifts in SST of 1°C and more, over widespread areas, within 100 years. If, however, the spinup was run for 20,000 years and a different convection scheme used that reduced intermittence in convection, the residual drifts were very small and, in their overall magnitude, independent of whether the model was coupled, driven by Newtonian damping, or driven by fixed surface heat flux (corresponding to models #4, #2, and #1, respectively).

In the remainder of this section, the connection between flux adjustments and the choice of atmospheric model (in the sense of section 3) will be made explicit. Assume an incorrect atmospheric energy balance in the coupled model #5 (MS; atmospheric transports proportional to the meridional temperature gradient), specifically that χ is incorrect. Equations (18)-(20) then state that both λ and T_E are false (denoted by a subscript F), but that λT_E is correct. Then introduce flux adjustment such that at the steady state, $\bar{T}$, the ocean model still receives the right amount of heat. Define the adjustment by

$$H_T^{adj}(\bar{T}) = \lambda_F(T_{E,F} - \bar{T}) + \Delta H_T \equiv \lambda(T_E - \bar{T}), \tag{68}$$

so

$$\Delta H_T \equiv (\lambda_F - \lambda)\bar{T} \tag{69}$$

and

$$H_T^{adj}(\bar{T}) = \lambda_F(T_{E,F} - \bar{T}) + (\lambda_F - \lambda)\bar{T}. \tag{70}$$

For arbitrary T, it follows that

$$H_T^{adj}(T) = \lambda(T_E - \bar{T}) + \lambda_F(\bar{T} - T) = \lambda_F\left\{\left[\bar{T} + \frac{\lambda}{\lambda_F}(T_E - \bar{T})\right] - T\right\}, \tag{71}$$

or

$$H_T^{adj}(T) = \lambda_F(T_E^{adj} - T) \tag{72}$$

with

$$T_E^{adj} \equiv \bar{T} + \frac{\lambda}{\lambda_F}(T_E - \bar{T}). \tag{73}$$

A comparison between eqs. (72) for the surface flux anomalies and (37) for linearised atmospheric heat transport anomalies shows that flux adjustment is equivalent to choosing a different power of the heat transport law while ensuring that the steady state is unchanged. Equation (73) states that for $\lambda_F < \lambda$, the $\dot{T} = 0$ curve is steeper than is correct, while for $\lambda_F > \lambda$, it is flatter (compare Fig. 2). For restoring weaker than correct, a salinity gradient anomaly causes too large temperature gradient changes; the converse is true for too strong restoring. This behavior is illustrated

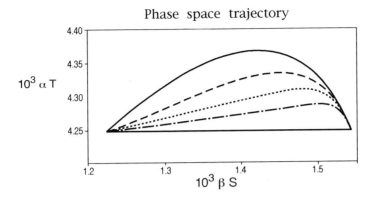

Figure 8: *Phase space trajectories of 5 models of section 3, after perturbing the high-latitude sinking equilibrium by -0.2 psu in the polar box and +0.2 psu in the equatorial box. The propagation on the trajectories is from larger to smaller salinity gradients (right to left). Heavy solid: model #1, fixed heat flux; thin solid: #2, fixed temperatures; dashed: #3, radiative restoring; dotted: #5, linear atmospheric transports; dash-dotted: #8, cubic atmospheric transports. The trajectories of the models not shown are nearly identical to the ones of the models with the same atmospheric heat transport laws.*

in Fig. 8, showing the phase space trajectories of 5 models of section 3, in response to an initial perturbation in salinity gradient of 0.4. The crucial parameter is the exponent of the heat transport law; weaker responses in atmospheric heat transport lead to larger excursions in temperature gradient. The anomalous temperature gradients increase from their starting value of zero and approach, on the respective Newtonian damping timescales, the respective $\dot{T} = 0$ curves, which in their final approach to equilibrium are well approximated by straight lines (with the exception of model #1, fixed fluxes, which is discussed in detail in section 6).

What happens if the spinup is too short? The tuning then occurs away from equilibrium, and each model continues its evolution on its own, separate $\dot{T} = 0$ curve - the curves intersect at the point of tuning only, as indicated by eqs. (72) and (73), with $\bar{T}$ replaced by the temperature gradient at the end of the spinup. Fig. 8 suggests the following interpretation of a typical coupled model experiment preceded by an uncoupled ocean spinup; confirmation through numerical experiment is readily obtained. The spinup has very tightly controlled surface temperatures, which corresponds to slowly moving to the left, towards the equilibrium, on the horizontal line marking model #2 (fixed temperatures). Assume now that, in contrast to the case depicted in Fig. 8, the spinup is stopped at a significant distance from the equilibrium point (measured according to the salinity gradient - the temperature gradient has already equilibrated), and a switch to a different atmospheric transport model occurs. All models then further decrease their salinity gradients while moving on their separate $\dot{T} = 0$ curves, which continue (with the slope indicated in Fig. 8) to the left of their focal point, so model temperature gradients diverge from each other. All steady states are different, and the steady state SST of a coupled GCM is likely to be further away from the observations than the SST of the spinup, which is often forced towards climatology.

A spinup may have to be considerably longer than often practiced, if it is used for flux adjustment and subsequent coupled runs. If the experience of Sausen *et al.* (1987) can be generalised, even 3000 years are too short, which is likely to have affected both the 150 year and the 475 year spinup runs discussed by Murphy (1995), and possibly even the experiments of Rahmstorf (1995a). The 30,000 year spinup of Sausen *et al.* (1988) virtually eliminated the drift, but this could have been aided by restoring salinity boundary conditions, as assumed by Rahmstorf (1995a). In a GCM, a shift in convection patterns might be able to amplify drifts that would be very small in a purely advective model as the one considered here. It is not possible to point at the single most important cause of climate drift after flux adjustment, but a very well equilibrated spinup is a necessary condition for its avoidance.

6   The 'neutrally buoyant mode'

Saravanan and McWilliams (1995) see a mode of variability in T and S in an idealised coupled model that is nearly compensated in density and hence not felt by the flow field; they call it the *neutrally buoyant mode*. It is associated with the portion of the surface fluxes that is unaffected by feedbacks; if atmospheric transports change due to the temperature variability of the neutrally buoyant mode, they are likely to modify T and S such that density is no longer compensated.

The neutrally buoyant (or neutral) mode manifests itself in the model used here, and is another contender for causing drifts in a coupled model that are difficult to control. A *low-latitude* sinking equilibrium is obtained with model #5 under the standard parameter set of Table 1. Figure 9 shows the response of this steady state to a salinity perturbation of 0.2 psu to the high-latitude box and -0.2 psu to the low-latitude box. Shown are the first 1000 years of 4 models (fixed fluxes, #1; fixed T, #2; radiative restoring, #3, and NSM, #8), represented by the time series of meridional temperature gradient (Fig. 9a), salinity gradient (Fig. 9b), the flow strength (and hence density gradient, Fig. 9c), and, in Fig. 9d, the combination of T and S orthogonal to density, denoted ν here and called spiciness by Munk (1981, p. 282; Olbers *et al.*, 1985, used the term veronicity). The models with strong Newtonian relaxation (#2, #8) return to the equilibrium monotonically. Model #3 shows slight overshoot in all quantities, while model #1 amplifies the 0.2 psu initial anomaly to excursions of 3.5°C and 0.7 psu. Density gradient and flow rate return to equilibrium monotonically, but spiciness changes sign and reaches quadrupled magnitude of the original perturbation.

That the flow anomaly in model #1 decays more slowly than in any other model is plausible from the feedback analysis of section 3: All feedbacks change sign when the low-latitude sinking equilibrium is considered, so model #1 lacks both *negative* feedbacks associated with changes in air-sea fluxes. The *growth* in spiciness gradient, however, which by far outweighs the decay in density gradient, is less expected since model #1 has only this one steady state, with the given set of surface fluxes, and all eigenvalues are negative (see below). The temporary growth in anomalies is caused by the interference between the eigensolutions of the non-normal operator (only normal matrices, *i.e.*, for which $AA^T = A^T A$, have orthogonal eigenvectors; see Trefethen *et al.*, 1993, for a general discussion of non-normality in hydrodynamic stability problems); it is readily shown that the original perturbation is small enough to ensure linear behavior so nonlinear effects play no role.

The perturbation solution of model #1 is now given explicitly. Define density and spiciness gradients and their forcing as

$$\rho = \alpha T - \beta S \tag{74}$$
$$\nu = \alpha T + \beta S \tag{75}$$

$$H_\rho = \alpha H_T - \beta H_S \tag{76}$$
$$H_\nu = \alpha H_T + \beta H_S \tag{77}$$

Notice that the density difference is defined as high-latitude minus low-latitude, in contrast to all other quantities. Hence,

$$q = k\rho. \tag{78}$$

With fixed differential surface fluxes, H_T and H_S, we derive from the equations for the temperature and salinity gradients, eqs. (29) and (30),

$$\dot{\rho} = 2H_\rho - 2k|\rho|\rho \tag{79}$$
$$\dot{\nu} = 2H_\nu - 2k|\rho|\nu, \tag{80}$$

with fixed forcing terms. If thermal forcing is stronger than haline forcing, $H_\rho > 0$, and one obtains high-latitude sinking, $\bar{q} > 0$. Linearization about the equilibrium leads to

$$\begin{pmatrix} \dot{\rho}' \\ \dot{\nu}' \end{pmatrix} = \mathbf{A} \begin{pmatrix} \rho' \\ \nu' \end{pmatrix} \equiv \mp \begin{pmatrix} 4k\bar{\rho} & 0 \\ 2k\bar{\nu} & 2k\bar{\rho} \end{pmatrix} \begin{pmatrix} \rho' \\ \nu' \end{pmatrix}, \tag{81}$$

where the minus sign holds if $\bar{q} > 0$, and the plus sign if $\bar{q} < 0$. The matrix $\mathbf{A}$ does not commute with its transpose. Its eigenvalues and associated eigenvectors are

$$\lambda_1 = -4k|\bar{\rho}|, \quad \begin{pmatrix} \rho' \\ \nu' \end{pmatrix}^{(1)} = (1 + \bar{\nu}^2/\bar{\rho}^2)^{-1} \begin{pmatrix} 1 \\ \bar{\nu}/\bar{\rho} \end{pmatrix}, \tag{82}$$

$$\lambda_2 = -2k|\bar{\rho}|, \quad \begin{pmatrix} \rho' \\ \nu' \end{pmatrix}^{(2)} = \begin{pmatrix} 0 \\ 1 \end{pmatrix} \tag{83}$$

Density anomalies evolve according to the first eigenfunction only; in contrast, spiciness anomalies evolve according to the superposition of two exponentials. With zero initial temperature anomalies,

$$\nu'(0) = -\rho'(0) = \beta S'(0), \tag{84}$$

the solution of the problem linearised about the low-latitude sinking state is readily verified as

$$\rho'(t) = -\beta S'(0) exp(-4k|\bar{\rho}|t), \tag{85}$$
$$\nu'(t) = \beta S'(0) exp(-2k|\bar{\rho}|t) \left\{ 1 + \frac{\bar{\nu}}{\bar{\rho}}[1 - exp(-2k|\bar{\rho}|t)] \right\}. \tag{86}$$

Density and hence circulation anomalies decay exponentially from their initial values, governed by the higher decay rate, λ_1 (Fig. 9c). The spiciness

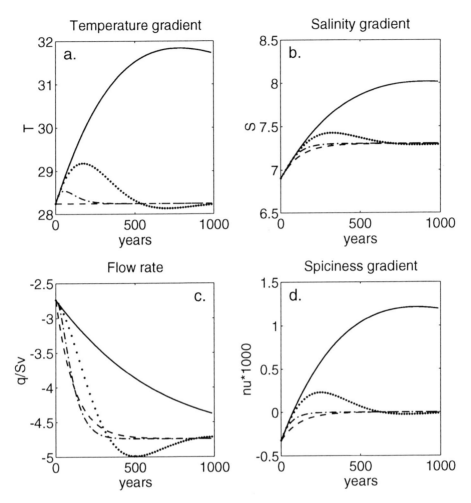

Figure 9: *The first 1000 years of the time histories of 4 models after the low-latitude sinking equilibrium is perturbed by 0.2 psu in the high-latitude box and -0.2 psu in the low-latitude box. Solid: #1, fixed fluxes; dashed: #2, fixed temperatures; dotted: #3, radiative restoring; dash-dotted: #8, cubic atmospheric transports. (a) Temperature gradient, (b) salinity gradient, (c) flow strength (and hence density gradient), (d) spiciness gradient anomaly.*

gradient anomaly shows a more interesting behavior (Fig. 9d), which can be understood as follows. Since T and S are both positive for all cases considered here, $|\bar{\nu}| > |\bar{\rho}|$ always and hence $\bar{\nu}/\bar{\rho} < -1$ in eq. (86). At $t = 0$, the square bracket is zero also, and the initial spiciness anomaly has the same sign as the initial salinity anomaly. As the exponential in the square bracket decays, however, $\bar{\nu}/\bar{\rho}$ is multiplied by a number approaching unity, and the term in curly brackets has a zero crossing, which is readily shown to occur at $t_{\nu,1}$, given by

$$t_{\nu,1} = \frac{1}{2|\bar{q}|} \ln \frac{\alpha \bar{T} + \beta \bar{S}}{2\alpha \bar{T}}. \tag{87}$$

There must be a maximum in spiciness deviation from equilibrium, of sign opposite to the initial anomaly, which occurs at

$$t_{\nu,2} = \frac{1}{2|\bar{q}|} \ln \frac{\alpha \bar{T} + \beta \bar{S}}{\alpha \bar{T}}. \tag{88}$$

The time for the zero crossing is positive only if $\alpha \bar{T} < \beta \bar{S}$, i.e., for low-latitude sinking: The spiciness anomaly decays nearly linearly and overshoots; the zero crossing occurs after 75 years for the parameters from Table 1, according to eq. (87) and, consistently, Fig. 9d. An extremum is reached after 800 years [again consistently from eq. (88) and Fig. 9d], and finally the spiciness anomaly decays exponentially (not shown). For high-latitude sinking, the initial perturbation grows linearly and reaches an extremum after typically 50-100 years, before it decays.

It is not clear how relevant the large excursions of model #1 about its low-latitude sinking state are. The assumption of fixed surface fluxes is very strong; in addition, Marotzke (1990) and Marotzke and Willebrand (1991) discussed how wind forcing in GCMs tends to eliminate the counterparts of low-latitude sinking states found in box models. On the other hand, Zorita and Frankignoul (1996) have recently identified modes of North Atlantic variability in a coupled GCM, on a timescale of decades, with no significant associated response of surface heat flux to SST anomalies. Model #1 might thus be applicable in some cases. Also, the Atlantic overturning circulation can be viewed as two back-to-back Stommel box models (Welander, 1986), with the Northern Hemisphere showing thermally direct circulation and the Southern Hemisphere thermally indirect circulation. One can then speculate that the neutral model contributes significantly to drift and variability in the Southern Hemisphere and hence the entire Atlantic thermohaline circulation.

7 Concluding remarks

This paper analyzes, in a systematic and within limits rigorous way, feedbacks between the thermohaline circulation and meridional transports in the atmosphere that can be represented in a coupled ocean-atmosphere box model. The approach is austere: The model can be analyzed to considerable depth, but many processes acting in more complex coupled models or in the real climate are not represented at all, or possibly distorted in their relative importance.

Chief among the processes in more realistic models of the thermohaline circulation that are missing here is perhaps the feedback between convective activity, surface heat flux, and the strength of the deep circulation. Lenderink and Haarsma (1994) first suggested a mechanism by which convection might be stopped, locally, by a local perturbation, without initially affecting the vicinity much, but eventually changing the meridional overturning. Rahmstorf (1995b) later expanded this picture, showing that significant changes in overturning strength of an idealised global model can be triggered by the cessation of convection in very few gridpoints. Rahmstorf and Willebrand (1995) discuss how the convective feedback depends on the surface heat flux parameterization, and demonstrate a stabilising effect of weaker SST restoring just like in the models with fixed E-P in section 3 here. For a detailed discussion of the convective feedback and its relationship with advective feedbacks, see the recent review by Rahmstorf et al. (1996).

What then can be learned from this simple model, with its potential to overestimate the importance of the processes it *can* represent (as, for example, emphasised by Lohmann et al., 1996a)? A prime virtue of the model is that it is fully coupled and allows a self-consistent representation of extratropical, large-scale ocean-atmosphere interactions. It makes possible the investigation of a surprisingly rich variety of effects even beyond the ones presented here (for example, one can analyze the heat transport capability of the total climate system, depending on a few fundamental parameters). The limits of applicability are as yet unclear; for example, the difference between the two statistically steady states in the coupled GCM of Manabe and Stouffer (1988) shows the important effects of differing thermohaline circulation patterns in the Atlantic, on meridional heat transport and air temperatures, which cannot be fully understood by the more local analysis of convective effects. An analysis of a coupled GCM can be aided considerably by the conceptual framework developed in this paper. The effects described here must act in the complex model as well; if they are unimportant in the GCM, this begs for an explanation of what processes do govern the GCM dynamics. We see the recurrent theme in dynamical meteorology and oceanography that simple models are needed to build up a complete understanding.

The sensitivity experiments of section 3 have made extensive use of a general strategy to analyze various feedback processes independently. By spinning up a coupled model and keeping some transports of the equilibrium state fixed, feedbacks can be turned off at will, providing the most rigorous way of assessing the importance of competing effects. The same strategy can be used in a coupled GCM; there is, however, the possibility that upon a switch in degree of coupling, convection patterns do become unconditionally unstable, as is the assumption underlying Rahmstorf's (1995a) analysis. Whether our approach is feasible for GCMs can only be determined by applying it.

The analysis of land effects suggests the need for idealised coupled GCM studies, in which the zonal transports are modelled explicitly rather than parameterized. The simple model offers guidance for how to conduct these experiments; for example, it can be analyzed how transport efficiencies depend on the size of the ocean basin with active thermohaline circulation; in particular, whether and how a relatively small ocean like the North Atlantic can influence the overall strength of the Jet Stream and the atmospheric eddies it produces through baroclinic instability. Again, the results of Manabe and Stouffer (1988) indicate that, at least on very long timescales, the North Atlantic does play a major active role.

To conclude, I firmly believe that a simple model as the one used here can provide important insights into the dynamics of the coupled climate system. It provides a starting point for further analyses or experimentation, which might then corroborate or reject the ideas formulated here.

Acknowledgements

I wish to thank David Anderson and Jürgen Willebrand for organising this exciting School, and for giving me the opportunity to present this material. A number of people read drafts of the manuscript and made useful suggestions; I am particularly indebted to Stefan Rahmstorf, Jeff

Scott, and Detlef Stammer. Sixty-five per cent ($10,000) of this work was supported jointly by the Northeast Regional Center of the National Institute for Global Environmental Change and by the Program for Computer Hardware, Applied Mathematics, and Model Physics (both with funding from the U.S. Department of Energy). Financial support was also contributed by the Tokyo Electric Power Company through the TEPCO/MIT Environmental Research Program.

Glossary of symbols

Index F	Erroneous quantity	S	Meridional ocean salinity gradient
Index i	High-latitude (i=1) or low-latitude (i=2)	S_0	Reference salinity
Overbar	Time average	S_{crit}	Critical salinity gradient ($S_{crit} \leq \bar{S}$)
Prime $'$	Perturbation from time average	S_i	Salinity
Index $*$	Meridional average	T	Meridional ocean temperature gradient
[]	Zonal average	T_E	Atmospheric equilibrium temperature
A_i	Surface radiative forcing		gradient
B	Longwave radiation coefficient	T_i	Ocean temperature
$c\rho_0$	Heat capacity per unit volume of water	T_L	Meridional land temperature gradient
D	Ocean depth	$T_{L,i}$	Land temperature
E	Evaporation minus precipitation	α	Thermal expansion coefficient
F_{0i}	Total area	β	Haline expansion coefficient
F_i	Ocean area	γ_m	Atmospheric meridional moisture
F_W	Meridional moisture transport		transport efficiency; power m
H_{0i}	Downward radiation, top of atmosphere	γ	γ_1
H_d	Meridional atmospheric heat transport	$\tilde{\gamma}$	γF_{01}
$\tilde{H}_i$	Downward surface heat flux	ε	Area ratio, ocean over total (F_i/F_{0i})
H_i	$\tilde{H}_i/(c\rho_0 D)$	ε_L	Effective area ratio (land effects)
$H_{L,i}$	Zonal heat transport, land-to-ocean	ε_W	Area ratio, ocean over catchment
$H_{R,i}$	Zonal heat transport, ocean-to-land	χ_n	Atmospheric meridional heat transport
H_S	Virtual surface salinity flux		efficiency; power n
H_T	$H_2 - H_1$	χ	χ_1
H_ν	Differential surface 'spiciness' forcing	$\tilde{\chi}$	χF_{01}
H_ρ	Differential surface density forcing	λ	Newtonian damping coefficient
k	Hydraulic constant	μ	Atmospheric zonal heat transport
m	Power, atmospheric heat transport		efficiency
n	Power, atmospheric moisture transport	ν	Spiciness gradient ($\alpha T + \beta S$)
q	Flow strength	ρ	Density gradient ($\alpha T - \beta S$)

376

References

BOYLE, E. A., 1990, Quaternary deepwater paleoceanography. *Science*, **249**, 863-870.

BAUMGARTNER, A., AND E.REICHEL, 1975, Die Weltwasserbilanz. *Oldenbourg Verlag, Müchen, 179pp.*

BRETHERTON, F.P., 1982, Ocean climate modeling. *Progr. Oceanogr.*, **11**, 93-129.

BROECKER, W.S., T-P.PENG, J.JOUZEL, AND G.RUSSELL, 1990, The magnitude of global fresh-water transports of importance to ocean circulation. *Clim. Dyn.*, **4**, 73-79.

CUBASCH, U., K.HASSELMANN, H.HÖCH, E.MAIER-REIMER, U.MIKOLAJEWICZ, B.D.SANTER, AND R.SAUSEN, 1992, Time-dependent greenhouse warming computations with a coupled ocean-atmosphere model. *Climate Dynamics*, **8**, 55-69.

GILL, A.E., 1982, Atmosphere-Ocean Dynamics. *Academic Press, London, 662 pp.*

HALL, M.M., AND H.L.BRYDEN, 1982, Direct estimates and mechanisms of ocean heat transport. *Deep-Sea Res.*, **29**, 339-359.

HANEY, R.L. 1971, Surface thermal boundary condition for ocean circulation models. *J. Phys. Oceanogr.*, **1**, 241-248.

HUANG, R.X., J.R.LUYTEN, AND H.M.STOMMEL, 1992, Multiple equilibrium states in combined thermal and saline circulation. *J. Phys. Oceanogr.*, **22**, 231-246.

HUGHES, T.C.M., AND A.J.WEAVER, 1994, Multiple equilibria of an asymmetric two-basin model. *J. Phys. Oceanogr.*, **24**, 619-637.

LENDERINK, G., AND R.J.HAARSMA, 1994, Variability and multiple equilibria of the thermohaline circulation, associated with deep water formation. *J. Phys. Oceanogr.*, **24**, 1480-1493.

LOHMANN, G., R.GERDES, AND D.CHEN, 1996A, Sensitivity of the thermohaline circulation in coupled oceanic GCM atmospheric EBM experiments. *Clim. Dyn., in press.*

LOHMANN, G., R.GERDES, AND D.CHEN, 1996B, Stability of the thermohaline circulation in an analytical investigation. *Submitted for publication.*

MANABE, S., AND R.J.STOUFFER, 1988, Two stable equilibria of a coupled ocean-atmosphere model. *J. Climate*, **1**, 841-866.

MANABE, S., AND R.J.STOUFFER, 1994, Multiple century response of a coupled ocean-atmosphere model to an increase of atmospheric carbon dioxide. *J. Climate*, **7**, 5-23.

MAROTZKE, J., 1990, Instabilities and multiple equilibria of the thermohaline circulation. *Ph.D. thesis. Ber. Inst. Meeresk. Kiel, 194, 126pp.*

MAROTZKE, J., 1994, Ocean models in climate problems. Ocean Processes in Climate Dynamics: Global and Mediterranean Examples, *P. Malanotte-Rizzoli and A.R. Robinson, eds., Kluwer, 79-109.*

MAROTZKE, J., AND P.H.STONE, 1995., Atmospheric transports, the thermohaline circulation, and flux adjustments in a simple coupled model. *J. Phys. Oceanogr.*, **25**, 1350-1364.

MAROTZKE, J., AND J.WILLEBRAND, 1991, Multiple equilibria of the global thermohaline circulation. *J. Phys. Oceanogr.*, **21**, 1372-1385.

MIKOLAJEWICZ, U., AND E.MAIER-REIMER, 1994, Mixed boundary conditions in ocean general circulation models and their influence on the stability of the model's conveyor belt. *J. Geophys. Res.*, **99**, 22,633-22,644.

MUNK, W., 1981, Internal waves. Evolution of Physical Oceanography, Scientific Surveys in Honor of Henry Stommel. *B. A. Warren and C. Wunsch (eds.), The MIT press, Cambridge, MA, 264-291.*

MURPHY, J.M., 1995, Transient response of the Hadley Centre coupled ocean-atmosphere model to increasing carbon dioxide. Part I: Control climate and flux adjustment. *J. Climate*, **8**, 36-56.

NAKAMURA, M., P.H.STONE, AND J.MAROTZKE, 1994, Destabilization of the thermohaline circulation by atmospheric eddy transports. *J. Climate*, **7**, 1870-1882.

NEELIN, J.D., AND H.A.DIJKSTRA, 1995, Ocean-atmosphere interaction and the tropical climatology. Part I: The dangers of flux correction. *J. Climate*, **8**, 1325-1342.

OLBERS, D.J., M.WENZEL, AND J.WILLEBRAND, 1985, The inference of North Atlantic circulation patterns from climatological hydrographic data. *Rev. Geophys.*, **23**, 313-356.

PIERCE, D.W., K-Y.KIM, AND T.P.BARNETT, 1996, Variability of the thermohaline circulation in an ocean general circulation model coupled to an atmospheric energy balance model. *J. Phys. Oceanogr., in press.*

POWER, S.B., AND R.KLEEMAN, 1994, Surface heat flux parameterization and the response of ocean general circulation models to high latitude freshening. *Tellus*, **46A**, 86-95.

RAHMSTORF, S., 1995A, Climate drift in an ocean model coupled to a simple, perfectly matched atmosphere. *Climate Dyn.*, **11**, 447-458.

RAHMSTORF, S., 1995B, Multiple convection patterns and thermohaline flow in an idealised OGCM. *J. Climate*, **8**, 3028-3039.

RAHMSTORF, S., AND J.WILLEBRAND, 1995, The role of temperature feedback in stabilising the thermohaline circulation. *J. Phys. Oceanogr.*, **25**, 787-805.

RAHMSTORF, S., J.MAROTZKE, AND J.WILLEBRAND, 1996, Stability of the thermohaline circulation. The Warm Water Sphere of the North Atlantic Ocean, W. Krauss, ed., *in press.*

ROEMMICH, D.H., AND C.WUNSCH, 1985, Two transatlantic sections: Meridional circulation and heat flux in the subtropical North Atlantic Ocean. *Deep-Sea Res.*, **32**, 619- 664.

SANTER, B.D., W.BRÜGGEMANN, U.CUBASCH, K.HASSELMANN, H.HÖCK, E.MAIER-REIMER, AND U.MIKOLAJEWICZ, 1994, Signal-to-noise analysis of time-dependent greenhouse warming experiments. Part I: Pattern analysis. *Climate Dyn.*, **9**, 267-285.

SARAVANAN, R. AND J.C.MCWILLIAMS, 1995, Multiple equilibria, natural variability, and climate transitions in an idealised ocean-atmosphere model. *J. Climate*, **8**, 2296-2323.

SARNTHEIN, M., K.WINN, S.J.A.JUNG, J.C.DUPLESSY, L.LABEYRIE, H.ERLENKEUSER, AND G.GANSSEN, 1994, Changes in East Atlantic deepwater circulation over the last 30,000 years: Eight time slice reconstructions. *Paleoceanography*, **9**, 209-267.

SAUSEN, R., K.BARTHEL, AND K.HASSELMANN, 1987, A flux correction method for removing the climate drift of coupled atmosphere-ocean models. Max-Planck-Institut für Meteorologie Report No. 1, *Hamburg, 39pp.*

SAUSEN, R., K.BARTHEL, AND K.HASSELMANN, 1988, Coupled ocean-atmosphere models with flux correction. *Climate Dyn.*, **2**, 145-163.

SCHNEIDER, E.K., 1996, Flux correction and the simulation of changing climate. *Submitted for publication.*

STOMMEL, H., 1961, Thermohaline convection with two stable regimes of flow. *Tellus*, **13**, 224-230.

STONE, P.H., AND D.A.MILLER, 1980, Empirical relations between seasonal changes in meridional temperature gradients and meridional fluxes of heat. *J. Atmos. Sci.*, **37**, 1708- 1721.

TREFETHEN, L.N., A.E.TREFETHEN, S.C.REDDY, AND T.A.DRISCOLL, 1993, Hydrodynamic stability without eigenvalues. *Science*, **261**, 578-584.

WALIN, G., 1985, The thermohaline circulation and the control of ice ages. *Palaeogeogr., Palaeoclimatol., Palaeoecol.*, **50**, 323-332.

WANG, W-C., AND P.H.STONE, 1980, Effect of ice-albedo feedback on global sensitivity in a one-dimensional radiative-convective climate model. *J. Atmos. Sci.*, **37**, 545-552.

WELANDER, P., 1986, Thermohaline effects in the ocean circulation and related simple models. Large-Scale Transport Processes in Oceans and Atmosphere, *J. Willebrand and D.L.T. Anderson, eds., NATO ASI series, D. Reidel, 163-200.*

WILLEBRAND, J., 1993, Forcing the ocean with heat and freshwater fluxes. Energy and Water Cycles in the Climate System, *E. Raschke, ed., Springer Verlag, 215-233.*

ZHANG, S., R.J.GREATBATCH, AND C.A.LIN, 1993, A re-examination of the polar halocline catastrophe and implications for coupled ocean-atmosphere modelling. *J. Phys. Oceanogr.*, **23**, 287-299.

ZORITA, E., AND C.FRANKIGNOUL, 1996, Modes of North Atlantic decadal variability in the ECHAM1/LSG coupled ocean-atmosphere general circulation model. *Submitted for publication.*

AN OVERVIEW OF CENTURY TIME-SCALE VARIABILITY IN THE CLIMATE SYSTEM: OBSERVATIONS AND MODELS

THOMAS F. STOCKER
Climate and Environmental Physics
Bern, Switzerland

Contents

Abstract

Estimates of the development of the Earth's climate subject to anthropogenic forcing depend critically on our knowledge of natural climate variability on time scales of decades to centuries. Time scales extracted from high-resolution proxy records and observations indicate that the spectrum of climate variability exhibits significant power in the range of decades to centuries superimposed on a red-noise continuum. The classical view of climate variability is based on the concept that observed fluctuations have their origin in periodic forcings on the same time scale, i.e. that the climate system behaves like a linear system that is externally forced. The present sensitivity of the climate system, however, would require strong positive feedback mechanisms to translate the weak forcing signals (e.g. variability of solar irradiation) into detectable fluctuations in observed and proxy variables. Instead, it is proposed that these fluctuations are linked to interactions within and between the different climate system components. An overview of recent modeling results and the discussion of mechanisms involved show that such interactions internal to the climate system cannot only exhibit the correct time scales but also easily account for the amplitudes observed.

NATO ASI Series, Vol. I 44
Decadal Climate Variability
Dynamics and Predictability
Edited by David L. T. Anderson and Jürgen Willebrand
© Springer-Verlag Berlin Heidelberg 1996

1   Introduction

Variability is a fundamental property of our climate system. Two decades ago Mitchell (1976) proposed a theoretical framework of climate variability based on a schematic spectrum of climate variations spanning time scales from one hour to some billions of years. He distinguishes two different types of processes in the climate system: (i) internal stochastic mechanisms; and (ii), external forcing mechanisms including their resonant amplification of internal modes. The spectrum in Mitchell's original figure (Fig. 1) thus consists of a background continuum on which is superimposed a series of spectral peaks. The power of the background continuum is stronger for lower frequencies and is the result of a number of red-noise spectra with increasing characteristic time scales. According to Mitchell this is the consequence of the stochastic aspect of the different climate system components for which first-order autoregressive processes are good conceptual models.

The purpose of this paper, which is an extended and updated version of Stocker (1995), is to give an overview on recent results that indicate the importance of these processes on time scales of many decades to centuries. Apart from a general interest this time scale is of particular importance, because detection of anthropogenic climate change depends on our knowledge of the time scales and patterns associated with the natural level of variability on the decadal to century time scale. A second purpose is to summarize those mechanisms of interdecadal-to-century variability that are quantitatively capable of causing detectable fluctuations. Up to now, our knowledge about these processes comes mostly from a hierarchy of dynamical models of the climate system and still only marginally from an observational or proxy network. The development and expansion of the latter is an important task of future research.

Many aspects of Mitchell's schematic picture can be found in records of climatic data both, directly observed and proxy. For instance, the red noise background of the spectrum (i.e. longer time scales exhibit stronger spectral power) can be seen in observed data from the atmosphere and ocean, and model simulations exhibit very similar spectral characteristics (Delworth *et al.* (1993), their Fig. 2, reproduced here as Fig. 5a). The spectra of longer time series (oxygen isotopes on planktonic foraminifera shells found in sea sediments) also confirm the red-noise continuum background and the presence of preferred time scales of variability (see e.g. Imbrie *et al.* (1992,

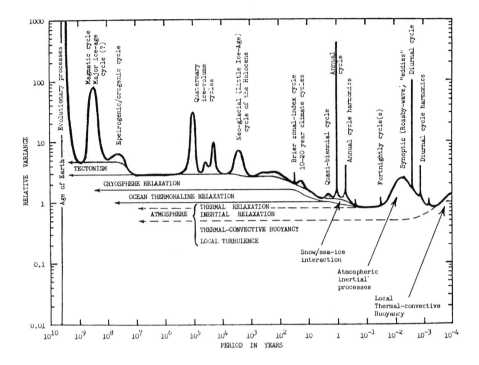

Figure 1: The original Figure of Mitchell (1976) showing a schematic spectrum of climate variability which is a composite of a red noise background due to stochastic fluctuations and distinct peaks resulting from various external forcings. Note that no peaks are in the El Niño and decadal-to-century band [From Mitchell (1976)].

Imbrie *et al.* (1993)). Of more interest due their predictive potential are the external mechanisms that produce variability on distinct time scales such as the diurnal and seasonal cycles, cyclic processes without a single distinct time scale such as ENSO, the Milankovic cycles and even slower, tectonic processes.

While this concept is a useful starting point, recent high-resolution paleo-climatic archives (e.g. sea sediment cores by Lehman and Keigwin (1992, Bond *et al.* (1993) and the two recent ice cores from Summit, Greenland) have clearly demonstrated that additional aspects of climate variability must be taken into account. Cyclic or periodic fluctuations in a linear, dissipative system are due only to external forcing. While the climate system is dissipative, it is certainly not a linear system.

Therefore, processes additional to those mentioned above can generate variability. First, a non-linear system can exhibit *self-sustained oscillations* in response to external forcing. Their characteristic time scale is determined by feedback mechanisms internal to the system. Second, external forcing can generate *abrupt reorganisation* of the system, i.e. switches from one equilibrium state to another. While these switches are caused by 'external' perturbations (external for the ocean, i.e. for example atmosphere or ice sheets), the evolution of the system occurs on a time scale determined, again, by internal mechanisms.

The four features of climate change are shown in Fig. 2:

1. EXTERNALLY FORCED CHANGES,

2. SELF-SUSTAINED OSCILLATIONS,

3. NON-DETERMINISTIC (CHAOTIC) VARIABILITY,

4. ABRUPT REORGANISATIONS.

Various proxy and direct climatic data from the ice, sea and lake sediment archive illustrate the climatic evolution during the last 40,000 years. The transition from the last glacial to the Holocene is an **externally forced change** due to changes in the distribution of solar radiation and the operation of a number of feedback mechanisms (greenhouse gases, albedo). This is manifested in the gradual increase of $\delta^{18}O$ starting around 20,000 BP and in many proxy records from various archives (Imbrie *et al.* 1992; Imbrie *et al.* 1993) and strong support for the Milankovic theory of climate change.

Self-sustained oscillations of the ocean-atmosphere-terrestrial ice sheet system are likely to be responsible for the climatic swings between milder and colder phases (Dansgaard/Oeschger cycles) during the glacial and it appears that they involve large changes of ice sheet volume which, when melting, influence significantly ocean circulation (Bond *et al.* 1992; Bond *et al.* 1993; Paillard and Labeyrie 1994).

Non-deterministic (chaotic) variability is visible indirectly in the $\delta^{18}O$ record of the last 10,000 years (Holocene) which fluctuates about a well defined mean value; variations are likely due to changes in the hydrological cycle. Classically, this dynamical behaviour is thought to be most relevant for mesoscale processes (and hence comparatively short time scales) in atmosphere and ocean but there are first indications that also on the decadal-to-century time scale the concept of non-deterministic evolution is

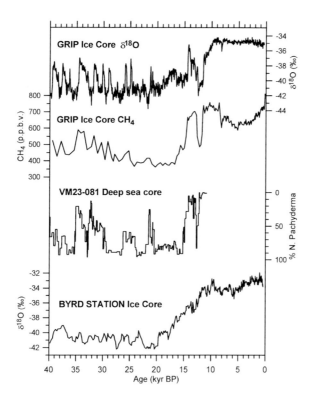

Figure 2: *Climatic change over the last 40'000 years as obtained from the measurement of $\delta^{18}O$ on water (from Johnsen et al. (1992), Dansgaard et al. (1993) and Hammer et al. (1994)) and CH_4 on air of bubbles trapped in the ice core (from Chappellaz et al. (1993) and Blunier et al. (1995). Four different features of climate variability are evident in the different time series: (i) slow, astronomically forced transition from the glacial to the interglacial ($\delta^{18}O$ records and CH_4); (ii) self-sustained oscillations during the glacial (foraminifera assemblages in a sea sediment core (from Bond et al. (1993)) and ice cores before 20 kyr BP); (iii) non-deterministic (chaotic) variability during the Holocene ($\delta^{18}O$ records); and (iv) abrupt reorganisations during the Bølling/Allerød/Younger Dryas Period (all but Byrd Station core) [Figure compiled by T. Blunier]*

required to describe the system (Saltzman 1983; James and James 1992; Mysak *et al.* 1993; Roebber 1995). Methane, on the other hand, exhibits surprising systematic changes during the Holocene (Blunier *et al.* 1995) which would rather hint at an externally forced cause (receding continental ice sheet give way to changing and evolving biosphere which influences methane production, see Blunier (1995)). This clearly indicates the need for a variety of climate proxies in order to characterize the dynamics and variability during a given period.

Three events of **abrupt reorganisations** recorded in sea sediments and ice cores, are superimposed on the longer term glacial-interglacial transition. The abrupt warming into the Bølling/Allerød, the cooling initiating the Younger Dryas and its termination all occur on time scales of a few decades to a few years (Dansgaard *et al.* 1989; Taylor *et al.* 1993). Model simulations have shown that such changes can be understood in terms of abrupt reorganisations, i.e. switches from one mode of operation to another, initiated by perturbations such as melting terrestrial ice sheets (Wright and Stocker 1993). It is important to distinguish between abrupt reorganisations and chaotic variability where the system resides for some time in one dynamical regime and then switches to another (Lorenz 1963; Lorenz 1990). Abrupt reorganisations are rapid changes between different *equilibrium* states when system parameters are slowly changing (Stocker and Wright 1991; Mikolajewicz and Maier-Reimer 1994; Rahmstorf 1995). In the search for mechanisms of periodic climatic variability it is often tempting to look for periodic processes in some forcing variables and then postulate enhancing feedback mechanisms which cause a response of sufficient amplitude in the climate variable under consideration. Examples are the various solar cycles such as the sunspot (10–11 yr), Hale (22 yr), and Gleissberg (84 yr) cycles. Although these time scales do occur in spectra of various climate proxies abundantly (see below), it seems unlikely that such cycles are due to direct solar forcing. The global sensitivity of the climate system to changes in the shortwave irradiation is estimated at about $0.14\,K/Wm^{-2}$ (based on an AGCM simulation by Lean and Rind (1994)) with a spatially rather uniform response. The only effect on solar irradiation that has been directly measured is that of the 11-year solar cycle whose peak-to-peak amplitude has been determined at about $2.5\,Wm^{-2}$ (ERBE 1990). This would result in a temperature variation of about 0.35 K which would have to be detected as a globally uniform signal. Significant positive feedback mechanisms would have to operate in the climate system to amplify such a signal to the amplitude seen in many climate records (order of 1 K) and to impose regional patterns such as observed in the climatic reconstruction (e.g. Briffa *et al.* (1992)).

2 Observations and Proxy Indicators

2.1 Direct Observations

Direct observations begin to exhibit large-scale changes in the climate system on the interdecadal time scale. From the comparison of systematic temperature and salinity measurements of the last 40 years a distinct warming in depths between 700 m and 3000 m could be identified in the North Atlantic (Roemmich and Wunsch 1984; Levitus 1989c). Moreover, Parilla *et al.* (1994) find, superimposed on this warming, zonally alternating regions of cooling at 24°N and attribute this to large-scale decadal variability in the North Atlantic. There is also evidence of long-term variability and even step-like changes of properties in the marginal seas of the North Atlantic. Schlosser *et al.* (1991) found a significant reduction of ventilation of the deepwater in the Greenland Sea on the basis of tracers, while Lazier (1996) shows a notable decrease in salinity in the Labrador Sea over the last thirty years. Although these latter two studies do not show cycles they provide potentially important information on mechanisms and focal areas of variability.

Variability on decadal-to-century time scales is also found in atmospheric variables such as air temperature, wind speed and sea level pressure and can be related to concurrent changes in the underlying ocean. Deser and Blackmon (1993) find a cycle of 9–12 years in their analysis of 90 years of winter time means of sea surface temperature (SST), air temperature, wind speed and sea level pressure in the North Atlantic region. They identify two modes in the North Atlantic. The first is a north-south aligned dipole pattern with a negative anomaly of sea level pressure to the north. This results in a westerly wind anomaly enhancing the prevailing wind field and increasing the sea-air fluxes leading to negative SST anomalies. The second pattern is associated with the Gulf Stream where strong positive SST anomalies are found between the period 1939–1968 and 1900–1929. It is probably a response to the global surface warming trend in the 1920s and resulted in a general weakening of the basin-scale atmospheric circulation over the North Atlantic.

Kushnir (1994) has also analysed the above data set with an emphasis to identify the difference patterns between cold (1900–14, and 70–84) and warm periods (1925–39 and 50–64) during this century. He found that the difference between warm and cold years consists of positive SST anomalies along the east coast of Greenland, the entrance of the Labrador Sea and off-shore of Newfoundland/Nova Scotia and negative SST anomalies along the northern part of the US east coast (Fig. 3). This is associated with a basin-wide cyclonic sea level pressure anomaly very similar to that reported by Deser and Blackmon (1993). The SST anomaly pattern is present in both the warm and the cold season (Fig. 3a, b), whereas the pressure anomaly is primarily a winter time phenomenon. While in the southern region of this SST anomaly the atmospheric circulation anomalies tend to attenuate the SST anomalies, they are maintained by them in the northern region. Hydrographic changes in the shallow and intermediate North Atlantic reported by Levitus (1989a), Levitus (1989b) are consistent with these findings. A similar structure (cooling at the Labrador Sea entrance and east coast of Greenland and warming is again found in the subsurface temperature (125 m) trends from 1966 to 1990 (Levitus *et al.* 1994)).

It is very important to address the question whether the North Atlantic is a region of enhanced variability as suggested by some proxy records or whether this is solely a result of the sparsity of data elsewhere. Schlesinger and Ramankutty (1994) analyze global data sets of surface air temperature covering the period 1858–1992 in 11 different geographical regions. Using singular spectrum analysis, they found a 50–80 year cycle in this variable in various sub-regions of the globe with the largest amplitudes in the North Atlantic and North American regions. This is an important hint towards an underlying mechanism. Similar decadal-to-century cycles at 24 and 100 years have been found in the Central England temperature time series, the longest instrumental record available, covering 318 years (Stocker and Mysak 1992). They are statistically significant at the 99% level.

2.2   Proxy Data

A review of long-term cyclic fluctuations on the century time scale is given in Stocker and Mysak (1992). Cycles of 50 years and longer are abundant in high-resolution proxy records (Fig. 4), but a clear and unequivocal time scale is missing. Stuiver (1980) tested the hypothesis whether such cycles found in proxy data could be due to solar variability for which $\Delta^{14}C$ from tree rings was used as a proxy. A statistically significant relationship could not be established.

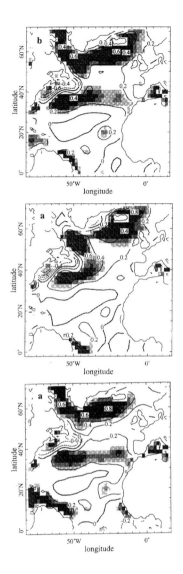

Figure 3: Differences of observed sea surface temperatures during 15 consecutive years of above average (1950–64) and below average (1970–84) temperatures in annual means (top), winter months (centre, Dec-Apr) and summer months (bottom, June-Oct). Large-scale, coherent anomalies can cause anomalous circulation patterns in ocean and atmosphere as independent numerical models also demonstrate. The SST patterns are better developed during the winter season [From Kushnir (1994)].

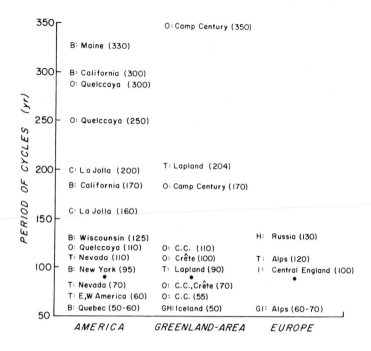

Figure 4: Summary of climatic variations observed on the century time scale. The findings are spatially and temporally ordered. The capital letters denote evidence in biological (B), radio carbon (C), glaciological (G), historical (H), instrumental (I), oxygen isotope and other ice core parameters (O) and tree ring records (T). The observed cycle period is given in brackets. C.C. denotes the Camp Century $\delta^{18}O$ record and $*$ indicates the 83-yr cycle in marine air and sea surface temperatures of Folland et al., (1984). [From Stocker et al. (1992)]

Briffa *et al.* (1992) have spectrally analyzed their 1480-year long tree ring record of Fennoscandia from which summer temperatures are reconstructed. They find significant power in the band of 30–40 years. An interesting feature is the fact that the periods are not stable: they vary by several years depending on the sub-interval considered. Such behavior is not expected if the oscillation is due to external forcing at a fixed period. It rather suggests that an internal mechanism may be responsible for these cycles.

Only recently researchers have begun to synthesize information derived from *different* paleoclimatic archives. The study of Mann *et al.* (1995) combines 35 different proxy records (tree rings, ice and coral cores, lake varves, historical records) distributed mainly in the northern hemisphere with a few "stations" in the south and covering the last 500 years. This allows them not only to determine time scales of variability but is a first estimate of spatial correlation and patterns of natural variability as recorded in high-resolution proxy data time series of sufficient length. Mann *et al.* (1995) find cycles in the interdecadal (15-35 years) and century (50-150 years) range that are significant. As with the summer temperatures derived from tree ring widths of Fennoscandinavia, these variations are not stable throughout time. Century scale cycles are particularly strong during the mid 17 and 18 century, a time where fluctuations to colder conditions have been reported extensively and are referred to as "Little Ice Age" (Bradley and Jones 1995). The spatial distribution of phase relations indicates that the variability is mostly confined to the North Atlantic region with a clear phase difference of 45°-135° between the eastern and western side of the Atlantic basin. Mann *et al.* (1995) interpret this as the result of in-phase and out-of-phase variability. The former may be due to the basin-wide transport of heat by the meridional overturning circulation and its century scale variability (Mikolajewicz and Maier-Reimer 1990; Mysak *et al.* 1993) while the latter is consistent with a mechanism described by Delworth *et al.* (1993) (see below).

It appears from this that the North Atlantic region is one of the pace makers of climate variability. However, one should note that still only few proxy data come from regions other than Europe or North America, and that therefore our view may be biased. Future emphasis must be given to the retrieval and analysis of high-resolution paleorecords of the tropical regions and the southern hemisphere. In the meantime, models are the only tool that can help us understand mechanisms of variability on the decadal-to-century time scale. Also, they are suitable in pointing to locations of increased variability where proxy records, if made available, should most likely exhibit variability.

3   Models and Mechanisms

As mentioned above, the weakness of the hypothesis of a purely solar origin of decadal-to-century variability lies in the fact that the sensitivity of the climate system to changes in the shortwave irradiance is about a factor of 5 smaller than that for longwave emission changes including water vapour feedback (Cess *et al.* (1989) give a mean sensitivity of $0.68 \, \mathrm{K/Wm^{-2}}$). By monitoring the most recent solar cycle it was found that corresponding solar irradiance changes have an amplitude of less than $2\,Wm^{-2}$ (ERBE 1990) which would yield an amplitude of about $0.3\,K$ using the above sensitivity. It is therefore important to look for possible alternative explanations and mechanisms of climate variability.

Modeling has become an important branch of climate research because only with physically based models is it possible to *quantitatively* test and verify hypotheses on climate change. Ocean, atmosphere and coupled models are successfully simulating the large-scale fields of the climate system (Trenberth 1992), and it is now timely to assess these models' capability of simulating also the natural variability.

During the last decade, oceanic circulation models have made significant progress due to the implementation of mixed boundary conditions that take into account feedback mechanisms between atmosphere and ocean. SST anomalies generate local heat flux anomalies that operate to remove the SST anomalies within a few weeks. Sea surface salinity (SSS) anomalies, on the other hand, do not influence the hydrological cycle, i.e. the surface freshwater balance, and hence can have a longer lasting impact on the surface buoyancy distribution. By relaxing SST to a fixed surface air temperature and keeping the surface freshwater fluxes constant, one arrives at a first approximation of the important difference between the feedback character of ocean-to-atmosphere heat and freshwater fluxes (Stommel 1961; Rooth 1982). However, local heat flux anomalies are bound to also change surface air temperature, an effect which is explicitly excluded when using mixed boundary conditions. Recognizing these limitations several studies have proposed improved parameterizations of the surface fluxes by formulating various types of energy balance models coupled to the ocean circulation models (e.g. Stocker *et al.* (1992), Zhang *et al.* (1993), Rahmstorf and Willebrand (1995), Lohmann *et al.* (1996)). Two effects are of importance: (i) SST anomalies cause heat flux anomalies which modify the surface air temperature locally; (ii) due to the possibility of meridional heat flux in the atmospheric part of the coupled model far-field effects can occur.

The basic mechanism for oscillations due to different feedback processes of SST and SSS anomalies was summarized by Welander (1986). He showed that self-sustained oscillations and different equilibrium states can be realized in a circular convection loop in which one side is heated and salted while the other side is cooled and freshened. This is reminiscent of the low and high latitudes where the surface ocean is heated and evaporation causes an increase in salinity whereas the opposite happens in the high latitudes. He showed that if the time scales characteristic for temperature and salinity anomalies are different, multiple steady states as well as self-sustained oscillations are possible.

In more complex, 2- and 3-dimensional models mechanisms generating natural variability are more difficult to understand, and the range of mechanisms and time scales is quite broad. At present, we do not have a unifying theory explaining natural variability on the century time scale but we are in the stage of collecting evidence for such fluctuations both from the observational and the modeling side. It is hoped that over the coming years the representation of simple atmospheres used in driving ocean models becomes more realistic and that with this improvement, a more consistent picture will emerge. For now, a list of the type and time scales of variability found in numerical models (ocean, atmosphere, coupled) helps us in discussing physically plausible mechanisms of variability.

Table 1 gives a summary of self-sustained oscillations found in a number of ocean, atmosphere and coupled models. We focus only on the most robust cycles in these models. The decadal-to-century time scale does in most cases include the ocean circulation, in particular the thermohaline part of it. Mechanisms are connected with mainly with the thermohaline but also with the wind-driven circulation as well as with the hydrological cycle. Note that long-term variability is also found in atmospheric GCMs suggesting interesting possibilities of interaction between the atmosphere and the ocean also on interdecadal time scales.

Model Type	Geometry	Perturbation to Forcing	Surface Flux Parameterisation

Gyre–Thermohaline Circulation

Model Type	Geometry	Perturbation to Forcing	Surface Flux Parameterisation
B&C 3D OGCM	box, flat bottom, hemispheric	–	mixed
B&C 3D OGCM	1 sector ocean, flat bottom, fixed ACC	var induced by mismatch of surface heat flux and heat storage	no-heat-capacity atm (i.e. effective restoring time of 1-2 yr), salt flux constant
3D B&C OGCM	box, flat bottom, hemispheric	increase of freshwater flux	freshwater flux only, no thermal forcing
B&C 3D OGCM	1 sector ocean, flat bottom, fixed ACC	var induced by mismatch of surface fluxes	mixed, constant heat or freshwater fluxes
3D OGCM	hemispheric box	–	mixed
3D OGCM	hemispheric box	–	mixed

SST Anomalies in the Northwest Atlantic

Model Type	Geometry	Perturbation to Forcing	Surface Flux Parameterisation
3D A/OGCM	global	–	coupled, flux correction
3D OGCM	hemispheric box	–	constant heat flux, mixed or zero heat capacity atm.

Marginal Seas

Model Type	Geometry	Perturbation to Forcing	Surface Flux Parameterisation
B&C 3D OGCM	flat N Atlantic and Labrador Sea	stochastic freshwater flux	mixed
3D LSG OGCM	global, topography	stochastic freshwater flux	mixed

Basin-Scale Thermohaline Circulation

Model Type	Geometry	Perturbation to Forcing	Surface Flux Parameterisation
2D, zonally av. OCM	1 sector ocean	stochastic freshwater flux	mixed
3D LSG OGCM	global, topography	stochastic freshwater flux	mixed

The Southern Ocean-North Atlantic Connection

Model Type	Geometry	Perturbation to Forcing	Surface Flux Parameterisation
3D LSG OGCM	global, topography	modified constant or stochastic freshwater flux	mixed

Other Mechanisms

Model Type	Geometry	Perturbation to Forcing	Surface Flux Parameterisation
2D, zonally averaged OCM, ice	1 sector ocean	–	mixed
3D OGCM	hemispheric box	–	mixed
3D AGCM, coupled to LSG OGCM	global	–	coupled, flux correction
3D AGCM, dry	global, flat surface	–	surface energy balance

Table 1: Internal variability found in various numerical models ordered ...

Period	Mechanism	Reference

Gyre–Thermohaline Circulation

Period	Mechanism	Reference
9	advection of SSS anomaly by subtrop. and subpolar gyres	(Weaver and Sarachik 1991a)
24	advection of SST anomalies by subtropical gyre, overturning influences exposure time of SST anomalies	(Cai and Godfrey 1995)
20–350	advection of SSS anomalies influencing overturning (transition to chaos)	(Huang and Chou 1994)
20–100	advection of SSS or SST anomalies	(Cai 1995)
250–500	succession of haloclines and flushes triggered by gyre salt transport	(Winton 1993)
> 500	advection of SSS anomalies by gyres and vertical diffusion	(Winton 1993)

SST Anomalies in the Northwest Atlantic

Period	Mechanism	Reference
40–60	baroclinic vortex in West Atlantic, meridional heat flux and local heat storiage	(Delworth et al. 1993)
50–70	meridional heat transport and local heat storage	(Greatbatch and Zhang 1995)

Marginal Seas

Period	Mechanism	Reference
20	Labrador Sea, zonal and meridional overturning	(Weaver et al. 1994)
10–40	Labrador Sea operates as stochastic integrator sending SSS anomalies into N Atlantic	(Weisse et al. 1993)

Basin-Scale Thermohaline Circulation

Period	Mechanism	Reference
200–300	large-scale SSS advection	(Mysak et al. 1993)
320	large-scale SSS advection in the Atlantic around entire deep circulation loop	(Mikolajewicz and Maier-Reimer 1990)

The Southern Ocean-North Atlantic Connection

Period	Mechanism	Reference
320	halocline in ACC influences NADW, its outflow causes deep heating in ACC and triggers convection	(Pierce et al. 1995)

Other Mechanisms

Period	Mechanism	Reference
13.5	brine release–deep water formation–meridional heat flux feedback	(Yang and Neelin 1993)
17	ice cover–thermal insulation feedback	(Zhang et al. 1995)
10–20	advection of T anomalies in the Pacific	(Von Storch 1994)
5–40	chaotic nature, subtropical and mid-latitude atmospheric jets	(James and James 1992)

...according to time scale and mechanisms discussed in the text.

The table is grouped into six categories of mechanisms that will be discussed below and ordered according to associated time scales. The latter is not rigorous, of course, since periods depend on model parameters. Nevertheless, it contains information which processes in the climate system might be important on a given time scale. This should contribute to the identification of the most robust and fundamental modes of variability and change and distinguish them from more model-specific fluctuations.

3.1   Gyre–Thermohaline Circulation

Weaver and Sarachik (1991a) report self-sustained oscillations in a hemispheric 3-D OGCM under mixed boundary conditions. Formed in the western boundary current, warm and saline anomalies travel eastward and are picked up by the sub-polar gyre which transports them into the region of deep water formation; this journey takes about 8 to 9 years. There, they influence the basin-scale overturning and so feedback to the surface advection of these anomalies. The role of the latitudinal structure of the freshwater forcing was also studied (Weaver *et al.* 1993). When precipitation in the high latitude is increased internal variability on the decadal (as before) and the interdecadal (15–20 yr) time scales is generated. For sufficiently strong forcing sequences of violent overturnings and little deep water formation could be excited. As these are connected to diffusive processes in the ocean interior during the time of reduced ventilation, time scales are on the order of 500 years and are decreasing with increasing amplitude of the stochasting forcing.

Similar oscillations were found by Winton (1993) using a frictional-geostrophic model. Important in these models is the fact that anomalies are advected by the gyre circulation near the surface which determines the decadal time scale. These oscillations affect the entire water column in the high latitudes by turning on and off deep water formation, and the amplitude of the changes of the meridional heat flux are of order $0.2 \times 10^{15} \, W$. While these models allow us to isolate and investigate various mechanisms of internal variability, immediate application to the real world is limited because of the simple parameterization of ocean-atmosphere interaction, simplified geometry, and their still fairly coarse resolution.

3.2   SST Anomalies in the Northwest Atlantic

Coupled climate models are also beginning to exhibit natural variability. Delworth *et al.* (1993) integrate the GFDL climate model for 600 years and find natural variability whose spectral properties are remarkably similar to observations (Fig. 5a). Superimposed on a red-noise spectrum a number of spectral peaks are visible. Interdecadal oscillations of 40–60 yr are evident in the maximum meridional overturning in the North Atlantic reaching amplitudes of about 2 Sv (Fig. 5b). When the thermohaline circulation is weak, decreased advection of lower-latitude warm and saline waters into the central regions of the North Atlantic generates a pool of anomalously cold and fresh water. The thermal anomaly dominates and hence generates a geostrophically controlled cyclonic circulation at the surface (baroclinic vortex). The western half of this anomalous circulation enhances the mean northward current of warm and saline waters located in the center of the Atlantic which is part of the large-scale conveyor belt circulation. The strengthened conveyor then carries more saline and warm low-latitude waters into this region. Again, the thermal contribution is stronger and creates an anomalous warm pool which is associated with anti-cyclonic circulation. The latter weakens the mean flow again, and the cycle begins anew.

The oscillation is distinctly irregular, a common feature of nonlinear dynamical systems. Although the first 200 years of the integration show a quasi-periodic cycle of 40–50 years, the periods are longer in the following 400 years. This is indicative of some preconditioning of the ocean independent of the feedback mechanism. One plausible possibility is the volume, i.e. the heat and salt content, of the anomalous pool. The bigger the pool, the longer it takes a certain mass transport anomaly to erode the SST anomaly. While the amplitude of the anomaly governs the strength of this anomalous mass flux via the pressure gradients, the spatial extent itself does not influence it but determines the time scale on which the anomaly can be removed. For an estimate of orders of magnitude we assume a typical extent of this pool (65°W–45°W, 35°N–50°N, (Delworth *et al.* 1993)) and a depth of the anomaly of roughly 300m (Greatbatch and Zhang 1995). The corresponding volume of $8.2 \cdot 10^{14} m^3$ is renewed once in 26 years by a flux anomaly of 1 Sv. Changing pool sizes are thus likely to be responsible for the changes of the period lengths during different segments of the 600-year run of Delworth *et al.* (1993). It is intriguing that a similar in-

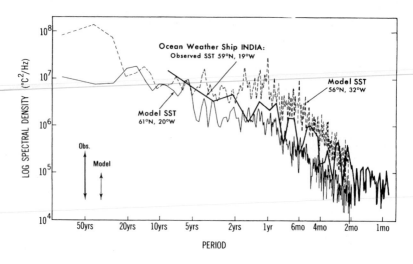

Figure 5a: Comparison of the spectral properties of a coupled A-OGCM with observations. The total amount of variability in the displayed frequency band is remarkably similar to observations [From Delworth et al. (1993)]

stationarity of interdecadal cycles is clearly present in proxy data (Briffa *et al.* 1992; Mann *et al.* 1995).

There are still limitations of such models. In order to achieve a stable climate consistent with the observations spatially dependent flux corrections have to be applied. Although these corrections do not impose a time scale on the model, they act as an additional forcing which, in a nonlinear system, could generate additional variability. Also, the region where the oscillation is observed (northwest Atlantic) is one of the areas of large flux correction (Manabe and Stouffer 1988). On the other hand, the model does suggest a deterministic mechanism with time scales and amplitudes similar to those of proxy records and also indicates the focal regions of oscillatory activity. Moreover, the spatial patterns and depth structures of temperature and salinity in the ocean and sea level pressure that are suggested by the model can be directly compared to long-term observations. The basic mechanism is also present in an ocean-only model driven by *constant* surface heat flux only, suggesting that the phenomenon is primarily thermally driven (Greatbatch and Zhang 1995). Occurring in two entirely different models is good evidence that this mechanism is a very robust feature. Greatbatch and Zhang (1995) have also found that the inclusion

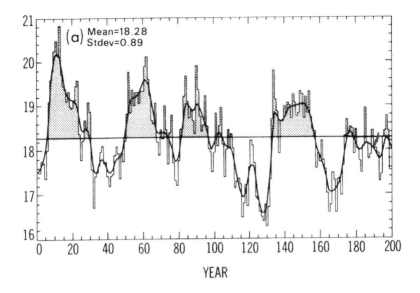

Figure 5b: *Time series of the maximum overturning streamfunction and running average in the North Atlantic for the first 200 years of their 600-year integration. [From Delworth et al. (1993)]*

of freshwater flux forcing increases the period and decreases the amplitude of the oscillation. A changing hydrological cycle in the atmosphere thus represents a second mechanism for instationarity of periods.

3.3 Marginal Seas

Recent modeling results indicate that marginal seas are potentially very important pacemakers for self-sustained variability of the ocean circulation. Weaver *et al.* (1994) found oscillations of about 20 years in the Labrador Sea of their model. The mechanism is linked to an interaction between the meridional (and zonal) pressure gradients and the geostrophically controlled zonal (and meridional) overturning in the Labrador Sea. In contrast to the mechanism discussed in the previous section, this variability is neither dependent on the surface freshwater budget nor on the wind-driven gyral circulation. While the resolution of this model in the most important area is admittedly very low (only 3 tracer grid points), it is encouraging to note that SST anomalies are very similar to those reported by Kushnir (1994) based on observations.

A complementary study with a global model was done by Weisse *et al.*
(1993). A stochastic freshwater flux perturbation excites century (320
yr) and decadal scale oscillations. The latter are localized, again, in the
Labrador Sea area, have periods in the range of 10–40 years, but the
mechanism is distinctly different from that above. The relatively isolated
marginal sea integrates the stochastic freshwater flux perturbations and
sends salinity anomalies into the North Atlantic on a time scale of about
10–40 years. This time scale is determined by the flushing time of the
upper 250 m of the basin, since the stratification is quite stable. Once the
perturbations arrive in the North Atlantic, where the stratification between
50°N and 60°N is weak, they strongly influence the deep water formation
rates and create the variability observed.

These two mechanisms differ distinctly from each other in that in the first
study, the Labrador Sea itself generates the variability by changing rates
of local deep water formation, whereas in the other case the same region
merely appears as a storage of perturbations which, once accumulated, act
outside the basin. An increased resolution and, with it, a better representation of the water masses in the Labrador Sea will refine our understanding
of its role in controlling the natural variability in the North Atlantic region.

3.4   Basin-Scale Thermohaline Circulation

Basin-scale variations of the Atlantic meridional overturning and hence
meridional heat flux are possible mechanisms for century time scale variability. The Hamburg global OGCM was run under mixed boundary conditions including a stochastic freshwater flux perturbation (Mikolajewicz
and Maier-Reimer 1990). Large fluctuations with a time scale of 320 years
are found in the mass transport through Drake Passage and the meridional heat flux and overturning in the North Atlantic. Amplitudes would
be large enough to be detected in paleoclimatic archives (the heat budget
in the Southern Ocean shows peak-to-peak amplitudes of up to $3 \times 10^{15}\,W$);
events are not regular cycles but appear in the time series as distinct events.
The mechanism is associated with the long residence time of SSS anomalies.
The random freshwater flux perturbations create local salinity anomalies
that are advected northward by the near-surface circulation in the Atlantic.
Depending on the spatial structure of the mean freshwater fluxes they are
enhanced or removed before they reach the deep water formation area in
the high-latitude Atlantic. If they are enhanced, they tend to accelerate

meridional overturning. The authors trace these anomalies along the entire loop of overturning so that they eventually surface again in the South Atlantic/Southern Ocean region. Although much diffused and diminished by that time, they are the seed of new anomalies that start their northbound travel at the sea surface.

Mysak *et al.* (1993) forced a zonally averaged one-basin ocean circulation model with random freshwater fluxes and found oscillations with time scales between 200 and 300 years over a wide range of parameters (Fig. 6). In contrast to the previous model, salinity anomalies could not be traced around a complete loop of the circulation but became well mixed once they were in the deep ocean. This indicates that these cycles are primarily due to an interaction between the detailed spatial structure of the surface freshwater flux and the strength of the thermohaline overturning which determines the preferred time scale.

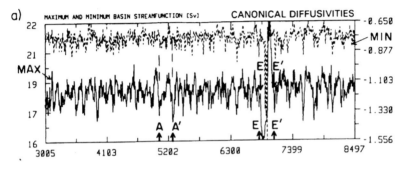

Figure 6a: Time series of the minimum (dashed) and maximum (solid) overturning streamfunction in a one-basin 2-d ocean model. [From Mysak et al. (1993)]

3.5   The Southern Ocean-North Atlantic Connection

Very recently two studies have re-examined natural variability on the century time scale found in the Hamburg ocean model (Pierce *et al.* 1995; Osborn 1995). Although the cycles have the same period they identify and demonstrate a mechanism that is significantly different from that proposed by Mikolajewicz and Maier-Reimer (1990). Also, variability need not be generated by a stochastic perturbation of the freshwater flux but can be initiated by a slight change ($\pm 10\%$) of the magnitude of the freshwater flux, i.e. an imposed incompatibility of the fluxes with the steady state (a

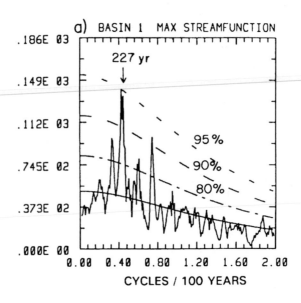

Figure 6b: *Spectrum of the time series in Fig. 6a. Significant power is found on the century time scale. [From Mysak et al. (1993)]*

similar way of inducing variability was presented by Cai (1994)). This is an indication that models which use the classical mixed boundary conditions, where the salt flux is fully consistent with the steady-state circulation field, probably underestimate natural variability.

The mechanism of variability is based on a newly found coupling between the hemispheres by the Atlantic thermohaline circulation. The model oscillates between two extreme states during one cycle: (i) strong deep water formation in the Southern Ocean with significant influx of Antarctic Bottom Water into the Atlantic where the thermohaline circulation is slightly weakened; (ii) halocline around Antarctica, no AABW in the Atlantic with slightly increased Atlantic overturning. The development and destruction of the Circum-Antarctic halocline has – via the JEBAR effect (see Cai (1994) for an example in a 3D model) – a strong impact on barotropic transport through the Drake Passage which changes between 60 and 160 Sv!

A cycle evolves as follows. When a strong halocline is present in the Southern Ocean, surface heat exchange is strongly reduced (due to stable stratification and ice cover). The warmer waters exiting from the Atlantic ocean at a depth of 2–3 km manage to slowly heat up the subsurface Southern Ocean until it is destabilised. Convection then establishes fairly quickly all around Antarctica and reaches full strength within less than 100 years. During the following 200 years convection slowly decreases again due to the action of net precipitation in that area. The result is an asymmetric oscillation pattern. Pierce *et al.* (1995) show that this oscillation is only possible when a nonlinear equation of state is used.

The above mechanism is similar to the flushes found in many other models (Marotzke 1989; Wright and Stocker 1991; Weaver and Sarachik 1991b; Winton and Sarachik 1993) whose characteristic feature is a decoupling of the lower ocean from the surface. In contrast to these earlier studies where an entire basin had to be destabilized by *diffusion* and hence evolved on much longer time scales of o($10^3 - 10^4$yrs), destabilisation here is due to *advection* of warmer NADW, i.e. an "efficient" process with a time scale of a few hundred years.

3.6   Other Mechanisms

Internal variability on the interdecadal time scale is found in simple ocean models when a thermodynamic sea ice component is included. Formation of a sea ice cover releases salt into the water column and induces convection. By altering density gradients this increases the thermohaline circulation advecting more heat northward which then melts the sea ice cover. Melting increases the surface density which tends to decrease the thermohaline circulation leading to a net cooling. This feedback loop was found in a 2-dimensional ocean circulation model coupled to a thermodynamic sea ice model (Yang and Neelin 1993). Zhang *et al.* (1995) showed that this process is also present in a 3-dimensional OGCM coupled to an ice model and found the destabilizing influence of brine rejection less important than that of slow heating below the ice cover due to thermal insulation.

In the previous section mainly oceanic processes localized in or related to the Atlantic have been discussed. However, there are also recent model examples of interdecadal cycles in the Pacific and in the atmosphere alone. James and James (1992) find in their atmospheric circulation model variability on time scales of 10 to 40 years and associate them with changes in the structure of the subtropical and mid-latitude jets.

Von Storch (1994), on the other hand, uses a coupled A/OGCM and identifies two types of low-frequency variability. The atmospheric fluctuations are essentially red noise, and there appears an out-of-phase relationship between the stratosphere and the troposphere. In the ocean there is an irregular cycle of about 17 years located in the Pacific Ocean. Note that no distinct variability is found in the Atlantic basin neither in marginal seas nor basins-wide.

4 Conclusions

The study and quantitative understanding of internal variability in the climate system is still in its infancy. Focal regions where internal variability is enhanced or generated have been identified and preliminary results have shed light on a palette of mechanisms. However, we are still at the stage where each model tends to produce its characteristic set of fluctuations with the associated time scales. Important results have been found regarding the role of the ocean, the influence of atmosphere-ocean exchange fluxes and the dynamics in marginal seas, but a consistent *quantitative theory* on climatic cycles on the decadal-to-century time scale is still missing. Further work is urgently needed both with respect of models and observations. We need to improve the formulation of surface exchange processes (boundary conditions) especially in the ocean-only models, the parameterisation of convection and deep water formation; the resolution must be increased in order to better represent topography and marginal but important regions of the ocean basins must be included.

More important, however, are efforts to obtain and analyze high-quality terrestrial and oceanic proxy data which allow annual or seasonal resolution. Special attention must be focused on the transfer function, i.e. how is the climate signal transferred into the archive and how is the possible temporal evolution of such a transfer function. Ongoing European and US efforts to retrieve two high-resolution ice cores from different locations in Antarctica will provide vital information about natural variability centered in the Southern Ocean. Finally, spatial networks of homogenized paleoclimatic data will certainly allow us to find some of the keys to better understand natural variability in the climate system.

Acknowledgment: I thank Jürgen Willebrand and David Anderson for this stimulating NATO ASI workshop held at Les Houches in February 1995. Thomas Blunier compiled Figure 2. I thank Thomas Tschannen for his expert help with LaTeX and an anonymous reviewer for comments which clarified the distinction between different types of climate variability. This work is supported by the Swiss National Science Foundation.

References

Blunier, T. (1995). *Methanmessungen aus Arktis, Antarktis und den Walliser Alpen, Inter-hemisphärischer Gradient und Quellenverteilung*. Ph. D. thesis, Physics Institute, University of Bern, Switzerland.

Blunier, T., J. Chappellaz, J. Schwander, B. Stauffer, and D. Raynaud (1995). Variations in the atmospheric methane concentration during the Holocene. *Nature 374*, 46–49.

Bond, G. *et al.* (1992). Evidence for massive discharges of icebergs into the North Atlantic ocean during the last glacial period. *Nature 360*, 245–249.

Bond, G., W. Broecker, S. Johnsen, J. McManus, L. Labeyrie, J. Jouzel, and G. Bonani (1993). Correlations between climate records from North Atlantic sediments and Greenland ice. *Nature 365*, 143–147.

Bradley, R. S. and P. D. Jones (Eds.) (1995). *Climate since AD 1500*. Routledge. 706 pp.

Briffa, K. R. *et al.* (1992). Fennoscandian summers from ad 500: temperature changes on short and long timescales. *Clim. Dyn. 7*, 111–119.

Briffa, K. R., P. D. Jones, and F. H. Schweingruber (1992). Tree-ring density reconstructions of summer temperature patterns across Western North America since 1600. *J. Climate 7*, 735–754.

Cai, W. (1994). Circulation driven by observed surface thermohaline fields in a coarse resolution ocean general circulation model. *J. Geophys. Res. 99*, 10163–10181.

Cai, W. (1995). Interdecadal variability driven by mismatch between surface flux forcing and oceanic freshwater/heat transport. *J. Phys. Oceanogr. 25*, 2643–2666.

Cai, W. and S. J. Godfrey (1995). Surface heat flux parameterizeations and the variability of the thermohaline circulation. *J. Geophys. Res. 100*, 10679–10692.

Cess, R. D. *et al.* (1989). Interpretation of cloud-climate feedback as produced by 14 atmospheric general circulation models. *Science 245*, 513–516.

Chappellaz, J., T. Blunier, D. Raynaud, J. M. Barnola, J. Schwander, and B. Stauffer (1993). Synchronous changes in atmospheric CH_4 and greenland climate between 40 and 8 kyr BP. *Nature 366*, 443–445.

Dansgaard, W., S. J. Johnsen, H. B. Clausen, D. Dahl-Jensen, N. S. Gundestrup, C. U. Hammer, C. S. Hvidberg, J. P. Steffensen, A. E. Sveinbjornsdottir, J. Jouzel, and G. Bond (1993). Evidence for general instability of past climate from a 250-kyr ice-core record. *Nature 364*, 218–220.

Dansgaard, W., J. W. C. White, and S. J. Johnsen (1989). The abrupt termination of the Younger Dryas climate event. *Nature 339*, 532–534.

Delworth, T., S. Manabe, and R. J. Stouffer (1993). Interdecadal variations of the thermohaline circulation in a coupled ocean-atmosphere model. *J. Climate 6*, 1993–2011.

Deser, C. and M. L. Blackmon (1993). Surface climate variations over the North Atlantic ocean during winter: 1900–1989. *J. Climate 6*, 1743–1753.

ERBE (1990). Earth radiation budget experiment. *EOS, Trans. Am. Geophys. Union 71*, 297–305.

Greatbatch, R. J. and S. Zhang (1995). An interdecadal oscillation in an idealized ocean basin forced by constatn heat flux. *J. Climate 8*, 81–91.

Hammer, C. U., H. B. Clausen, and C. C. Langway Jr. (1994). Electrical conductivity method (ECM) stratigraphic dating of the Byrd Station ice core, Antarctica,. *Ann. Glaciology 20*, 115–120.

Huang, R. X. and R. L. Chou (1994). Parameter sensitivity study of the saline circulation. *Clim. Dyn. 9*, 391–409.

Imbrie, J. *et al.* (1992). On the structure and origin of major glaciation cycles, 1. linear responses to Milankovitch forcing. *Paleoceanogr. 7*, 701–738.

Imbrie, J. *et al.* (1993). On the structure and origin of major glaciation cycles, 1. the 100,000 years cycle. *Paleoceanogr. 8*, 699–735.

James, I. N. and P. M. James (1992). Spatial structure of ultra-low frequency variability of the flow in a simple atmospheric circulation model. *Q. J. Roy. Met. Soc. 118*, 1211–1233.

Johnsen, S. J., H. B. Clausen, W. Dansgaard, K. Fuhrer, N. Gundestrup, C. U. Hammer, P. Iversen, J. Jouzel, B. Stauffer, and J. P. Steffensen (1992). Irregular glacial interstadials recorded in a new Greenland ice core. *Nature 359*, 311–313.

Kushnir, J. (1994). Interdecadal variations in north atlantic sea surface temperature and associated atmospheric conditions. *J. Climate 7*, 141–157.

Lazier, J. R. N. (1996). The salinity decrease in the Labraodr Sea over the past thirty years. In *Climate Variability on Decade-to-Century Time Scales*. National Research Council. (in press).

Lean, J. and D. Rind (1994). Solar variability: Implications for global change. *EOS, Trans. Am. Geophys. Union 75*, 1–7.

Lehman, S. J. and L. D. Keigwin (1992). Sudden changes in North Atlantic circulation during the last deglaciation. *Nature 356*, 757–762.

Levitus, S. (1989a). Interpentadal variability of salinity in the upper 150 m of the North Atlantic ocean, 1970-1974 versus 1955-1959. *J. Geophys. Res. 94*, 9679–9685.

Levitus, S. (1989b). Interpentadal variability of temperature and salinity at intermediate depths of the North Atlantic ocean, 1970-1974 versus 1955-1959. *J. Geophys. Res. 94*, 6091–6131.

Levitus, S. (1989c). Interpentadal variability of temperature and salinity in the deep North Atlantic, 1970-1974 versus 1955-1959. *J. Geophys. Res. 94*, 16125–16131.

Levitus, S., J. I. Antonov, and T. P. Boyer (1994). Interannual variability of temperature at a depth of 125 meters in the North Atlantic ocean. *Science 266*, 96–99.

Lohmann, G., R. Gerdes, and D. Chen (1996). Sensitivity of the thermohaline circulation in coupled ocean GCM – atmospheric EBM experiments. *Clim. Dyn. xx*, yy. (in press).

Lorenz, E. N. (1963). Deterministic non-periodic flow. *J. Atm. Sci. 20*, 130–141.

Lorenz, E. N. (1990). Can chaos and intransitivity lead to interannual variability ? *Tellus 42A*, 378–389.

Manabe, S. and R. J. Stouffer (1988). Two stable equilibria of a coupled ocean-atmosphere model. *J. Climate 1*, 841–866.

Mann, E., J. Park, and R. S. Bradley (1995). Global interdecadal and century-scale oscillations during the past five centuries. *Nature 378*, 266–270.

Marotzke, J. (1989). Instabilities and multiple steady states of the thermohaline circulation. In D. L. T. Anderson and J. Willebrand (Eds.), *Ocean Circulation Models: Combining Data and Dynamics*, NATO ASI, pp. 501–511. Kluwer.

Mikolajewicz, U. and E. Maier-Reimer (1990). Internal secular variability in an ocean general circulation model. *Clim. Dyn. 4*, 145–156.

Mikolajewicz, U. and E. Maier-Reimer (1994). Mixed boundary conditions in ocean general circulation models and their influence on the stability of the model's conveyor belt. *J. Geophys. Res. 99*, 22633–22644.

Mitchell, J. M. (1976). An overview of climatic variability and its causal mechanisms. *Quat. Res. 6*, 481–493.

Mysak, L. A., T. F. Stocker, and F. Huang (1993). Century-scale variability in a randomly forced, two-dimensional thermohaline ocean circulation model. *Clim. Dyn. 8*, 103–116.

Osborn, T. J. (1995). *Internally-generated variability in some ocean models on decadal to millennial timescales*. Ph. D. thesis, Climatic Research Unit, School of Environmental Sciences, University of East Anglia.

Paillard, D. and L. Labeyrie (1994). Role of the thermohaline circulation in the abrupt warming after Heinrich events. *Nature 372*, 162–164.

Parilla, G., A. Lavin, H. Bryden, M. Garcia, and R. Millard (1994). Rising temperatures in the subtropical North Atlantic Oocean over the past 35 years. *Nature 369*, 48–51.

Pierce, D. W., T. P. Barnett, and U. Mikolajewicz (1995). Competing roles of heat and freshwater flux in forcing thermohaline oscillations. *J. Phys. Oceanogr. 25*, 2046–2064.

Rahmstorf, S. (1995). Bifurcations of the Atlantic thermohaline circulation in response to changes in the hydrological cycle. *Nature 378*, 145–149.

Rahmstorf, S. and J. Willebrand (1995). The role of temperature feedback in stabilizing the thermohaline circulation. *J. Phys. Oceanogr. 25*, 787–805.

Roebber, P. J. (1995). Climate variability in a low-order coupled atmosphere-ocean model. *Tellus 47A*, 473–494.

Roemmich, D. and C. Wunsch (1984). Apparent changes in the climatic state of the deep North Atlantic. *Nature 307*, 447–450.

Rooth, C. (1982). Hydrology and ocean circulation. *Prog. Oceanogr. 11*, 131–149.

Saltzman, B. (1983). Climatic systems analysis. *Adv. Geophys. 25*, 173–233.

Schlesinger, M. E. and N. Ramankutty (1994). An oscillation in the global climate system of period 65-70 years. *Nature 367*, 723–726.

Schlosser, P., G. Bönisch, M. Rhein, and R. Bayer (1991). Reduction of deepwater formation in the Greenland Sea during the 1980s: Evidence from tracer data. *Science 251*, 1054–1056.

Stocker, T. F. (1995). An overview of decadal to century time-scale variability in the climate system. In C. M. Isaacs and V. L. Tharp (Eds.), *Proc. 11th Annual Pacific Climate (PACLIM) Workshop*, Number 40 in Tech. Rep., pp. 35–46. Interagency Ecological Program for the Sacramento-San Joaquin Estuary: Calif. Dept. of Water Resources.

Stocker, T. F. and L. A. Mysak (1992). Climatic fluctuations on the century time scale: a review of high-resolution proxy-data. *Clim. Change 20*, 227–250.

Stocker, T. F. and D. G. Wright (1991). Rapid transitions of the ocean's deep circulation induced by changes in surface water fluxes. *Nature 351*, 729–732.

Stocker, T. F., D. G. Wright, and L. A. Mysak (1992). A zonally averaged, coupled ocean-atmosphere model for paleoclimate studies. *J. Climate 5*, 773–797.

Stommel, H. (1961). Thermohaline convection with two stable regimes of flow. *Tellus 13*, 224–241.

Stuiver, M. (1980). Solar variability and climate change during the current millennium. *Nature 286*, 868–871.

Taylor, K. C., G. W. Lamorey, G. A. Doyle, R. B. Alley, P. M. Grootes, P. A. Mayewski, J. W. C. White, and L. K. Barlow (1993). The 'flickering switch' of late Pleistocene climate change. *Nature 361*, 432–436.

Trenberth, K. E. (Ed.) (1992). *Climate System Modeling.* Cambridge.

Von Storch, J. (1994). Interdecadal variability in a global coupled model. *Tellus 46A*, 419–432.

Weaver, A. J., S. M. Aura, and P. G. Myers (1994). Interdecadal variability in an idealized model of the North Atlantic. *J. Geophys. Res. 99*, 12423–12441.

Weaver, A. J., J. Marotzke, P. F. Cummins, and E. S. Sarachik (1993). Stability and variability of the thermohaline circulation. *J. Phys. Oceanogr. 23*, 39–60.

Weaver, A. J. and E. S. Sarachik (1991a). Evidence for decadal variability in an ocean general circulation model: an advective mechanism. *Atmosphere-Ocean 29*, 197–231.

Weaver, A. J. and E. S. Sarachik (1991b). The role of mixed boundary conditions in numerical models of the ocean's climate. *J. Phys. Oceanogr. 21*, 1470–1493.

Weisse, R., U. Mikolajewicz, and E. Maier-Reimer (1993). Decadal variability of the north atlantic in an ocean general circulation model. *J. Geophys. Res. 99*, 12411–12422.

Welander, P. (1986). Thermohaline effects in the ocean circulation and related simple models. In J. Willebrand and D.L.T.Anderson (Eds.), *Large-Scale Transport Processes in Oceans and Atmosphere*, pp. 163–200. D. Reidel.

Winton, M. (1993). Deep decoupling oscillations of the oceanic thermohaline circulation. In W. Peltier (Ed.), *Ice in the climate system*, Volume I 12 of *NATO ASI*, pp. 417–432. Springer.

Winton, M. and E. S. Sarachik (1993). Thermohaline oscillations induced by strong steady salinity forcing of ocean general circulation models. *J. Phys. Oceanogr. 23*, 1389–1410.

Wright, D. G. and T. F. Stocker (1991). A zonally averaged ocean model for the thermohaline circulation. Part I: Model development and flow dynamics. *J. Phys. Oceanogr. 21*(12), 1713–1724.

Wright, D. G. and T. F. Stocker (1993). Younger Dryas experiments. In W. R. Peltier (Ed.), *Ice in the Climate System*, Volume I 12 of *NATO ASI*, pp. 395–416. Springer Verlag.

Yang, J. and J. D. Neelin (1993). Sea-ice interaction with the thermohaline circulation. *Geophys. Res. Lett. 20*, 217–220.

Zhang, S., R. Greatbatch, and C. A. Lin (1993). A reexamination of the polar halocline catastrophe and implications for coupled ocean-atmosphere modeling. *J. Phys. Oceanogr. 23*, 287–299.

Zhang, S., C. A. Lin, and R. Greatbatch (1995). A decadal oscillation due to the coupling between an ocean circulation model and a thermodynamic sea-ice model. *J. Marine Res. 53*, 79–106.

STEADY STATES AND VARIABILITY IN OCEANIC ZONAL FLOWS

DIRK OLBERS and CHRISTOPH VÖLKER
Alfred-Wegener-Institute for Polar and Marine Research
Bremerhaven, Germany

Contents

1 Introduction

The Antarctic Circumpolar Current (ACC) is the only oceanic flow system of large scale bearing similarity to the atmospheric zonal circulation. There is not only the obvious geometrical similarity – the zonal unboundedness of a current reaching all around the earth – , there are also deeper dynamical correspondences. Though the forcing is different, the dynamical balance of the zonal atmospheric flow and the ACC resides substantially on the excitation of and interaction with synoptic-scale eddies and the intricate correlation of the large-scale pressure field with respect to the underlying topography in shaping what is known as mountain drag in the atmospheric system and bottom form drag (or stress) in the oceanic case.

In fact, the only sound explanation of the momentum budget of the ACC, which has recently been confirmed by numerical experiments with eddy resolving models, is merely a confirmation of a hypothesis which was

NATO ASI Series, Vol. I 44
Decadal Climate Variability
Dynamics and Predictability
Edited by David L. T. Anderson and Jürgen Willebrand
© Springer-Verlag Berlin Heidelberg 1996

put forward almost half a century ago as a narrow parallel to the balance of zonal atmospheric flow by an oceanographer and a meteorologist (Munk and Palmén 1951). In view of the apparent inability of friction at the bottom or lateral coasts to extract the eastward momentum which is generously imparted to the ocean in the ACC belt by the strong surface wind, they suggested that it could flux out of the fluid system into the solid earth by the bottom form stress mechanism. This stress arises from a systematically higher pressure amplitude on the luv compared to the lee side of the submarine ridges blocking the flow at great depths in the Drake Passage and the other oceans which the ACC has to cross. The mechanism can hardly be measured: the pressure difference across a ridge would correspond to only a few centimeters of surface displacement (the role of the barotropic and baroclinic pressure fields in balancing the ACC is however much more complicated, see e. g. Olbers et al. 1992). Experiments by numerical models of the ACC (McWilliams et al. 1978, Wolff and Olbers 1989, Wolff et al. 1991, Marshall et al. 1993, The FRAM Group 1991 and others) have clearly shown that eddies do not transport eastward momentum out of the current system to enable lateral frictional loss. They rather concentrate the jets and – by interfacial form stress – they also help to transport momentum downward to the bottom. Furthermore, the entire flow arranges a pressure field such that the bottom form stress is by far more effective than friction to extract momentum.

This strong resemblance in the balance of forces is encouragement to look for more correspondences of the ACC physics with the physical features discovered for zonal atmospheric flows. Outstanding theoretical issues in this field of atmospheric science are the occurrence of multiple steady states, their stability and their conjectured role in temporal variability. This problem is addressed in the present paper for a simple system of fluid flow which in most of its physical aspects resembles the ACC but is still understandable with analytical treatment.

The concept of multiple equilibria in a severely truncated 'low-order' image (the CdV model) of the atmospheric circulation was put forward by Egger, Charney, Wijn-Nielsen and others in a series of papers (Egger 1978, Charney and DeVore 1979, Wijn-Nielsen 1979, Charney et al. 1981) to explain the variability of atmospheric large-scale flow, in particular the occurrence and transition between blocked and unblocked situations in the midlatitudes of the northern hemisphere. It is appealing to connect these 'Großwetterlagen' with the steady regimes of a low-order subsystem of the atmospheric dynamics such as the CdV model and explain transitions by interaction with shorter waves simply acting as white noise (see e. g. Egger 1981, De Swart and Grassman 1987). The observational evidence for dynamically disjunct multiple states, particularly with features of the CdV model, in the atmospheric circulation is however sparse (see the collection of papers in Benzi et al. 1986) and the applicability has correctly been questioned (see e. g. Tung and Rosenthal 1985). Indications of a double peak distribution of large-scale wave amplitudes have been found, however with no counterpart in the distribution of the mean flow amplitudes. Moreover, on the theoretical side it was found that relaxing the severe truncation and allowing for more modes and wave-wave interactions the multiple equilibria in barotropic models disappeared. They also disappeared for realistic forcing and friction parameters in the original CdV model.

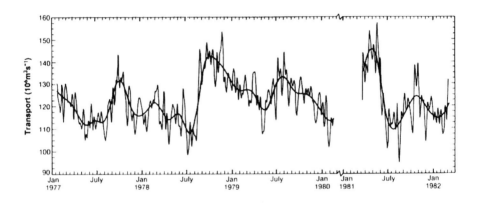

Figure 1: *Net transport through the Drake Passage from January 1977 till February 1980 and March 1981 till March 1982, obtained from bottom pressure measurements (after Whitworth and Peterson (1985)).*

Again, as with the dynamical balance, observing multiple equilibria or variability in the Southern Ocean circulation is far beyond the capability of existing data. There are some hints of interannual variability of the transport of the ACC through the Drake Passage obtained from difference pressure recording across the passage (see Figure 1). The data are however by far not adequate to decide between the different possible causes of temporal variability of a dynamical system: is it driven by variations in the forcing (windstress or buoyancy flux in case of the ACC), is it due to internal instability of the system (structural instability of the large-scale components) or is it caused by the perturbation of phase space trajectories around the stable and unstable steady states by some internally or externally generated noise (e. g. unresolved waves of smaller scale or meso-scale eddies)? These questions must be considered open as well for the anomalous atmospheric circulation in mid-latitudes: the extensive collection of observational examples and theoretical investigations of externally forced and internally generated variability in the atmospheric circulation in Benzi et al. (1986) give the strong impression of lack of adequate data and lack of dynamical concepts to find the causes of variability.

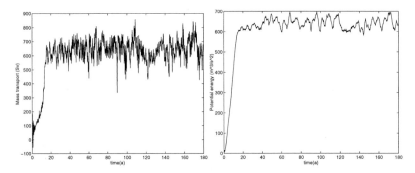

Figure 2: *Time series of transport and potential energy of a simulation of wind forced channel flow with an eddy resolving quasigeostrophic numerical model.*

The same dilemma is present in case of a numerical model where one could 'measure' all data which are needed to decide between the different causes of variability and at least part of the system can be controlled. Figure 2 shows fluctuations of transport and potential energy in a two-layer β-plane channel driven by steady zonal windstress (the model is described in section 4). Externally induced variability is here excluded but to decide between the other causes would certainly require more information than just a few time series, in fact a rather detailed dynamical concept must elaborated to perform statistical tests on these different hypotheses. An attempt along this way is made in the present paper.

The CdV model describes the steady response of a barotropic flow driven by an external stress over a terrain with simple sinusoidal topography. In the linearized system the forced zonal flow excites a Rossby wave with the wavelength of the topography and this wave then feeds back on the flow by bottom form stress. Depending on the structural parameters (forcing amplitude, friction, wavelength and height of topography) the system may bifurcate (the mean flow and the Rossby wave are in resonance) and besides an unstable state two stable states emerge one of which is dominated by friction and the other one is in a balance dominated by bottom form stress. For low forcing or high friction or low topography there is no bifurcation and only the stable frictional state survives. The system has thus the elements of the dynamical balance – forcing, friction and form stress – which we have described above for the ACC. However, barotropic Rossby waves have the same phase speeds in the atmosphere and the ocean and in the later medium the mismatch of the sizes of mean currents and barotropic phase speeds is even more severe: there is no way how such waves can be locked into resonance by the much smaller observed mean flow. A 'barotropic ACC' is thus necessarily in a frictional state, in contrast to high-resolution numerical experiments and – very likely – to reality. This context is outlined in the next section.

We thus proceed to investigate the resonance mechanism in a baroclinic generalization of the CdV system by considering the flow in a two layer quasigeostrophic channel on a β-plane. In contrast to previous baroclinic models (e. g. Charney and Straus 1980) studied in atmospheric context the oceanic parameter range allows simplifications for planetary scale dynamics. The simplest of the low-order models which we have derived has nine degrees of freedom (compared to three in the CdV model or six in the model of Charney and Straus). The model is nonlinear, it includes wave-wave interactions, there is form stress arising from the barotropic and baroclinic waves, but despite of these complexities the model can still be completely solved by analytical means. We are thus able to determine the equilibria and analyse in detail the momentum balance of these states as well as the resonance mechanism of the zonal flow with the baroclinic Rossby waves. We find that the system may have up to three steady states which in certain parts of the parameter domain may all become unstable and generate quite complex temporal behavior. Finally, we compare the structural and temporal features of the low-order model with a more realistic model of zonal channel flow, an eddy resolving quasigeostrophic β-plane model.

2 Topographic resonance in the CdV model

The effect of large-scale topographic obstacles on a zonal flow is investigated in a zonal periodic channel of length X and width Y on a β-plane with Coriolis parameter $f = f_0 + \beta y$. The meridional boundaries at $y = 0, Y$ are rigid and to begin with we are considering a layer of fluid with thickness $H + \zeta - B$ where H is the undisturbed layer height, B is the elevation of the bottom topography and ζ is the elevation of the upper surface. We assume quasigeostrophic (QG) dynamics: the horizontal velocities in the layer and the surface elevation are expressed in terms of the streamfunction Ψ which equals the geostrophic pressure devided by f_0. Thus, in particular, the surface elevation relates to the streamfunction by $\zeta = f_0\Psi/g$. The time evolution is governed by the balance of the quasigeostrophic potential vorticity (QPV)

$$\frac{\partial Q}{\partial t} + J(\Psi, Q) = F \tag{1}$$

where J is the Jacobian and

$$Q = \nabla^2\Psi - \frac{f_0}{H}(\zeta - B) + \beta y \tag{2}$$

is the QPV of the layer (multiplied by H). The forcing and dissipation term F on the rhs of (1) arises from the stress τ at the top and the frictional bottom stress modelled here simply by Newtonian friction. The horizontal momentum transport by unresolved eddies is neglected altogether, so we end up with

$$F = \frac{\partial F^{(y)}}{\partial x} - \frac{\partial F^{(x)}}{\partial y} = \frac{\mathrm{curl}\tau}{H} - \epsilon\nabla^2\Psi \tag{3}$$

The boundary conditions appropriate for the channel geometry require vanishing velocities normal to the walls, i. e. for the geostrophic part $\partial\Psi/\partial x = 0$ at $y = 0, Y$. In addition, a constraint must be considered which determines the difference of the streamfunction values (i. e. the transport) between the northern and southern walls. This is achieved by the physically motivated requirement that the ageostrophic pressure field is uniquely defined. This leads to the condition

$$-\frac{\partial}{\partial t}<\frac{\partial \Psi}{\partial y}>-\frac{\partial}{\partial y}<\frac{\partial \Psi}{\partial x}\frac{\partial \Psi}{\partial y}>=<F^{(x)}> \qquad \text{at} \qquad y=0 \qquad (4)$$

where the angle brackets denote the zonal average. Vanishing of the ageostrophic velocity normal to the wall has been implemented.

This simple QG setting will be generalized to two layers in the next section. The topographic resonance of perturbations on a mean (meridionally constant) geostrophic zonal flow U is however most easily exemplified in this barotropic model.

We separate the flow into a time and zonal mean and a perturbation about this state in the form $(u+U, v), \zeta = \xi + Z$ with $Z = -f_0 U y/g$ (which is summarized by $\Psi = \phi - Uy$). The QPV becomes

$$Q = q + \left(\beta + \frac{U}{R^2}\right) + \frac{f_0}{H}B$$

$$q = \nabla^2 \phi - \frac{f_0}{H}\xi = \left[\nabla^2 - \frac{1}{R^2}\right]\phi \tag{5}$$

where q is the wave part and $R = \sqrt{gH}/f_0$ is the external Rossby radius of deformation. The QPV balance

$$\left(\frac{\partial}{\partial t}+U\frac{\partial}{\partial x}\right)q + \mathcal{J}(\phi, q) + \mathcal{J}\left(\phi, \frac{f_0 B}{H}\right) + \left(\beta + \frac{U}{R^2}\right)\frac{\partial \phi}{\partial x} + \frac{f_0 U}{H}\frac{\partial B}{\partial x} = F \tag{6}$$

yields the dispersion relation

$$\omega = -k\frac{\beta - K^2 U}{K^2 + R^{-2}} = kU - k\frac{\beta + UR^{-2}}{K^2 + R^{-2}} \tag{7}$$

for free linear waves with the wave vector $\boldsymbol{k} = (k, l)$ and $K^2 = k^2 + l^2$. Notice the so-called *non-Doppler* effect: the advection of the perturbation stretching vorticity by the mean flow cancels the advection of the mean stretching vorticity by the perturbed flow. Stationary waves do exist for eastward currents ($U > 0$) and due to the non-Doppler effect they occur for $U = \beta/K^2$ rather than for U equal to the phase velocity of the Rossby wave in the reference frame of the mean flow. If the short-wave limit $(KR)^2 \gg 1$ is applicable this distinction disappears.

The stationary response in the wave field to zonal flow over topography is governed in the linearized regime by

$$U\frac{\partial}{\partial x}\nabla^2\phi + \beta\frac{\partial \phi}{\partial x} + \epsilon\nabla^2\phi = -\frac{f_0 U}{H}\frac{\partial B}{\partial x} + \frac{\text{curl}\tau}{H} \qquad (8)$$

Here the linear term $\mathcal{J}(\phi, f_0B/H)$ describing the effect of the anomalous flow over the topography is neglected to keep the analysis simple (it is formally $O(B^2)$). The stationary wave response is actually not influenced by the free surface, hence one could use the rigid lid condition $1/R^2 \to 0$ to yield identical results. The amplitude of the wave component $\phi = \phi_{\boldsymbol{k}}\exp i(\boldsymbol{k}\cdot\boldsymbol{x}) + c.c.$ is found to be

$$\phi_{\boldsymbol{k}} = \frac{f_0 U}{H}\frac{k b_{\boldsymbol{k}}/K^2}{k\left[U - \beta/K^2\right] - i\epsilon} \qquad (9)$$

provided the stress τ is zonally constant so that there is no directly forced contribution. For the sinusoidal topography elevation

$$B(x,y) = b_0 \sin\frac{2\pi x}{X}\sin\frac{\pi y}{Y} \qquad (10)$$

which we shall use in this investigation we have $\boldsymbol{k} = (2\pi/X, \pi/Y)$ and $b_{\boldsymbol{k}} = -ib_0/2$. The resonance in the wave amplitude (9) at $U = \beta/K^2$ was first pointed out in the classical paper of Charney and Eliassen (1949). It occurs when the mean advection of relative vorticity (to the east) cancels the advection of the planetary vorticity by the wave perturbation (the Rossby wave propagation to the west).

The feedback of the topographically induced wave pattern on the mean flow occurs via the stress exerted by the waves. The balance of the zonal flow is formally obtained by integration of the QPV balance (6) over the area bounded by the southern wall and a latitude y, utilizing the momentum constraint (4). This procedure recovers the zonally averaged zonal momentum balance, written here for the steady state

$$- < \frac{\partial \phi}{\partial x} \nabla^2 \phi > - \frac{f_0}{H} < \frac{\partial \phi}{\partial x} B > = < F^{(x)} > \tag{11}$$

Notice that again there is no effect of the free surface. The balance may alternatively be expressed in terms of the meridional QPV flux $< vq >$ and the bottom form stress $-f_0/H < vB >$ induced by the resonant waves

$$- < vq > = \frac{\partial}{\partial y} < uv > + \frac{f_0^2}{H} < v\xi > = \frac{f_0}{H} < vB > + < F^{(x)} > \tag{12}$$

Notice that the stretching term $< v\xi > = 0$ in this barotropic case. Finally, with

$$< F^{(x)} > = -\epsilon \left[U - < \frac{\partial \phi}{\partial y} > \right] + \frac{\tau^{(x)}}{H} \tag{13}$$

and integration across the channel to derive the net momentum balance we arrive at the barotropic CdV model

$$\frac{\tau^{(x)}}{H} - \epsilon U - S[U] = 0 \tag{14}$$

where $S[U]$ is the net form stress

$$
\begin{aligned}
S[U] &= -\frac{f_0}{HY} \int_0^Y dy < vB > = \frac{f_0}{H} 2k\Im < \phi_{\boldsymbol{k}} b_{\boldsymbol{k}}^* > = \\
&= \frac{1}{2} \left(\frac{f_0 b_0}{H} \right)^2 \frac{\epsilon U (k/K)^2}{k^2 [U - \beta/K^2]^2 + \epsilon^2}
\end{aligned}
\tag{15}
$$

acting in the channel. The sign convention is here such that $S > 0$ corresponds to a loss of eastward momentum. The existence of multiple solutions for U is evident from the graphical display of the two curves (14) and (15) as shown in Figure 3.

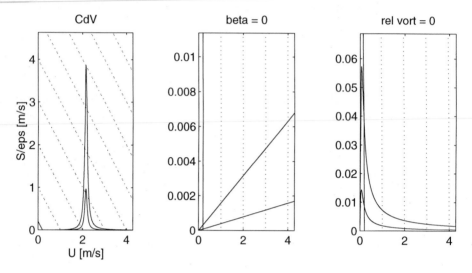

Figure 3: *Graphical display of the CdV model showing S/ϵ vs U for oceanic conditions. The parameter values are appropriate for $55°$ latitude and $\epsilon = 10^{-7}s^{-1}$, $H = 4000m$, $X = 4000km$, $Y = 1500km$. The form stress is plotted for $b_0 = 400m$ and $b_0 = 800m$. The left panel applies to the β-plane and various values for τ are used: the leftmost line represents the momentum balance for typical ACC conditions ($\tau = 10^{-4}m^2s^{-2}$, solid line), for the other lines τ is increased (to unrealistic values) by a factor of $5, 10 \cdots 40$ (dashed lines), respectively. The middle panel is for the f-plane ($\beta = 0$), the right panel for a β-plane with neglection of the relative vorticity. Notice the difference in the S/ϵ-axes.*

Several important properties are worth mentioning. At first v and B are exactly out of phase for zero friction and there is no form stress in this case. Of course, the flow would then not become steady. For zero friction the Reynolds stress $< uv >$ vanishes as well. This is a special case of the Eliassen-Palm theorem. Thus, friction plays a twofold role in the balance of the mean zonal flow: there is a direct frictional effect on U, manifested here by the bottom friction, and an indirect effect through the feedback by the topographically induced waves where friction enables to build up the phase shift and generate bottom form stress. The resonance of the wave amplitude (9) carries over to the form stress. Thus, if the relative vorticity is neglected (by a scale argument $L^2 \gg R^2$ which is however not appropriate for a barotropic model), the resonance disappears and the momentum balance of this barotropic model is linear in U. Then multiple steady states do not exist. The resonance disappears as well when the Rossby waves are filtered (f-plane approximation). In these cases the form stress is very small compared to the forcing and the flow is directly controlled by the bottom friction: we find $U \cong \tau/(\epsilon H)$ (notice that we have plotted $S[U]/\epsilon$ vs U in the Figure 3 so that both intersections of the straight lines with the axes are equal to the frictional solution $\tau/(\epsilon H)$).

Significant sizes of the form stress can thus only arise if the topography is sufficiently high, Rossby waves propagation is present and the forcing is sufficiently strong. The three possible steady states which then exist can be classified according to the size of the mean flow U compared to the wave amplitudes: the *high zonal index regime* is frictionally controlled, the flow is intense and and the wave amplitude is low; the *low zonal index regime* is controlled by form stress, the mean flow is weak and the wave is intense. The intermediate state is transitional, it is actually unstable to perturbations. This 'form drag instability' works obviously when the slope of the resonance curve is below the one associated with friction ($\epsilon + \partial S/\partial U > 0$, see Figure 3) so that a perturbation must run away from the steady state. Apparently the criterion of instability is always satisfied for the intermediate state whereas the other two states are always stable.

In realistic atmospheric applications of the CdV model the parameter window (topographic height, forcing and friction parameters) for multiple solutions is quite narrow, maybe even nonexisting for more complex topographies (see e. g. the discussion and results of Tung and Rosenthal 1985). The most important feature revealed in the oceanic parameter range, as displayed in Figure 3, is the absence of multiple states for realistic values of the forcing τ and damping rate ϵ. Oceanic velocities are always considerably smaller than the barotropic Rossby wave speed at relevant topographic wavelengths, hence the resonance cannot be met and a barotropic model is necessarily in the frictionally controlled state (see e. g. Völker 1991). A different regime can be found when the topography allows for blocked geostrophic contours (Krupitsky and Cane 1994) which however is questionable for simple geometrical reasons: f/H-contours connecting the walls of a channel are an artifact of the model setup.

3   A large scale baroclinic CdV model

The above comments are a motivation to transfer the CdV model to oceanic stratified conditions where baroclinic Rossby waves are present and a resonance could be build up at the baroclinic Rossby wave speed βR_{int}^2 which is of order of a few $cm\ s^{-1}$ (the internal Rossby radius R_{int} is typically $10km$ in the Southern Ocean). A baroclinic successor of CdV was constructed by Charney and Straus (1980) using a two-layer QG model. For the most part this model defies however analytical treatment and because of the mismatch in the parameter ranges for atmospheric and oceanic conditions it cannot readily be transferred to meet our purposes.

3.1   QG planetary scale dynamics

The CdV concept is readily extended to a baroclinic situation and – with some considerations of planetary scale dynamics – cast into a form which allows complete analytical treatment. We consider now a stable density stratification of the fluid given by two homogeneous layers with reduced gravity g', thicknesses $H_1 + \zeta - \eta$ and $H_2 + \eta - B$ and streamfunctions $\Psi_i\ (i = 1, 2)$. The QPVs are given by

$$Q_1 = \nabla^2 \Psi_1 - \frac{f_0}{H_1}(\zeta - \eta) + \beta y$$

$$Q_2 = \nabla^2 \Psi_2 - \frac{f_0}{H_2}(\eta - B) + \beta y$$

(16)

with $\zeta = f_0 \Psi_1/g$ as above and $\eta = f_0(\Psi_2 - \Psi_1)/g'$ for the interface elevation. The setup is indicated in Figure 4.

The QPV balance (1) is now formulated for two layers with forcing and dissipation terms F_i which must include interfacial friction parameterizing the vertical transfer of momentum by small scale processes (meso-scale eddies) that are not resolved in the analytical model. Neglecting the horizontal momentum transport by these eddies we have

$$F_1 = \frac{\mathrm{curl}\tau}{H_1} - \mu \nabla^2 (\Psi_1 - \Psi_2)$$

(17)

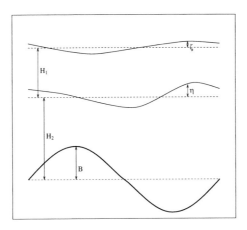

Figure 4: *Sketch of the two-layer model. There is a density jump $\rho_1 - \rho_2$ across the interface of the layers giving rise to a reduced gravity $g' = g(\rho_2 - \rho_1)/\rho_2$.*

$$F_2 = \mu h \nabla^2 \left(\Psi_1 - \Psi_2\right) - \epsilon \nabla^2 \Psi_2$$

where $h = H_1/H_2$ is the ratio of the mean layer heights. The coefficient μ describes the interfacial friction in the momentum balance of the two layers.

We will focus on scales L of motion which are large compared to the internal Rossby radius $R_{int} = (g' H_1 H_2 / f_0^2 (H_1 + H_2))^{\frac{1}{2}}$. With $L^2 \gg g' H_i / f_0^2$ the relative vorticity $\nabla^2 \Psi_i$ is much smaller than the stretching term proportional to η and, because of $R^2 \gg R_{int}^2$, the upper surface may be replaced by a rigid lid. Hence, we may further simplify the dynamics by neglecting the relative vorticity and the surface stretching term, which leads to the large scale QPVs

$$Q_1 = -\frac{f_0^2}{g' H_1} \left(\Psi_1 - \Psi_2\right) + \beta y$$

$$Q_2 = \frac{f_0^2}{g' H_2} \left(\Psi_1 - \Psi_2\right) + \frac{f_0}{H_2} B + \beta y$$

(18)

This approximation implies an infinite phase velocity of barotropic Rossby waves, hence the barotropic part of the flow reacts instantaneously to changes in the slowly varying baroclinic part. We anticipate therefore a diagnostic balance for the barotropic QPV. In addition, the short baroclinic Rossby waves are filtered out and only the long nondispersive waves with phase speed βR_{int}^2 are retained.

We transform to barotropic and baroclinic streamfunctions Φ and Θ, defined by

$$\Psi_1 = \Phi + \Theta \qquad \Phi = \frac{1}{1+h}(h\Psi_1 + \Psi_2)$$

$$\Psi_2 = \Phi - h\Theta \qquad \Theta = \frac{1}{1+h}(\Psi_1 - \Psi_2)$$

(19)

The corresponding barotropic and baroclinic QPV balances will be given in dimensionless form. We define a timescale $1/|f_0|$ and a lengthscale $L = Y/\pi$ and scale the dependent and independent variables according to

$$t' = t|f_0| \qquad (x', y') = \frac{(x, y)}{L} \qquad (\Phi', \Theta') = \frac{1}{f_0^2 L^2}(\Phi, \Theta)$$

The dimensionless width of the channel is then π and the length is $2\pi\ell$ where $\ell = X/2Y$ is a ratio of the channel dimensions. We also define dimensionless quantities by

$$B' = B/H \qquad \boldsymbol{\tau}' = \boldsymbol{\tau}/(f_0^2 LH) \qquad \beta' = L\beta/|f_0|$$

$$\epsilon' = \epsilon/|f_0| \qquad \mu' = \mu/|f_0| \qquad \sigma' = (R_{int}/L)^2$$

and omit the primes in the following for brevity. The barotropic and baroclinic QPV balances then take the form

$$\beta \frac{\partial \Phi}{\partial x} - J(\Phi - h\Theta, B) = \text{curl}\boldsymbol{\tau} - \frac{\epsilon}{1+h}\nabla^2(\Phi - h\Theta) \quad (20)$$

$$-\frac{1}{\sigma}\left(\frac{\partial \Theta}{\partial t} + J(\Phi, \Theta)\right) + \beta\frac{\partial}{\partial x}(\Phi + \Theta) = \frac{1+h}{h}\text{curl}\boldsymbol{\tau} - (1+h)\mu\nabla^2\Theta \quad (21)$$

The barotropic balance (20) is diagnostic and linear. These noteworthy properties are a consequence of the neglection of the relative vorticity.

The sign of the topography term is here given for the southern hemisphere where f_0 is negative. Equation (21) is the vorticity balance for the upper layer, expressed in terms of the barotropic and baroclinic streamfunctions.

Kinematic boundary conditions appropriate to the channel geometry are

$$\Phi(x,y) = \Phi(x + 2\pi\ell, y) \qquad \left.\frac{\partial \Phi}{\partial x}\right|_{y=0,\pi} = 0 \qquad (22)$$

and correspondingly for Θ. Notice that with neglection of lateral Reynolds stresses there are no dynamical boundary conditions but because there are two layers and a multiply connected domain, four auxiliary conditions are needed to determine the values of the streamfunctions at the northern and southern channel walls. An appropriate choice is

$$\Phi|_{y=0} = 0 \qquad (23)$$

$$\int_0^{2\pi\ell} \int_0^\pi \Theta \, dy \, dx = 0 \qquad (24)$$

$$\oint_0^{2\pi\ell} \left[\frac{\epsilon}{1+h}\frac{\partial}{\partial y}(\Phi - h\Theta) + \tau^{(x)}\right]_{y=0} dx = 0 \qquad (25)$$

$$\oint_0^{2\pi\ell} \left[\mu\frac{\partial \Theta}{\partial y} + \frac{\tau^{(x)}}{h}\right]_{y=0} dx = 0 \qquad (26)$$

The first of these conditions is an arbitrary setting using the freedom to add a constant to all streamfunctions without changes in the physical variables. The second condition guarantees that the layers do not exchange mass. The remaining two conditions correspond to (4), they are integrals of the zonal ageostrophic momentum balance along the southern boundary and guarantee the existence of a continuous ageostrophic pressure field.

3.2 The low-order model

We now derive the equations of a low-order-approximation of (20) and (21) for the case of a sinusoidal bottom topography (10) and a zonal windstress of the form

$$\tau^{(x)} = \tau_0 \sin y \qquad (27)$$

We follow the general procedure for deriving low-order-models and expand the streamfunctions into an orthogonal system of basis functions and then truncate the resulting system of equations by setting most of the expansion coefficients to zero. In channel geometry the appropriate basis functions are

$$F_{mn}(x, y) = \exp(imx/\ell) \sin(ny) \tag{28}$$

where m is an integer and n a positive integer. They are eigenfunctions of the operators ∇^2, $\frac{\partial}{\partial x}$ and $\frac{\partial}{\partial y}$ which appear in the QPV balance, in particular

$$\nabla^2 F_{mn} = -\left(\frac{m^2}{\ell^2} + n^2\right) F_{mn} =: -a_{mn} F_{mn}$$

The F_{mn} vanish on the channel boundaries but the streamfunctions do not vanish. Due to the kinematic boundary conditions they are equal to time dependent constants along the channel walls. Therefore we must add the two functions $F_0(x, y) = 1$ and $F_1(x, y) = y$ that are not orthogonal to the F_{mn}. This guarantees an uniform convergence of the expansion

$$\Phi(x, y, t) = -U(t)y + \sum_{n=1}^{\infty} \sum_{m=-\infty}^{\infty} \phi_{mn}(t) F_{mn}(x, y)$$

$$\tag{29}$$

$$\Theta(x, y, t) = -V(t)y + c(t) + \sum_{n=1}^{\infty} \sum_{m=-\infty}^{\infty} \theta_{mn}(t) F_{mn}(x, y)$$

The U and V parts define the constant (in y) shear (in z) flow. Due to reality of the streamfunctions the wave amplitudes satisfy

$$\phi_{-m,n} = \phi_{m,n}^* \quad \text{and} \quad \theta_{-m,n} = \theta_{m,n}^*.$$

The expansion automatically satisfies the boundary conditions (22) and the auxiliary condition (23).

The smallest possible truncation of the expansion (29) that still retains some nonlinear interactions between the modes and coupling to the topography is a triangular one that includes only the wavenumbers $(m, n) = (0, 1), (0, 2), (1, 1), (-1, 1)$. The dynamics of the mode $(0, 1)$ is trivial and uncoupled with the other modes, so we end up with a smallest model that contains the nine degrees of freedom $\phi_{0,2}, \phi_{1,1}, \theta_{0,2}, \theta_{1,1}, U, V$ and c (remember that ϕ_{11} and θ_{11} are complex). An extended model with 21 degrees of freedom – which however is not accessible by analytical means – is presented in Völker (1995).

The model equations are now obtained by projection. We simplify the notation by defining some parameters

$$\epsilon_{mn} = \epsilon \frac{a_{mn}}{1 + h} \qquad \mu_{mn} = \mu(1 + h)a_{mn} \qquad \tau_{02} = \frac{2\tau_0}{3\pi} \qquad \gamma = \beta\sigma$$

The ϵ_{mn} and μ_{mn} are geometrically modified friction parameters and γ is the phase speed of the baroclinic Rossby waves. By projecting the barotropic balance (20) on the wavenumbers $(0, 2)$ and $(1, 1)$ we obtain three barotropic equations (the first is real, the second is complex)

$$0 = -\epsilon_{02}(\phi_{02} - h\theta_{02}) + \frac{b_0}{\ell}\Re[\phi_{11} - h\theta_{11}] + 4\tau_{02} \tag{30}$$

$$0 = -\epsilon_{11}(\phi_{11} - h\theta_{11}) + i\frac{\beta}{\ell}\phi_{11} - \frac{b_0}{2\ell}[\phi_{02} + U - h(\theta_{02} + V)] \tag{31}$$

where $b_{11} = -ib_0/2$ has been used for the expansion coefficient of the topography. In the same way we obtain from the baroclinic balance (21) the following three equations

$$\frac{1}{\sigma}\frac{d}{dt}(\theta_{02} + V) = -\mu_{02}\theta_{02} + \frac{2}{\ell\sigma}\Im[\phi_{11}\theta_{11}^*] + 4\frac{1 + h}{h}\tau_{02} \tag{32}$$

$$\frac{1}{\sigma}\frac{d}{dt}\theta_{11} = -\mu_{11}\theta_{11} + i\frac{\gamma}{\ell\sigma}(\phi_{11} + \theta_{11}) -$$

$$- \frac{i}{\ell\sigma}[(\phi_{02} + U)\theta_{11} - \phi_{11}(\theta_{02} + V)] \tag{33}$$

The time derivative of V in (32) appears due to the nonorthogonality of y and the F_{mn}. The missing three equations can be derived from the auxiliary conditions (24) to (26). The latter simplify due to vanishing of windstress at the channel boundaries. Insertion of the low-order-expansion results in

$$c - \frac{\pi}{2}V = 0 \tag{34}$$

$$U - 2\phi_{02} - \frac{\tau_{02}}{\epsilon}\left[1 + h + \frac{\epsilon}{\mu}\right] = 0 \tag{35}$$

$$V - 2\theta_{02} - \frac{\tau_{02}}{h\mu} = 0 \tag{36}$$

The last equation ensures the vanishing of the area integral of interface elevation η. The other two equations have been derived with an additional assumption: the forcing $curl\tau = -\tau_0 \cos y$ projects on all modes with $m = 0$ and even meridional wavenumber n. We assume, that for all modes with $n \geq 4$ this forcing is balanced by interfacial and bottom friction. This assumption leads in (35) and (36) to the term proportional to τ_{02}. This procedure yields a slightly more precise representation of the forcing than suggested by strict application of the truncation which otherwise would omit a substantial fraction of the wind forcing. If we assumed a windstress of the form $\tau^{(x)} = \tau_0 \sin^2 y$ which only projects on $(m, n) = (0, 2)$ the terms $\sim \tau_{02}$ would vanish.

In contrast to the barotropic case studied in part 2, the zonal mean velocity $U - \phi_{02} \cos 2y$ of the baroclinic low-order model varies with latitude. But (35) and (36) immediately show that the zonal mean fields are completely specified by U and V. The balance of momentum is easily derived from (30), (32), (35) and (36). The balances of the total momentum (both layers) and the momentum of the upper layer are

$$0 = 3\tau_0 - \frac{\epsilon}{1+h}(U - hV) + \frac{b_0}{2\ell}\Re\left[\phi_{11} - h\theta_{11}\right] \tag{37}$$

$$\frac{h}{(1+h)\sigma}\frac{3}{4}\frac{dV}{dt} = 3\tau_0 - \mu hV + \frac{h}{\ell(1+h)}\Im\left[\phi_{11}\theta_{11}^*\right] \tag{38}$$

These equations express the balance between the momentum input by windstress at the surface with friction either at the bottom (proportional to the lower layer zonal velocity $U - hV$) or at the interface (proportional to the baroclinic velocity V) and form stresses caused by the waves. There is a barotropic ($\sim \phi_{11}$) and a baroclinic ($\sim \theta_{11}$) bottom form stress arising from the $(1,1)$-component of the lower layer pressure. In the second equation the interfacial form stress ($\sim \phi_{11}\theta_{11}^*$) appears as correlation of the surface pressure and the elevation of the interface induced by the $(1,1)$-wave.

3.3 Steady states, stability and baroclinic resonance

A formal variable elimination in the steady versions of (30) to (36) leads to a third-order polynomial in any of the variables with coefficients which are complicated functions of the parameters. We will not present the tedious procedure to determine the analytical solution of the steady states. Rather we will try to highlight the analogies and differences to the barotropic CdV model. To keep the expressions as simple as possible we consider here only the special case $h = \ell = 1$, i. e. a channel with equal layer thicknesses and a ratio 2 of length to width.

As in the barotropic case we express the wave components ϕ_{11} and θ_{11} in terms of the zonal mean flow, using the projection of the barotropic and baroclinic vorticity balances (31) and (33) onto the wavenumber (1,1). This is most easily achieved by representing ϕ_{11} and θ_{11} in terms of amplitudes and phase shifts relative to the topography, i.e.

$$\phi_{11} = \mathcal{P}e^{-i(\varphi + \pi/2)} \qquad \theta_{11} = \mathcal{T}e^{-i(\vartheta + \pi/2)} \tag{39}$$

The phase shift of the barotropic streamfunction (pressure) with respect to the topography is thus given by φ and correspondingly ϑ is the shift of the the baroclinic streamfunction. The steady balances of the zonal barotropic and baroclinic momentum of the system then take the simple form

$$0 = 6\tau_{02} - \epsilon(U - V) + b_0(\mathcal{P}\sin\varphi - \mathcal{T}\sin\vartheta) \tag{40}$$

$$0 = 6\tau_{02} - 2\mu V + \frac{1}{\sigma}\mathcal{P}\mathcal{T}\sin(\vartheta - \varphi) \tag{41}$$

The equations reflect that nonzero bottom form stress can only be established if either of the phase shifts φ or ϑ are nonzero. Nonvanishing interfacial stress requires that the barotropic and baroclinic waves are out of phase, i. e. $\varphi \neq \vartheta$.

With the shorthand notations

$$\mathcal{U} = U + \phi_{02} \qquad\qquad \mathcal{V} = V + \theta_{02}$$
$$\mathcal{A} = 1 - \frac{(\mathcal{U} - \gamma)(\mathcal{V} + \gamma)}{(4\mu\sigma)^2 + (\mathcal{U} - \gamma)^2} \tag{42}$$
$$\mathcal{B} = \beta + \epsilon\frac{(4\mu\sigma)(\mathcal{V} + \gamma)}{(4\mu\sigma)^2 + (\mathcal{U} - \gamma)^2}$$

and ϕ_{02} and θ_{02} determined from (35) and (36) we find the wave phases and amplitudes

$$\tan\varphi \;=\; \epsilon\,\frac{\mathcal{A}}{\mathcal{B}} \tag{43}$$
$$\mathcal{P} \;=\; \frac{b_0|\mathcal{U} - \mathcal{V}|}{2\mathcal{B}}\,\cos\varphi \tag{44}$$

from the barotropic balance (31) and

$$\tan(\vartheta - \varphi) \;=\; -\frac{4\mu\sigma}{\mathcal{U} - \gamma} \tag{45}$$
$$\mathcal{T} \;=\; \mathcal{P}\,\frac{|\mathcal{V} + \gamma)|}{\sqrt{(4\mu\sigma)^2 + (\mathcal{U} - \gamma)^2}} \tag{46}$$

from the baroclinic balance (33). Without going into more details these relations obviously reveal similar properties of the wave induced stresses as the discussed in section 2 for the barotropic CdV model. There is no form stress for vanishing friction. The barotropic part $\mathcal{P}\sin\varphi$ of the bottom from stress vanishes for zero bottom friction: for $\epsilon \to 0$ we have $\varphi \to 0$ from (43) but $\mathcal{P}$ remains finite. The baroclinic part $\mathcal{T}\sin\vartheta$ vanishes if bottom and interfacial friction vanish whereas the interfacial form stress vanishes for zero interfacial friction. Hence, in accordance with the Eliassen–Palm theorem, it is only through frictional processes that the induced waves feed back on the mean flow.

Resonance of the zonal mean flow with the wave field occurs when $\mathcal{U}$ equals the baroclinic Rossby wave phase speed γ. This modification of the linear theory (where the resonance is at $U = \gamma$) is due to the inclusion of the nonlinear interaction with the mode $(0,2)$. Notice that there is only a resonant behavior in the baroclinic amplitude $\mathcal{T}$.

Expressing the interfacial and bottom form stresses as functions of U and V by use of (42) to (46), the two momentum equations (40) and (41) become implicit functions defining curves in (U,V)-space. Their intersection points determine the steady states. Figure 5 displays the two curves for various values of the topography height. There are five straight lines in these plots with a specific physical meaning. The upper horizontal line reflects a balance between the wind input and interfacial friction in the baroclinic momentum, i. e. $6\tau_{02} - 2\mu V = 0$. On the lower horizontal line we have $V + \gamma = 0$, hence the baroclinic wave amplitude $\mathcal{T}$ vanishes. In addition, there is a change of sign in the interfacial form stress when this line is crossed: below we have a source of eastward momentum and above we have a sink. Thus for eastward windstress the steady states must lie above. The right diagonal line characterizes the balance between wind input and bottom friction in the barotropic balance, i. e. $6\tau_{02} - \epsilon(U - V) = 0$. On the left diagonal line we have $\mathcal{U} = V$ so both wave amplitudes vanish. The resonance $\mathcal{U} = \gamma$ of the mean zonal flow with the Rossby waves is on the vertical line.

For vanishing topography both amplitudes $\mathcal{P}$ and $\mathcal{T}$ vanish by (44) and (46) and only frictional processes can balance the momentum input. The solution curves are both straight lines. With increasing topography height the resonance near $\mathcal{U} - \gamma$ leads to high amplitudes of $\mathcal{T}$ and more complicated curve forms appear: the curve associated with the barotropic balance (40) bends with increasing height and at a certain height (which depends on the other parameters) a closed second part of the curve appears near the resonance line $\mathcal{U} = \gamma$. The extent of this isolated curve increases with increasing topography until the two parts join. In parallel, with increasing height, the baroclinic momentum balance (41) develops a resonance reaching down on the line $\mathcal{U} = \gamma$. In this way, there is a succession from one to three to two steady states when the topography is increased. These properties are displayed in the Figures 5 and 6 for two sets of parameters. Figure 5, using the unrealistic parameters $\tau_0 = 3 \times 10^{-5} m^2 s^{-1}$, $\beta = 2 \times 1.471421 \times 10^{-11} m^{-1} s^{-1}$, makes the structure more obvious and Figure 6 applies the standard ACC parameters.

428

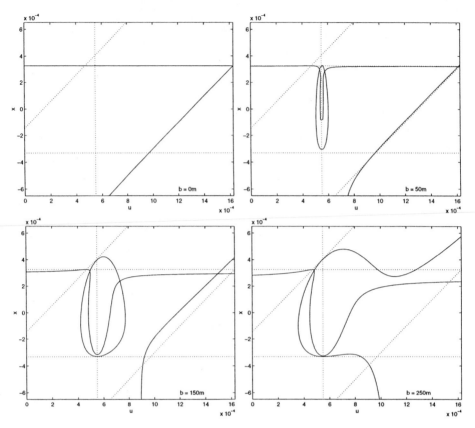

Figure 5: *The curves determined by the momentum balances (40) and (41) in U, V-space, for various values of the topography height. Parameters are $\tau_0 = 3 \times 10^{-5} m^2 s^{-2}$, $\beta = 2 \times 1.471421 \times 10^{-11} m^{-1} s^{-1}$.*

As in the barotropic model we may classify the steady states according to the role of form stress and frictional processes in the two momentum balances. Figure 7 gives the sizes of the frictional terms in (40) and (41) with varying topography height. The *low-index regime* arises from the intersection of the curves in Figure 5 in the upper right corner. It is mainly frictionally controlled, with high zonal velocities in both layers and low amplitudes of the wave disturbances. The *high index regime* has a barotropic zonal velocity near the baroclinic resonance condition $\mathcal{U} = \gamma$ and consequently a high amplitude of the baroclinic wave amplitude. In their balance, interfacial and bottom friction have only a minor role. The intermediate state is also resonant but always unstable.

The baroclinic resonance mechanism differs from the barotropic resonance in some aspects. Two important differences are:

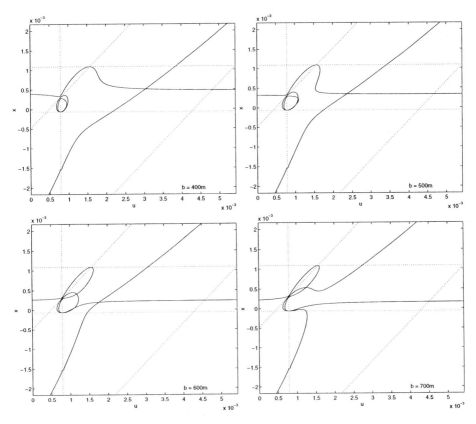

Figure 6: *Same as Figure 5 but the parameters are* $\tau_0 = 10^{-4}m^2s^{-1}$, $\beta = 1.471421 \times$ $10^{-11}m^{-1}s^{-2}$, *which are relevant to ACC conditions.*

- In the CdV model the bifurcation diagram of the steady states consists of the two branches of high- and low-index states which are always stable, joined by a branch of an intermediate state which is always unstable. In the baroclinic case the high- and low-index states may also become unstable by an oscillatory instability (i. e. by a Hopf-bifurcation). This situation is indicated in Figure 7 by a dotted line. While the intermediate branch is always unstable, the other two branches loose their stability well before the saddle-node bifurcation joining the branches. There is a window in the parameter space where all of the three steady states are in fact unstable.

- In the barotropic case the resonance depends on the lengthscale of the topography since the barotropic Rossby waves are dispersive. As a consequence, realistic topography with a continuum of Fourier amplitudes tends to produce very broad resonance curves which may inhibit the existence of multiple steady states (see e. g. Tung and Rosenthal

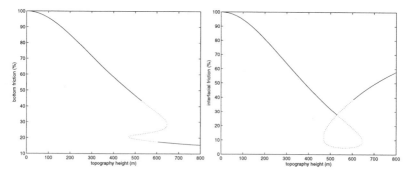

Figure 7: *Size of the frictional terms in the two momentum balances (37) and (38) with varying of topography height. The frictional terms are normalized by the momentum input by windstress and are given in %. Parameters are $h = 1, \ell = 1, g'/g = 0.001$ and otherwise the standard ACC values. The figures display at the same time the other terms of the momentum balances: the wind input is the 100% line and the form stress is the difference between wind and friction.*

1985). In contrast, the large-scale baroclinic Rossby waves are non-dispersive and hence the resonance velocities for different lengthscales coincide. We have calculated the resonance for a topography with a Gaussian shape in the zonal direction and found almost no difference to the sinusoidal topography concerning the existence of multiple steady states and their dependence on topography height.

3.4   Time dependence

As mentioned above, there exists a connected region in parameter space where none of the three coexisting steady states is stable. Parameter values in this region lead to a complex temporal behavior of the system. A numerical integration of the equations (30) to (36) is presented in the Figures 8 and 9. The phase plane picture of θ_{02} against $\Re\theta_{11}$ exemplifies the typical form of the attractor existing in this parameter region.

The orbits on this attractor are almost periodic, i. e. a typical succession of events is repeated. In a first phase the system is near an unstable steady state which is located in the center of the outward spiralling orbits in the left part of the phase plane picture. This behavior is caused by instability of the steady state exciting a baroclinic wave which propagates several times through the channel with increasing amplitude. This wave modifies the momentum balance via the form stress mechanism and thus

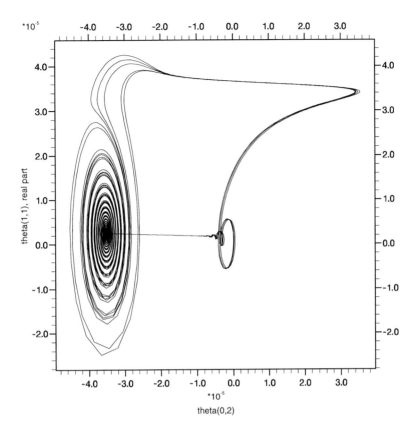

Figure 8: *Phase plane picture of the time-dependent behavior in a parameter region where none of the three steady states is stable. The topography amplitude is* $b_0 = 538\,m$, *the two layers have equal mean heights* $2500\,m$, *the reduced gravity is* $g'/g = 0.001$ *and the amplitude of the windstress is* $\tau_0 = 10^{-4} m^2 s^{-1}$.

leads to periodic oscillations of the zonal velocities. The amplitude of this oscillation then reaches a level where the excitation of the Rossby wave is stopped because the zonal velocity in the lower layer becomes to small. The time span of this Rossby wave phase of the attractor is thus few propagation times, of order $\gamma\ell$. The orbits then get attracted toward another unstable steady state located in the upper right corner of the phase plane picture. But this steady state is only halfway circled, the orbits finally get attracted by first unstable steady state to begin a new cycle. This transition phase is the longest lasting stage of the whole cycle; its timescale is determined by the baroclinic spin-down time $(\sigma\mu)^{-1}$.

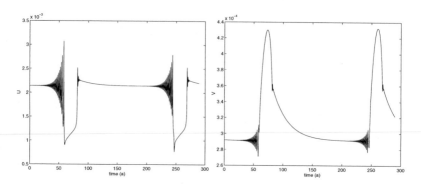

Figure 9: *Time series of two of the variables for the same parameter values as for Figure 8. The variables shown are the dimensionless values of the barotropic and baroclinic mean zonal velocities U and V.*

The existence of this complex attractor is linked to a homoclinic bifurcation of a type described first by L.P. Shilnikov. For certain parameter values a homoclinic orbit exists which connects the stable and the unstable manifold of the unstable steady state in the center of the oscillatory phase. This homoclinic orbit is the limiting case of an unstable periodic orbit that bifurcates in a Hopf-bifurcation from the low-index or from the high-index state. It can be shown (Guckenheimer and Holmes 1993) that depending on the ratio of the eigenvalues of the Jacobi-matrix of the steady state this homoclinic orbit can be accompanied by quite complex dynamical behavior, including the existence of horseshoe-maps in the flow and of infinitely many periodic orbits of arbitrary high period.

4 Comparison with high-resolution numerical models of zonal channel flow

Finally we want a comparison of the low-order model with a system which is closer to reality than this model with extremely reduced physics. Since there are no adequate observational data to determine possible multiple equilibria or temporal variability of the ACC we have to build a less realistic 'strawman' to compare with, i. e. we will use another model. Of course, the eddy resolving models recently constructed for the Southern Ocean, namely the FRAM experiment (The FRAM Group 1991) and the QG-Southern-Ocean-Model (Olbers 1993, Marshall et al. 1993), offer their service but a detailed comparison is certainly doomed to fail: these models with realistic geometry, topography and elaborate physics are far too complex. Instead we have performed numerical experiments with quasi-geostrophic dynamics in the same geometry as the low-order image. The same forcing, topography and stratification were chosen, the code is essentially the one of McWilliams et al. (1978). The numerical model solves the complete QPV equations (i. e. with rigid-lid approximation but inclusion of the relative vorticity) on a regular grid with a gridsize of $20km$. Since the model thus resolves most of the turbulent baroclinic activity of the flow there is no need for a interfacial friction-type parametrization. The forcing terms of the numerical model

$$\begin{aligned} F_1 &= \frac{curl\tau}{H_1} - A_H\nabla^6\Psi_1 \\ F_2 &= -\epsilon\nabla^2\Psi_2 - A_H\nabla^6\Psi_2 \end{aligned}$$

contain a weak biharmonic friction for numerical reasons ($A_H = 10^{10}\,m^4s^{-1}$). The channel has a length of $4000km$ and a width of $1500km$, corresponding to $\ell = 4/3$. This configuration has been investigated for isolated topography in various papers, mainly to study the elements of the balance of zonal momentum. We have performed a systematic study for the sinusoidal topography in a wide range of parameter values (see Völker 1995).

4.1 Flow patterns

The circulation of the numerical model displays a lot of variability which is mostly caused by meandering jets and eddy activity. Figure 10 shows an instantaneous streamfunction field in a model run with equal layer depths. There are two clearly distinguishable meandering jets shedding off an intensive eddy field. These jets can still be identified in the time mean field which otherwise is much smoother: Figure 11 shows a broad band of eastward flow in both layers that is deflected equatorward over the topographic elevation and poleward over the depression, in accordance with large-scale

layer 1, year 151

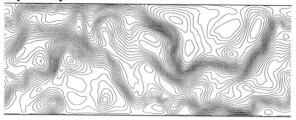

layer 2

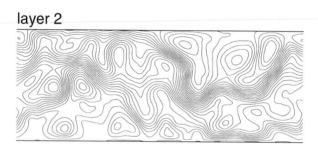

Figure 10: *Instantaneous streamfunctions in the two model layers, as calculated by the eddy-resolving quasigeostrophic model. The topography amplitude is $b_0 = 500$ m, the two layers have equal mean heights $H_i = 2500$ m, the reduced gravity is $g'/g = 0.001$ and the amplitude of the windstress is $\tau_0 = 10^{-4} m^2 s^{-2}$. The isolines are placed at regular intervals of $5000 m^2 s^{-1}$.*

barotropic vorticity constraints. Although the phase shift of this deflection relative to the topography is small and not visible in the figures, it is sufficient to induce a large topographic form stress.

Apart from the synoptic scale variability, the system has temporal variability on much longer timescales. Figure 2 (in the introduction) displays the transport and the potential energy of an integration over 180 years using the same parameter set as for the Figures 10 and (11). The potential energy gives a low-pass filtered view and we can clearly identify significant variations in the range from years to a decade.

The time mean balance of such experiments has been investigated by many authors (e. g. McWilliams et al. 1978, Wolff and Olbers 1989, Treguier and McWilliams 1990, Wolff et al. 1991). These studies have

layer 1, time mean

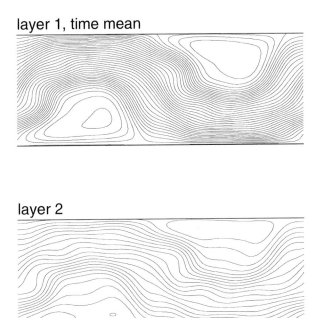

layer 2

Figure 11: *Time mean streamfunction in the same numerical experiment as Figure (10).*
Same isoline spacing.

manifested the momentum balance of the ACC as described in the introduction. In particular, the important role of the synoptic-scale eddies in
the transport of QPV and momentum has been pointed out. To make any
comparison between these eddy resolving experiments and the low-order
model we have to specify a representation of the eddy fluxes in the low-
order model. We have chosen the simple linear interfacial friction as a
low-order image of the interfacial form stress caused by the eddies and neglected the Reynolds stress contribution. In view of the extensive research
on the topic of parameterizing QG eddies in the general atmospheric circulation (see e. g. the review of Hoskins 1983) our representation of the
downward momentum flux is very crude and can only be justified by the
simplicity of the model.

The constant friction parameter μ was obtained by fitting the total
exchange of momentum by the eddies between the layers to the linear

friction law which we have used above, i. e.

$$\frac{\mu}{H_1} \left[\bar{\Psi}_1 - \bar{\Psi}_2 \right] \Big|_0^Y = \int_0^Y dy \frac{f_0^2}{g'} \overline{\Psi_1' \frac{\partial \Psi_2'}{\partial x}} \tag{47}$$

where the bar stands for the time mean and primes for deviations (the eddy part). With our parameter values we obtain values ranging from $\mu = 4 \cdot 10^{-7} s^{-1}$ for a flat bottom channel to $\mu = 1.4 \cdot 10^{-7} s^{-1}$ for a topography height of $500m$. Taking the low values in this range we find that the low-order model possesses only one steady state with a flow structure quite similar to the time mean field in Figure 11 but with transports that are about a factor of two to high. In this parameter range it is not possible to fit the momentum transfer between the layers to the numerical model using a linear frictional approach and in the same time obtain reasonable transports.

Using values from the high end of the range we can however find steady states of the low-order model with flow patterns that are very similar to Figure 11 without showing much difference in the transports. Figure 12 illustrates the steady states for $\mu = 2.9 \cdot 10^{-7} s^{-1}$ and $b_0 = 500m$. At this value of interfacial friction the model has three distinct steady states, of which only the low-index state is stable.

There is a remarkable similarity between the low-index state and the time mean field from the numerical model concerning the general flow structure, i. e. the direction of the deflection over topography, the strength of the band of flow and the weaker deflection in the lower layer than in the upper layer. There is also a qualitative agreement in the momentum balances, as can be seen from table 1. While the volume transport in the first layer of the numerical model and of the low-index state do not differ much, the transport in the lower layer differs roughly by a factor of 1.5.

While the agreement may be qualified as good at this comparatively low topography height, serious differences between low-order model and numerical simulation occur for higher topography amplitudes. This manifests itself mostly in the momentum balances. We therefore attribute this failure of the model to oversimplified parametrization of the interfacial friction.

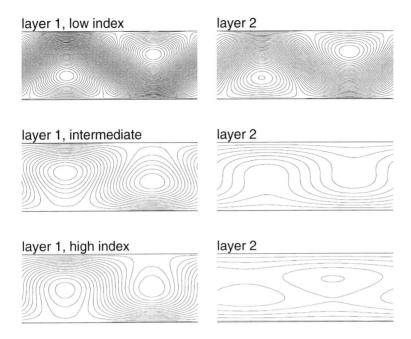

Figure 12: *The three steady states of the low-order-model for* $\mu = 2.9 \cdot 10^{-7} s^{-1}$, *all other parameters as in the numerical model. Isoline spacing as in the pictures above.*

	Transports [Sv]			Mom balance [%]			
	T	T_1	T_2	IS	IT	T	F
numerical	643	404	239	74	25	71	29
low-index	809	446	363	72	28	54	46
intermediate	224	106	118	104	-4	85	15
high index	250	165	85	74	26	89	11

Table 1: *Zonal volume transports and area integrated zonal momentum balance in the numerical model run and in the steady states of the low-order model for* $\mu = 2.9 \times 10^{-7} s^{-1}, b_0 = 500m$. *The volume transport is given for both and each of the two layers, the terms in the momentum balances are in percent of the total momentum input by the wind stress. The terms are: interfacial form stress by standing eddies (IS), interfacial form stress by transient eddies respectively interfacial friction (IT), bottom form stress (T) and bottom friction (F).*

4.2 Model tests

By projecting the streamfunction fields of the numerical model onto the basis functions of the low-order-model at regular time intervals, we obtain 'experimental' time series of the low-order-model variables. We cannot expect that these time series fulfill the equations of the low-order model because of the neglection of relative vorticity and because of the interaction with unresolved modes. But we can test whether the difference between model and data can reasonably well be represented as an uncorrelated white noise. We exemplify these test with the barotropic equations (30) and (31).

The residuum $r(t)$ between model and 'data' is calculated by inserting the 'experimental' time series. We consider this residuum as a realization of a random process with mean zero. We have assured that the mean of r is significantly lower than the other terms in the equation. Furthermore, the hypothesis is that the true spectrum C_{rr} of $r(t)$ is constant. The test of the hypothesis follows the general procedure for a multivariate least-square fit as described e. g. in Hasselmann (1979). The noise level is fitted by minimizing an error function

$$\delta^2 = \sum_{i,j} \sigma_{ij}^{-1}(C_{rr}(f_i) - \hat{C}_{rr}(f_i))(C_{rr}(f_j) - \hat{C}_{rr}(f_j)) \qquad (48)$$

where $C_{rr}(f_i)$ is the value of the spectrum at frequency f_i from the hypothesis, $\hat{C}_{rr}(f_i)$ is the spectrum as estimated from the data and σ_{ij} is the covariance matrix of the spectral estimator for the frequencies f_i, f_j. Formulas for the spectral estimator and its variance have been taken from Jenkins and Watts (1968) and Bendat and Piersol (1986). Under the hypothesis of a true model and approximately Gaussian distributed errors in the spectral estimation the error function δ^2 is a χ^2-distributed random variable. We reject the hypothesis if the value of δ^2 is above the $\chi^2_{n\alpha}$-quantile of the χ^2-distribution, where $\alpha = 0.05$ according to a significance level of 95% and the number of degrees of freedom n is the number of independent spectral estimates minus one.

Figure 13 shows as an example the estimated spectrum of the residuum of equation (30). The spectrum may be considered constant in the low frequency range but it is surely not white in the high frequency part. The reason is that the neglection of relative vorticity limits the applicability

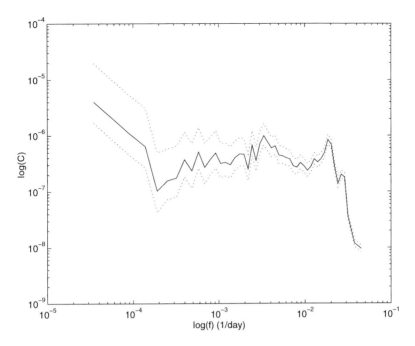

Figure 13: *Estimated spectrum of the residuum of equation (30) with 95% confidence intervals.*

of our model to timescales longer than the periods of typical barotropic Rossby waves. Restricting the test to a low-frequency range ($0 \leq f \leq 0.002$ $1/day$) we accept the hypothesis for the equations (30) and (31) for the $\alpha = 0.05$ quantile. The test have also been performed for the baroclinic equations with differing success: whenever the terms resulting from the linear friction parametrization are explicitly included in the test, the error bounds become large and the test is not very discriminating, i. e. the model must be accepted.

5   Conclusions

Compared to the research on the variability of the atmospheric circulation our knowledge of oceanic variability is still in its infancy, both in terms of the observational background as well as understanding the mechanisms. Direct observations of variations of the large-scale ocean circulation are very sparse, in particular outside the tropical regions where longer time scales are appropriate. In fact, it is fair to say that noteworthy theoretical studies of aperiodic variations of large-scale current systems have only been performed for the tropical processes associated with ENSO (see e. g. the review by Neelin et al. 1994). Here modelling efforts were strongly guided and verified by observations. On the contrary, other areas of recent research on temporal variability, such as the thermohaline circulation (e. g. Rahmstorf 1995) or coupled modes of the ocean–atmosphere system (see e. g. Latif et al. 1996, this volume), still lack observational support.

Our investigation on the existence of multiple equilibria and the temporal variability of zonally periodic ocean currents – with the Antarctic Circumpolar Current (ACC) as the only member of this species – has a similarly staggering foundation. There are virtually no observations to support our results, even numerical models which comprise the appropriate dynamical and thermohaline physics (i. e. eddy resolving primitive equation physics) have not been employed to study our topic. Therefore, apart from the pure investigation of a dynamical system, the results of this paper may serve to analyse and interpret future numerical experiments and observational data. In addition, the paper points out some benefits and limitations of low-order modelling.

To summarize the main results: we have developed a low-order image of wind forced flow in periodic zonal channel, governed by quasigeostrophic β-plane dynamics. The investigation supplements and extends previous studies in this area of Charney–DeVore type models in so far as it applies oceanic parameters and includes stratification and non-linear wave interactions. We have derived and discussed the dynamical balance and stability of the existing steady states and unraveled the complex aperiodic and chaotic temporal behavior. This analysis could be performed for the entire range of model parameters (topographic height, friction coefficients, channel dimensions and reduced gravity) since the model equations are completely solved by analytical techniques. The model described here has nine degrees of freedom, it is embedded in a more elaborate low-order model with 21 degrees of freedom (Völker 1995) which resolves a larger number of wave modes.

Finally, we have tested the balance and temporal behavior of this low-order model against 'observed data' obtained from a numerical simulation with an eddy resolving quasigeostrophic model in the identical configuration. We accept the hypothesis that the slow temporal variability of the numerical model is explained by low-order vacillations. This applies strictly to the barotropic components, the tests for baroclinic components are overshadowed by the necessity to parameterize the momentum and vorticity transports due to transient eddies by an extremely simplified representation in terms of a linear interfacial friction law in the low-order image.

Bendat, J.S., and A.G. Piersol (1986): *Random Data. Analysis and measurement procedures, 2nd ed.* John Wiley and sons, New York

Benzi, R., Saltzman, B., and A. Wijn-Nielsen (1986): *Anomalous atmospheric flows and blocking.* Advances in Geophysics Bd. 29, Academic Press, Orlando

Charney, J.G., and A. Eliassen (1949): A numerical method for predicting the perturbations of the middle latitude westerlies. *Tellus* 1: 38 – 54

Charney, J.G., and J.G. DeVore (1979): Multiple flow equilibra in the atmosphere and blocking. *Journal of the Atmospheric Sciences* 36: 1205 – 1216

Charney, J.G., and D.M. Straus (1980): Form drag instability, multiple equilibra, and propagating planetary waves in baroclinic, orographically forced, planetary wave systems. *Journal of the Atmospheric Sciences* 37: 1157 – 1176

Charney, J.G., Shukla, J., and K.C. Mo (1981): Comparison of a barotropic blocking theory with observation. *Journal of the Atmospheric Sciences* 38: 762 – 779

De Swart, H.E., and J. Grassman (1987): Effect of stochastic perturbations on a low-order spectral model of the atmospheric circulation. *Tellus* 39A: 10–24

Egger, J. (1978): Dynamics of blocking highs. *Journal of the Atmospheric Sciences* 35: 1788–1801

Egger, J. (1981): Stochastically driven large-scale circulations with multiple equilibra. *Journal of the Atmospheric Sciences* 38: 2606 – 2618

The FRAM-Group (1991): An eddy-resolving model of the southern ocean. *EOS, Transactions of the American Geophysical Union* 72: 169 – 175

Guckenheimer, J., and P. Holmes (1993): *Nonlinear oscillations, dynamical systems and bifurcations of vector fields, 4th ed.* Springer Verlag, New York

Hasselmann, K. (1979): Linear statistical models. *Dyn. Atmos. Oceans* 3: 501–521

Hoskins, B.J. (1983): Modelling of the transient eddies and their feedback on the mean flow. In: *Large-scale dynamical processes in the atmosphere*, Ed. Hoskins, B.J., and R.P. Pearce: Academic Press, New York

Jenkins, G.M. and D.G. Watts (1968): *Spectral analysis and its applications.* Holden-Day, San Francisco, Düsseldorf

Krupitsky, A., and M.A. Cane (1994): On topographic pressure drag in a zonal channel. *Journal of Marine Research* 52: 1 – 23

Marshall, J., Olbers, D., Ross, H., and D. Wolf-Gladrow (1993): Potential vorticity constraints on the dynamics and hydrography of the southern ocean. *Journal of Physical Oceanography* 23: 465 – 487

McWilliams, J.C., Holland, W.R., and J.H. Chow (1978): A description of numerical Antarctic Circumpolar Currents. *Dynamics of Atmospheres and Oceans* 2: 213 – 291

Munk, W.H., and E. Palmén (1951): Note on the dynamics of the Antarctic Circumpolar Current. *Tellus* 3: 53 – 55

Neelin, J.D., Latif, M., and F.-F. Yin (1994): Dynamics of coupled ocean–atmosphere models: the tropical problem. *Ann. Rev. Fluid Mechanics* 26: 617–659

Olbers, D., Wübber, C., and J.O. Wolff (1992): The dynamical balance of wind and buoyancy driven circumpolar currents. *Berichte aus dem Fachbereich Physik* 32. Alfred-Wegener-Institut für Polar- und Meeresforschung

Olbers, D. (1994): Links of the Southern Ocean to the global climate. In: *Modelling Oceanic Climate Interactions*, Ed. Willebrand J., Anderson D.L.T.:Springer-Verlag, Berlin

Rahmstorf, Stefan (1995): Bifurcations of the Atlantic thermohaline circulation in response to changes in the hydrological cycle. *Nature*, 378: 145 – 149

Treguier, A.M., and J.C. McWilliams (1990): Topographic influences on wind-driven, stratified flow in a β-plane channel: an idealized model for the Antarctic Circumpolar Current. *Journal of Physical Oceanography* 20: 321 – 343

Tung, K.K., and A.J. Rosenthal (1985): Theories of multiple equilibria – a critical reexamination. Part i: barotropic models. *Journal of the Atmospheric Sciences* 42: 2804 – 2819

Völker, C. (1991): *Ein niederdimensionales Modell einer turbulenten Kanalströmung.* Diplomarbeit, Universität Bremen.

Völker, C. (1995): *Barokline Strömung über periodischer Topographie: Untersuchungen an analytischen und numerischen Modellen.* PhD Thesis, Universität Bremen

Whitworth III, T., and R.G. Peterson (1985): The volume transport of the Antarctic Circumpolar Current from three-year bottom pressure measurements. *Journal of Physical Oceanography* 15: 810 – 816

Wijn-Nielsen, A. (1979): Steady states and stability properties of a low-order barotropic system with forcing and dissipation. *Tellus* 31: 375 – 386

Wolff, J.O., Maier-Reimer, E., and D.J. Olbers (1991): Wind-driven flow over topography in a zonal β-plane channel: a quasi-geostrophic model of the Antarctic Circumpolar Current. *Journal of Physical Oceanography* 21: 236 – 264

SPECTRAL METHODS: WHAT THEY CAN AND CANNOT DO FOR CLIMATIC TIME SERIES

MICHAEL GHIL[1] and PASCAL YIOU[2]
Ecole Normale Supérieure
Paris, France

Contents

Abstract

The analysis of time series — uni- or multivariate — is one of the high roads to our understanding of climatic variability. This classical field of study has recently been revitalized by the discovery and implementation of a number of new methodologies for extracting useful information from time series, as well as for interpreting the information so obtained in terms of dynamical systems theory. In this chapter, we describe the connections between time-series analysis and nonlinear dynamics, discuss signal-to-noise enhancement, and present some of the novel methods for spectral analysis. The various steps, as well as the advantages and disadvantages of these methods, are illustrated by their application to a well-known climatic time series, the Southern Oscillation Index. Open questions and further prospects conclude the chapter.

[1]also at: Institute of Geophysics and Planetary Physics, Los Angeles, USA
[2]Laboratoire de Modélisation du Climat et de l'Environnement, Saclay, France

NATO ASI Series, Vol. I 44
Decadal Climate Variability
Dynamics and Predictability
Edited by David L. T. Anderson and Jürgen Willebrand
© Springer-Verlag Berlin Heidelberg 1996

1 Wherefrom: Introduction and Motivation

The guiding idea of time-series analysis is that certain basic properties of a physical, biological or socio-economic system that generated the recorded time series can be detected by its analysis. These properties — subject to validation by using additional information, such as other (segments of) time series generated by the same system — can then be put to good use by helping predict the system's future behavior.

1.1 Analysis in the time vs. the spectral domain

Two basic approaches to time series analysis are associated with the time domain or spectral domain. Both are most easily understood in the linear context in which the physical sciences have operated for most of the last two centuries. In this context, the physical system can be described by a (system of) linear, ordinary or partial, differential or difference equation(s), subject to additive random forcing. The (constant) coefficients a_k of a scalar ordinary difference equation (OΔE), the simplest case of this type,

$$X(t+1) = \sum_{k=1}^{M} a_k X(t - M + k) + \xi(t), \tag{1}$$

determine its solutions $X(t)$ at discrete times $t = 0, 1, \ldots, j, \ldots$. In Eq. (1) the random forcing $\xi(t)$ is assumed to be white in time, i.e., uncorrelated from t to $t + 1$, and Gaussian at each t. Computing the coefficients a_k from a realization of X having length N, $\{X(t), 1 \le t \le N\}$, is the Yule (1927)–Walker (1931) method for the time-domain approach; this method is discussed further in Sec. 3.2 below, where (1) is treated as an auto-regressive (AR) process of order M.

The spectral-domain approach arises from the observation that — besides a quantity being constant — the most regular, and hence predictable, behavior of an observable is to be periodic. This approach then proceeds by determining the periodicities embedded in the time series. The classical implementation of this approach is based on the Bochner-Khinchin-Wiener theorem (Box and Jenkins 1970) which states that the lag-autocorrelation function of a time series and its spectral density are Fourier transforms of each other. Hannan (1960) provides an introduction to this approach and its implementation that excels by its brevity and clarity; the so-called Blackman-Tukey implementation is presented in Sec. 3.1 below.

The remainder of this chapter is organized, over all, as follows. Section 2 deals mainly with signal-to-noise ratio (S/N) enhancement and introduces singular-spectrum analysis (SSA) as an important and flexible tool for this enhancement. Statistical tests for the reliability of SSA results are also discussed. In Sec. 3, we present, in succession, three methods of spectral analysis: Blackman-Tukey, maximum entropy and multi-taper. Both Secs. 2 and 3 use the Southern Oscillation Index (SOI) for the purposes of illustrating the methods "in action". In Sec. 4, the multivariate extension of single-channel SSA is introduced and a few additional applications are mentioned or illustrated. The chapter concludes with a section on open questions, from the point of view of both the methodology and its applications.

1.2 Time series and nonlinear dynamics

Before proceeding with the technical details, we give in this subsection a quick perspective on the "nonlinear revolution" in time-series analysis. In the 1960s and '70s, the scientific community found out that much of the irregularity in observed time series, which had traditionally been attributed to the above-mentioned random "pumping" of a *linear* system by infinitely many (independent) degrees of freedom (*d-o-f*), could be generated by the *nonlinear* interaction of a few *d-o-f* (Lorenz 1963, Ruelle and Takens 1971, Smale 1967). This realization of the possibility of deterministic aperiodicity or "chaos" (Gleick 1987) created quite a stir and the purpose of this review is to describe briefly some of its implications for time-series analysis (Drazin and King 1992), with a special view to climatic time series.

A connection between deterministically chaotic time series and the nonlinear, possibly low-order, dynamics generating them was attempted fairly early in the young history of "chaos theory." The basic idea was to exploit specifically a univariate time series with apparently irregular behavior, generated by a deterministic or stochastic system, in order to (i) ascertain whether the underlying system has a finite number of *d-o-f*, by establishing an upper bound on this number; (ii) to verify that the observed irregularity arises from the fractal nature of the deterministic system's invariant set, as reflected by the fractional, rather than integer, character of (one of) its dimension(s); and (iii) reconstruct the invariant set or even the equations governing the dynamics from the data.

This ambitious program (Packard et al. 1980, Roux et al. 1980, Ruelle 1981) relied essentially on the *method of delays*, based in turn on the Whitney (1936) embedding lemma and the Mañé (1981)–Takens (1981) theorems. Heuristically, here is the sequence of steps involved. It is easy to transform a single nonlinear ordinary differential equation (ODE) of order n,

$$X^{(n)} = F(X^{(n-1)}, \ldots, X^{(j)}, \ldots, X), \tag{2}$$

into a system of first-order ODEs,

$$\dot{X}_i = f_i(X_1, \ldots, X_j, \ldots, X_n), \quad 1 \le i, j \le n; \tag{3}$$

here $X^{(n)} \equiv d^n X/dt^n$ and $\dot{X} \equiv dx/dt$. It suffices to write

$$X \equiv X_1, \ \dot{X}_1 = X_2, \ \ldots, \ \dot{X}_n = F(X_1, \ldots, X_n), \tag{4}$$

so that $f_1 = X_2$, $f_2 = X_3, \ldots, f_n = F$.

Equation (3) is considered as a fairly general description of a differentiable dynamical system in continuous time (Arnold 1973, 1983) and we are interested in the case in which, say, only a time series $X_j(t)$ were known. For the solutions of such a system to be irregular, i.e., other than (asymptotically) steady or periodic, three or more *d-o-f* are necessary. Can one then go from (3) to (2) just as easily as in the opposite direction? The answer, in general is "no"; hence, a slightly more sophisticated procedure needs to be applied. This procedure, in some sense, tries to imitate the Yule-Walker inference of (1) from $X(t)$, $t = 0, \ldots, N$. First of all, one acknowledges that the data $X(t_k) = X_k$ are typically given at discrete times $t_k = k\Delta t$ only. Next, one admits that, at first, it is hard to get actually the right-hand sides f_i; instead one attempts to reconstruct the invariant set on which the solutions of (3) that satisfy certain constraints lie.

In the case of conservative, Hamiltonian systems (Lichtenberg and Lieberman 1991), deterministically irregular motion is often called "stochastic." In this case, there are typically (unique) solutions through every point in phase space and the irregularity is associated with the intricate structure of *cantori* (Wiggins 1988), complicated sets of *folded* tori characterized by a given energy of the solutions lying on them. These cantori have, in particular, finite and fractional dimension, being *fractals* (Mandelbrot 1982).

Hamiltonian systems, however, are — mathematically speaking — structurally unstable (Smale 1967) in the (function) space of all differentiable dynamical systems, while — physically speaking — "open" systems in which energy is gained externally and dissipated internally — abound in nature. Therefore, climatic time series, as well as most other time series from nature or the laboratory, are more likely to be generated by forced dissipative systems (Lorenz 1963; Ghil and Childress 1987, Ch. 5). The invariant sets associated with irregularity here are "strange attractors" (Ruelle and Takens 1971), towards which all solutions tend asymptotically, i.e., long-term irregular behavior in such systems is associated with these attractors. These objects are also fractal, although rigorous proofs to this effect have been much harder to give than in the case of Hamiltonian cantori (Guckenheimer and Holmes 1983, Lasota and Mackey 1994).

The idea of Mañé (1981), Ruelle (1981) and Takens (1981) — as recently developed further by Sauer et al. (1991) — was that a single observed time series $X_j(t)$ or, more generally, $\phi(X_1(t), \ldots, X_n(t))$, could be used to reconstruct the attractor of a forced dissipative system, due essentially to the fact that such a solution covers the attractor densely, i.e., as time increases, it will pass arbitrarily close to any point on the attractor. The finite length and sampling rate, as well as the significant measurement noise associated with observed time series, have limited the applicability of this ingenious idea mostly to time series generated numerically or by laboratory experiments in which sufficiently long series could be obtained and noise was controlled better than in nature. In the climate context, Lorenz (1969) already had shown — by a judicious interpretation of results obtained while using more classical statistical methods — that the recurrence time of sufficiently good analogs for large-scale atmospheric motions was of the order of hundreds of years, at the spatial (and hence spectral) resolution of the observational network then available for the Northern Hemisphere.

The next-best target for demonstrating from an observed time series the deterministic cause of its irregularity was to show that the presumed system's attractor had a finite and fractional dimension. Of the various dimensions, metric and topological, that can be defined (Kaplan and Yorke 1979, Farmer et al. 1983), the one that became the most popular, since easiest to compute, was the correlation dimension (Grassberger and Procaccia 1983). While in other applications its computation proved rather reliable, and hence useful, climatic time series tended again to be rather too short and noisy for comfort (see for instance, Ghil et al. 1991 and Ruelle 1990 for a review of this controversial topic).

A more robust connection between classical spectral analysis and nonlinear dynamics seems to be provided by the concept of "ghost limit cycles." The road to chaos (Eckmann 1981) proceeds from stable equilibria, or *fixed points*, through stable periodic solutions, or *limit cycles*, and on through quasi-periodic solutions lying on *tori*, to *strange attractors*. The fixed points and limit cycles are road posts on this highway from the simple to the complex and, even after having lost their stability to successively more complex and realistic solutions, leave their imprint on the observed spatial patterns or time series generated by the system.

Consider the periodic solution shown in Fig. 1a as embedded in Euclidean three-dimensional phase space. It is neutrally stable in the direction tangent to itself, and in the plane perpendicular to this tangent it is asymptotically stable in one direction and unstable in the other, as shown in the Poincaré section of Fig. 1b. In a multi-dimensional phase space, it

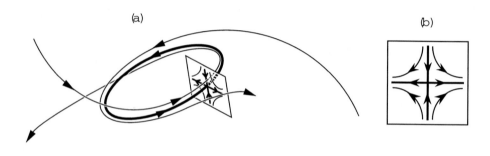

(a)　　　　　　　　　　　　　　　　　　　　　　　　(b)

Figure 1: *The schematic diagram of a ghost limit cycle. (a) A perspective sketch of the limit cycle in a three-dimensional Euclidean space. (b) Sketch of the flow lines in a Poincaré section (i.e., a plane intersecting transversally the limit cycle) in a neighborhood of the limit cycle. The shape of such a limit cycle need not be elliptic, as shown in the figure, except near the Hopf bifurcation point where it arises (courtesy of K. Ide).*

is plausible that the directions of stability are numerous or even infinite, while the directions of instability would still be few in number, for parameter values not too far from those at which the Hopf bifurcation that gave rise to the limit cycle in the first place occurs. Hence solutions of the full system would easily be attracted to this barely unstable limit cycle, follow it closely for one or a few turns, be ejected from its neighborhood, only to return later, again and again. The analogous picture for a "ghost fixed point" was illustrated in detail for an atmospheric model with 25 *d-o-f* by Legras and Ghil (1985); see also Ghil and Childress (1987; Figs. 6.12 and 6.18 there).

The episodes during which the solution circles near the ghost limit cycle result in nearly periodic parts of the time series and hence contribute to a spectral peak with that period. This concept was illustrated using 40 years of an atmospheric multivariate time series by Kimoto and Ghil (1993) for the so-called intraseasonal oscillations of the Northern Hemisphere (see also Ghil and Mo 1991a). We shall show in the subsequent sections of the present chapter how this concept can be generalized to associate multiple spectral peaks with a robust "skeleton" of the attractor, as proposed by Vautard and Ghil (1989).

2 S/N Enhancement: Singular Spectrum Analysis

2.1 Motivation

SSA is designed to extract information from short and noisy time series, and thus provides insight into the (unknown or partially known) dynamics of the underlying system that generated the series (Broomhead and King 1986a, Vautard and Ghil 1989). We outline here the method for univariate time series and generalize for multivariate ones in Sec. 4. The starting point is to *embed* a time series of observables $\{X(t), t = 1, \ldots, N\}$ in a vector space of dimension M, with M presumably larger than the effective but unknown dimension d of the underlying system. For topological reasons (Whitney 1936), the embedding dimension M must be larger than $2d + 1$ (Broomhead and King 1986a). The embedding procedure constructs a sequence $\{\tilde{X}(t)\}$ of M-dimensional vectors from the original time series X, by using lagged copies of the data $\{X(t)\}$,

$$\tilde{X}(t) = (X(t), X(t+1), \ldots, X(t+M-1)), \tag{5}$$

with $t = 1, \ldots, N - M + 1$. SSA allows one to unravel the information embedded in the delay-coordinate phase space by decomposing the sequence of augmented vectors thus obtained into elementary patterns of behavior in the time and spectral domains. It does so by providing data-adaptive filters that help separate the time series into statistically independent components, which can be classified essentially into (nonlinear) trend(s), deterministic oscillations, and noise.

SSA has been applied to the study of paleoclimatic time series (Vautard and Ghil 1989, Yiou et al. 1994, Yiou et al. 1995), interdecadal climate variability (Ghil and Vautard 1991, Allen and Smith 1994, Plaut et al. 1995), as well as interannual (Rasmusson et al. 1990, Keppenne and Ghil 1992) and intraseasonal (Ghil and Mo 1991a,b) oscillations. SSA algorithms and their properties have been investigated further by Penland et al. (1991), Allen (1992) and Vautard et al. (1992). The software of Dettinger et al. (1995a) is built, mainly but not exclusively, around this technique.

2.2   Decomposition and reconstruction

In this section, we illustrate the fundamental SSA formulae with a classical example of a climatic time series, the Southern Oscillation Index (SOI). SOI is a climatic index connected with the recurring El Niño conditions in the tropical Pacific; it is essentially the monthly mean difference in sea-level pressure between Darwin, Australia and Tahiti. In the data set we use here, the annual cycle was removed and the time series was normalized by its variance. The time interval considered goes from January 1942 to December 1990, during which no observations are missing at either station. The SOI so defined is shown in Fig. 2, and has $N = 588$ data points, one point for each month of record.

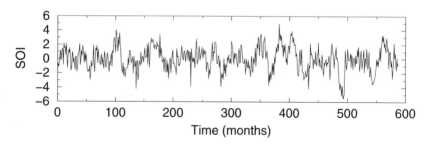

Figure 2: *Variations of the Southern Oscillation Index (SOI) between January 1942 and December 1990. Time on the abscissa in months and SOI on the ordinate normalized by its standard deviation.*

SSA is based on calculating the principal directions of extension of the sequence of augmented vectors $\{\tilde{X}(t), t = 1, \dots, N\}$ in phase space. The $M \times M$ covariance matrix C_X is computed, and its eigenelements $\{(\lambda_k, \rho_k), k = 1, \dots, M\}$ are obtained by solving

$$C_X \rho_k = \lambda_k \rho_k. \tag{6}$$

An equivalent formulation of Eq. (6), which will prove useful further on, is given by forming the $M \times M$ matrix E_X having the eigenvectors ρ_k as its columns and the diagonal matrix Λ_X composed of the eigenvalues λ_k, in deceasing order:

$$E_X^t C_X E_X = \Lambda_X, \qquad (7)$$

where E_X^t is the transpose of E_X. Each eigenvalue λ_k gives the degree of extension (and hence the variance) of the time series in the direction given by the orthogonal eigenvectors ρ_k. S/N separation is obtained by plotting the eigenvalue spectrum (Fig. 3) and distinguishing between an initial steep slope containing the signal, and noise characterized by lower values and a flat floor or very mild slope (Kumaresan and Tufts 1980, Pike et al. 1984, Vautard and Ghil 1989). As the matrix C_X is symmetric, standard

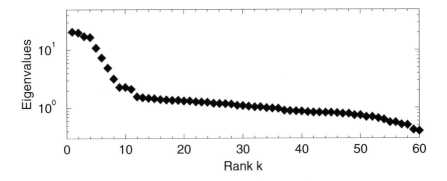

Figure 3: *Singular spectrum of the SOI time series with the eigenvalues plotted in decreasing order. The embedding dimension for Figs. 2.2 to 2.6 is $M = 60$, which represents a time window of 5 years.*

algorithms of decomposition (Press et al. 1988) will perform the numerical task efficiently. The method draws its name from the fact that often (Broomhead and King 1986a) singular-value decomposition (SVD: Golub and Van Loan 1983) is used to obtain the square roots of λ_k, called *singular values* of the *trajectory matrix* that has the $N - M + 1$ augmented vectors $\tilde{X}(t)$ as its columns. By analogy with the meteorological literature, the eigenvectors ρ_k are called empirical orthogonal functions [EOFs: Vautard and Ghil (1989)]. The EOFs corresponding to the first four eigenvalues are shown in Fig. 4. Note that the two pairs of EOFs, $(1, 2)$ and $(3, 4)$, are in quadrature and each corresponds in Fig. 3 to a pair of nearly equal eigenvalues. Vautard and Ghil (1989) argued that, subject to certain statistical significance tests, discussed further below, such pairs correspond to

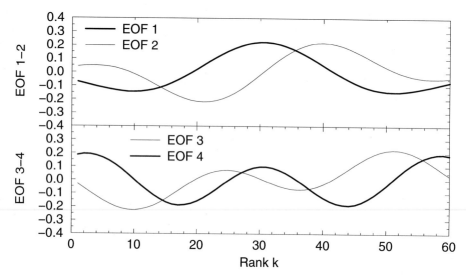

Figure 4: *First four EOFs of the SOI time series. The EOFs are grouped into two pairs:* $(1, 2)$ *and* $(3, 4)$.

the nonlinear counterpart of a sine-cosine pair in standard Fourier analysis of linear problems. In the terminology of our Sec. 1 here, such a pair gives a representation of a ghost limit cycle.

Projecting the time series onto each EOF yields the corresponding principal components (PCs) A_k:

$$A_k(t) = \sum_{j=1}^{M} X(t+j)\rho_k(j). \tag{8}$$

Figure 5 shows the variations of the four leading PCs. Again, the pairs of PCs $(1, 2)$ and $(3, 4)$ are in quadrature and strongly suggest periodic variability at two different periods, of about 4 and 2 years, respectively.

We can reconstruct that part of a time series that is associated with a single EOF or several by combining the associated PCs:

$$R_\mathcal{K}(t) = \frac{1}{M_t} \sum_{k \in \mathcal{K}} \sum_{j=1}^{M} A_k(t-j)\rho_k(j), \tag{9}$$

where $\mathcal{K}$ is the set of EOFs on which the reconstruction is based, and M_t is a normalization factor which is M for the central part of the time series and has slightly different values near its endpoints (Ghil and Vautard 1991, Vautard et al. 1992). The reconstructed components (RCs) have the

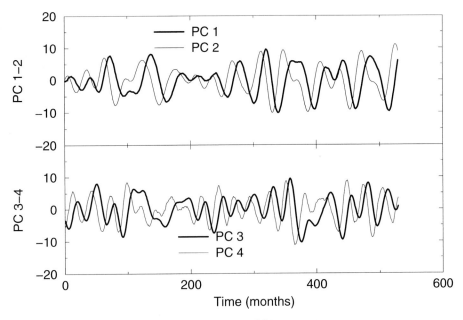

Figure 5: *First four PCs of the SOI time series.*

property of capturing the phase of the time series in a well-defined least-square sense, so that $X(t)$ and $R_\mathcal{K}(t)$ can be superimposed on the same time scale, $0 \le t \le N$, while the PCs have length $N - M$ and do not contain phase information within the window width M. No information is lost in the reconstruction because the sum of all individual reconstructed components gives back the original time series. Partial reconstruction is illustrated in Fig. 6 by summing the variability of PCs 1–4, associated with the two leading pairs of eigenelements and the quasi-periodic behavior isolated by them.

It is clear that the reconstruction (heavy solid curve in Fig. 6) is smooth and captures the essential part of interannual variability in the SOI: all large warm (El Niño) and cold (La Niña) events during the 49 years of record are captured as minima and maxima of the curve. Similar SSA results were obtained by Keppenne and Ghil (1992) for a slightly different treatment of the SOI, as well as by Jiang et al. (1995) for sea surface temperatures and by Unal and Ghil (1995) for sea level heights in the tropical Pacific.

Both the optimal selection of the S/N threshold and the reliable determi-

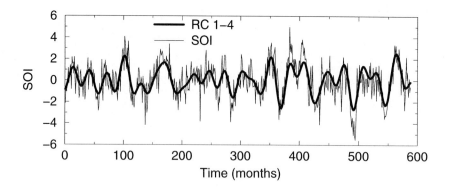

Figure 6: *Partial reconstruction of the SOI time series based on EOFs 1 to 4 (heavy solid). The raw SOI series is shown as the light solid curve.*

nation of oscillatory pairs depend on suitable criteria for statistical significance, which are treated in the next subsection. Subject to these caveats,
a clean signal — obtained by partial reconstruction over the correct leading set of indices $\mathcal{K}$ — can be analyzed, both visually and by using other
spectral-analysis tools. The Maximum Entropy Method (MEM), which we
describe in Sec. 3.2, works particularly well on signals so enhanced by SSA
(Penland et al. 1991).

2.3   Monte Carlo SSA

In the process of developing methodology for applying SSA to climatic time
series, a number of heuristic (Vautard and Ghil 1989, Ghil and Mo 1991*a*,
Unal and Ghil 1995) or Monte Carlo (Ghil and Vautard 1991, Vautard
et al. 1992) methods have been devised for S/N separation or the reliable
identification of oscillatory pairs. A selected subset of these methods have
been implemented in the current version of the SSA Toolkit (Dettinger et
al. 1995*a*). They are all essentially attempts to discriminate between the
significant signal as a whole, or individual pairs, and *white noise*, which
has a flat spectrum. A more stringent "null hypothesis" (Allen 1992) is
that of *red noise*, since most climatic and other geophysical time series
tend to have larger power at lower frequencies (Hasselmann 1976, Mitchell
1976, Ghil and Childress 1987).

In general, very straightforward tests can be devised to compare a given
time series with an "idealized" noise process: the spectrum of such a noise
process is known to have a particular shape, and if the data spectrum lies

above this theoretical noise spectrum, it is generally considered as "significant." This approach can be quite deceptive because a single realization of a noise process can have a spectrum that differs greatly from the theoretical one, especially if the number of data points is small; it is the (suitably weighted) average of such sample spectra over many realizations that will tend to the theoretical spectrum of the ideal noise process. Indeed, the Fourier transform of a single realization of a red-noise process can yield arbitrarily high peaks at arbitrarily low frequencies; such peaks could be attributed, quite erroneously, to periodic components. Therefore, more laborious tests have to be used, to determine error bars that are better adapted to each data set. Allen (1992) devised such tests comparing the statistics of simulated (or *surrogate*) red-noise time series with those of a given climatic time series. As mentioned in the previous paragraph, a number of Monte-Carlo-based tests against white-noise hypotheses have been published in the SSA literature. Still, Allen and Smith (1994, 1995) prefer to refer to their particular approach to such tests, against red noise, as "Monte Carlo SSA (MC-SSA)."

Red noise is a first-order auto-regressive, or AR(1), process u_t whose value at a time t depends on the value at time $t - 1$ only,

$$u_t = \gamma(u_{t-1} - u_0) + \alpha z_t + u_0; \qquad (10)$$

here z_t is a white-noise, or AR(0), process for which each value is independent of the previous one, while u_0 is the initial value and α and γ are deterministic coefficients.

The first step in MC-SSA is to determine the red-noise coefficients α and γ from the time series $X(t)$ using a maximum-likelihood criterion. Heuristic formulae with low bias are given by Allen (1992) and Allen and Smith (1996). Based on these coefficients, an ensemble of surrogate red-noise data can be simulated and, for each realization, a covariance matrix C_R is computed. These covariance matrices are then *projected* onto the eigenvector basis E_D of the original data by using Eq. (7) for their SVD,

$$\Lambda_R = E_D^t C_R E_D. \qquad (11)$$

Since (11) is not the SVD of that realization, the matrix Λ_R is not necessarily diagonal, like in Eq. (7), but it measures the resemblance of a given surrogate set with the data set. This resemblance can be quantified by computing the statistics of the diagonal elements of Λ_R. The statistical distribution of these elements, determined from the ensemble of Monte

Carlo simulations, gives confidence intervals outside which a time series can be considered to be significantly different from a generic red-noise simulation. For instance, if an eigenvalue λ_k lies above a 90% noise percentile, then the red-noise *null hypothesis* for the associated EOF (and PC) can be rejected with this confidence. Otherwise, that particular SSA component of the time series cannot be considered as significantly different from red noise.

As the next step in the analysis our SOI time series, we apply an MC-SSA noise test to it. In order to enhance the readability of the diagrams for the SSA spectra in the presence of MC-SSA error bars, we associate a dominant frequency with each EOF detected by SSA, as suggested by Vautard et al. (1992), and plot in Fig. 7 SSA eigenvalues (diamonds) vs. frequency, following Allen and Smith (1996). Such a plot is easier to interpret, with respect to the MC-SSA error bars, than plotting vs. SSA rank k as in Fig. 3. Care needs to be exercised, however, since the dominant-frequency estimate may be uncertain due to the possible anharmonicity of the EOFs, especially for low frequencies. The error bars in Fig. 7 represent 90% of the range of the eigenvalues obtained at that (dominant) frequency over the given ensemble of 1000 red-noise realizations (i.e., they denote the interval between the 5th and 95th percentile). Hence, the eigenvalues lying outside the bars have a 90% chance not to be due merely to red noise. The high values exhibit a significant quasi-biennial oscillation and an oscillatory component with a period between 5 and 6 years. The low values near 1 cycle yr^{-1} are probably associated with nonlinear side effects of the seasonal cycle (Jin et al. 1994, Tziperman et al. 1994) which have not been properly removed.

The MC-SSA algorithm described above can be adapted to eliminate *known* periodic components and test the residual against noise. This adaptation might provide a sharper insight into the dynamics captured by the data, since known periodicities (like orbital forcing on the Quaternary time scale or seasonal forcing on the intraseasonal-to-interannual one) often generate much of the variance at the lower frequencies manifest in a time series and alter the rest of the spectrum. Allen (1992) and Allen and Smith (1996) describe this refinement of MC-SSA which consists in restricting the projections given by Eq. (11) to the EOFs that do *not* account for known periodic behavior.

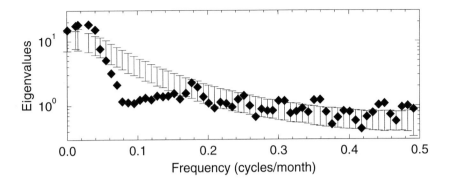

Figure 7: *Monte Carlo singular spectrum of the SOI time series. The diamonds indicate projections of the noise eigenvectors onto the data correlation matrix, cf. Eq. (11); the bars indicate the 5th and 95th noise percentiles. For each EOF, a characteristic frequency was estimated by maximizing its correlation with a sinusoid. Therefore, the frequency at which each diamond is plotted is just an average estimate.*

3 Spectral Analysis Methods

Both deterministic (Eckmann and Ruelle 1985) and stochastic (Hannan 1960) processes can, in principle, be characterized by a function of *frequency* (instead of *time*) called the power spectrum in the engineering or the spectral density in the mathematical literature. Thus, a very irregular motion (i.e., noise) possesses a smooth, continuous and rational spectrum, which indicates that *all* frequencies in a given band are excited by such a process. On the other hand, a purely periodic or quasi-periodic process is described by a single line or a (finite) number of lines in the frequency domain. Between these two extremes, nonlinear deterministic but "chaotic" processes can have spectral peaks superimposed on a continuous and wiggly background.

In theory, for a power spectrum to exist and be well defined, the dynamics generating the time series has to be ergodic and allow the definition of an invariant measure, with respect to which the spectrum is computed as an ensemble average. The reason for the spectral theory of linear random processes being more familiar is simply that the construction of such a measure is easier for them. In practice, the distinction between deterministically chaotic and truly random processes can be as tricky as the attempted distinctions based on the dimension of the invariant set (see

Sec. 1); here as there, the difficulty is due to the shortness and noisiness of climatic time series, but the estimation of "peaks" is intuitively more robust and easy to interpret than that of the continuous background.

The computation of the power spectrum of a random process is an ill-posed inverse problem (Jenkins and Watts 1968, Thomson 1982). For example, a straightforward calculation of the discrete Fourier transform of a random time series (which has a continuous spectral density) will provide a spectral estimate whose variance is equal to the estimate itself (Jenkins and Watts 1968, Box and Jenkins 1970). In the remainder of this section, we outline three techniques to reduce this variance that are commonly used for the spectral analysis of climatic time series and point out their respective properties, advantages and failings.

Each one of the three techniques outlined in Secs. 3.1–3.3 below — Blackman-Tukey, maximum entropy, and multi-taper — can provide error bars for the estimates it produces. Still, these bars are based on certain assumptions about the process generating the time series that is being analyzed. These assumptions are rarely, if ever, met in practice by the physical processes one wishes to examine. Therefore, we highly recommend to apply several independent techniques to any time series before drawing any conclusion about its spectrum.

3.1 Blackman–Tukey spectral estimate

A generic problem of time series analysis is the finiteness of the time interval on which the series is known. When the spectrum is estimated by a discrete Fourier transform, this corresponds to convoluting the true spectrum with the spectrum of a box-car function, which induces power leakage, due to the lobes of the Fourier transform of the box car. This systematic distortion of the spectrum adds to the problem of the variance of the spectral estimate. The Blackman-Tukey (1958) method gives an estimate of the power spectrum of a given time series $X(t)$ which reduces the estimate's variance and attenuates these leakage effects (Chatfield 1984). The starting point of this method is the so-called Bochner-Khinchin-Wiener identity, which states that the power spectrum P_X is equal to the Fourier transform of the auto-correlation function ϕ_X (Jenkins and Watts 1968; see also Sec. 1.1 here). Hence the power spectrum can be approximated based on a discretized version of this identity, by using the first $M + 1$

auto-correlation coefficients $\{\phi_X(k), k = 0, \ldots, M\}$:

$$P_X(\omega) \approx \sum_{k=0}^{M} w(k)\phi_X(k)e^{i\omega k}; \tag{12}$$

the truncation point $M < N$ as well as the weights $\{w(k), k = 0, \ldots, M\}$ have to be chosen carefully.

The result of (12) is smoother than the *periodogram*, i.e., the discrete Fourier transform of $X(t)$ itself, and provides a *consistent* estimate of the true spectrum, i.e., its variance converges to zero when $N \to \infty$ (which is not the case for the periodogram squared; see Hannan 1960 or Chatfield 1984). In (12), the choice of M is dictated by a trade-off between frequency resolution — the larger M the better — and the estimated variance, which is proportional to M/N — hence the smaller M the better (Kay 1988). Therefore, a rule of thumb is to take M no larger than $N/5$ or $N/10$, to avoid spurious results from high-variance estimates.

In addition, a *window* (or *taper*) $w(k)$ is applied to the data, so as to reduce the spectral leakage by lowering the side lobes of the box-car function. Such tapers are heuristically chosen as modified cosine functions, cubic functions or tent functions (Chatfield 1984). Classical Blackman-Tukey estimation deals largely with the choice of window shape, often called "window carpentry," and of the window width M, often called "opening and closing" the window. For example, choosing a truncation point $M = N - 1$ and a window $w(k)$ such that

$$w(k) = \frac{\sin(p\pi k/N)}{p\sin(\pi k/N)}, \quad k = 0, \ldots, N - 1, \tag{13}$$

with a given $p < N$, is equivalent to *smoothing* the periodogram over frequency "bins" of width $1/p$ (Chatfield 1984). We shall see in Sec. 3.3 how to choose, according to the *multi-taper* method (MTM), a set of optimal tapers and maximize therewith the resolution. Confidence levels relative to a pure red noise — or AR(1) process, which has a spectral slope of order $1/(1+\omega^2)$ — can be calculated for each window shape (Jenkins and Watts 1968), but the spectral resolution is generally poor if the number N of data points, and hence M, is low.

3.2 Maximum Entropy Method (MEM)

This method performs optimally when estimating line frequencies for a time series that is actually generated by a linear AR process (1) or order

M, AR(M). Exhaustive details can be found in Burg (1967) or Childers (1978).

Given a stationary time series $\{X(t), t = 1, \ldots, N\}$ of zero mean, $M'+1$ auto-correlation coefficients $\{\phi_X(k), k = 0, \ldots, M'\}$, are computed from it:

$$\phi_X(k) = \frac{1}{N+1-k} \sum_{t=0}^{N-k} X(t)X(t+k). \tag{14}$$

In the absence of the knowledge about the process that engenders the time series $X(t)$, M' is arbitrary and has to be optimized. The purpose of evaluating (14) is to determine the spectral density P_X, given by (12), which is equivalent to the most random, or least predictable process with the same auto-correlation coefficients ϕ. In terms of information theory (Shannon 1949), this corresponds to the concept of *maximal entropy*, hence the name of the method.

In practice, one determines the coefficients $\{a_k, k = 0, \ldots, M'\}$ from $X(t)$ by assuming that it is generated by an AR process and that its order $M = M'$. The autocorrelation coefficients $\phi_X(k)$ are computed and used to form the same Toeplitz matrix C_X as in SSA (Vautard and Ghil 1989, Penland et al. 1991, Vautard et al. 1992) and this matrix is then inverted using standard numerical schemes (Press et al. 1988) to yield the $\{a_k\}$. The spectral density P_X of the (true) AR process with coefficients $\{a_k, k = 0 \ldots, M\}$ is given by

$$P_X(\omega) = \frac{a_0}{\left| 1 + \sum_{k=1}^{M} a_k e^{ik\omega} \right|^2}, \tag{15}$$

where a_0 is the variance of the residual noise ξ in Eq. (1). Therefore, the knowledge of the $\{a_k, k = 0, \ldots, M'\}$ coefficients, determined from the time series $X(t)$, also yields an estimate of the power spectrum P_X.

An example of MEM estimates is given for the SOI time series in Fig. 8, using a number of lags $M = 10$, 20 and 40. It is clear that the number of peaks increases with M. A separation of two interannual peaks only occurs for $M = 40$, but is accompanied, unfortunately, by many spurious peaks at higher frequencies.

In general, if the time series is not stationary or otherwise not close to auto-regressive, great care in applying MEM — as well as cross-testing with the application of other techniques — is necessary. As the number of peaks in the spectrum increases with M, regardless of the spectral content of the

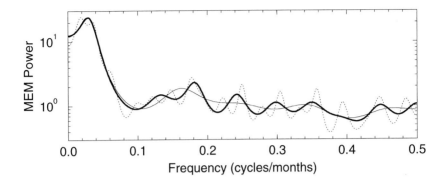

Figure 8: *MEM spectral estimates of the SOI time series. The frequency in this figure and subsequent ones is given in cycles per month. The autocorrelation orders are $M = 10, 20, 40$ (light solid, heavy solid, and dashed curves). The Akaike criterion (Haykin and Kesler 1983) predicts an order of $M \approx 10$, which is obviously too low.*

time series, an upper bound for M is generally taken as $N/2$. Heuristic criteria have been devised to refine the choice of a reasonable M (Haykin and Kesler 1983, Benoist 1986), based on minimizing the residual of a least-square fit between the AR approximation and the original time series (Akaike 1969, 1974; Haykin and Kesler 1983). The use of such criteria can be tricky because they tend to under- (Benoist 1986) or over-estimate (Penland et al. 1991) the order of regression of a time series, depending on its intrinsic characteristics.

The effects of the S/N enhancement performed by SSA decomposition (cf. Sec. 2) on MEM analysis are illustrated in Fig. 9, where the noise components identified by SSA were filtered out prior to MEM analysis. In this example, the power spectrum is much smoother than in Fig. 8. The regression order $M = 10$ suffices to separate a quasi-biennial (≈ 2.2 years) and a quasi-quadriennial (≈ 4.2 years) peak, while no spurious peaks at all appear at high frequencies. By contrast, Blackman-Tukey spectra of the same time series [not shown here, but see Rasmusson et al. (1990)] fail to separate these two peaks with sufficient statistical confidence.

3.3 Multi-Taper Method (MTM)

This method, like that of Blackman and Tukey (1958) in Sec. 3.1, is non-parametric, i.e. — unlike MEM — it does not use a specific, parameter-

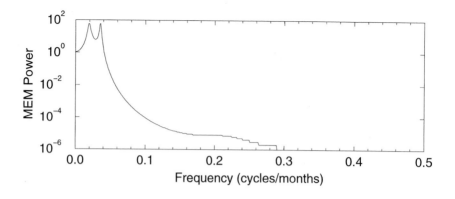

Figure 9: *MEM analysis of the SOI time series after removal of the noise identified by SSA. The auto-regression order is $M = 10$.*

dependent model of the process generating the time series. It attempts to reduce the variance of spectral estimates by using a (small) set of tapers (Thomson 1982, Percival and Walden 1993), rather than the unique data taper or spectral window used by Blackman-Tukey methods. A set of independent estimates of the power spectrum is computed, by pre-multiplying the data by orthogonal tapers which are built to minimize the spectral leakage due to the finite length of the data set. The optimal tapers, called discrete prolate spheroidal sequences (DPSS) and defined as the eigenvectors of a suitable Rayleigh-Ritz minimization problem, were extensively studied by Slepian (1978). Averaging over this (small) ensemble of spectra yields a better and more stable estimate — i.e., one with lower variance — than do single-taper methods (Thomson 1990b). As the tapers are explicitly designed to minimize the leakage outside a given bandwidth 2Ω, this method is less heuristic than traditional nonparametric techniques (Box and Jenkins 1970, Jenkins and Watts 1968).

Detailed algorithms for the calculation of these (DPSS or Slepian) tapers are given by Thomson (1990b), Percival and Walden (1993) and Rögnvaldsson (1993). In practice, only the most efficient tapers are retained: it turns out that only the first $\lfloor 2N\Omega \rfloor$ tapers have a close-to-minimal spectral leakage (Slepian 1978); thus the number of tapers K should always be less than $2N\Omega$. The choice of the bandwidth and number of tapers is therefore, again, a trade-off between frequency resolution and stability of the estimate (Thomson 1982). We show in Fig. 10 the first four Slepian tapers for a bandwidth parameter of $\Omega N = 4$; this means that (in this example) the effective spectral resolution is $4/N$, where N is the number of data points and 2Ω is the predetermined bandwidth, centered on a given line frequency outside which leakage is to be minimized. The maximum K is, accordingly, $K = 7$.

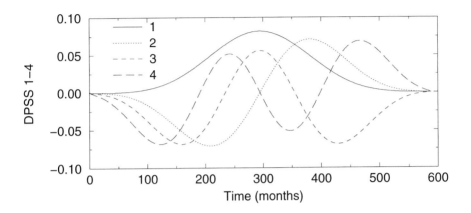

Figure 10: *The first four Slepian (1978) tapers for* $\Omega N = 4$, *computed for the* $N = 588$ *available points.*

Once the DPSS tapers $w_k(t)$ are computed for a chosen frequency bandwidth 2Ω, the total power spectrum P_X can be estimated by averaging the individual spectra given by each tapered version of the data set. We call $\hat{S}_k(\omega) \equiv |y_k(\omega)|^2$ the spectrum of the kth tapered time series $X(t)w_k(t)$, where y_k is the discrete Fourier transform (DFT) of Xw_k. The spectral estimate based on the first K tapers is, therewith,

$$\hat{S}(\omega) = \frac{1}{K} \sum_{k=0}^{K-1} \hat{S}_k(\omega). \tag{16}$$

Its resolution in frequency is $\pm\Omega$, which means that "line" components will be detected as peaks or bumps of width 2Ω. The multi-taper calculation (16) smoothes out, in principle, spurious irregularities and hence reduces the variance of the estimate.

What exactly is meant by a *line component* in the previous paragraph? The purpose of *harmonic analysis* is to determine the lines of a periodic or quasi-periodic signal, i.e., their frequency and amplitude. The Fourier transform of a clean periodic signal of infinite length yields a *Dirac function* at the frequency of the signal, viz., a line (or peak of zero width) with infinite magnitude. A spectral estimate based on the methods of Secs. 3.1 or 3.2 gives indirect information on the amplitude of the signal at a given frequency, through the area under the peak centered at that frequency and whose width is, roughly speaking, inversely proportional to the length N of the time series; this area is nearly constant, since the height of the peak is also proportional to N. Harmonic analysis attempts, instead, to determine directly the (finite) amplitude of a (pure) line in the spectrum of a time series of finite length. We explain next how this is done within MTM. Other approaches to this problem, closer in spirit to the periodogram mentioned in Sec. 3.1, are described by MacDonald (1989).

Let the time series $X(t)$ be the sum of a sinusoid of angular frequency $\hat{\omega}$ and amplitude μ, plus a "noise" $\xi(t)$ which is the sum of other sinusoids and white noise. One can then write

$$X(t) = \mu e^{i\hat{\omega}t} + \xi(t). \tag{17}$$

If $\{w_k(t), \ k = 0, \ldots, K-1\}$ are the first K tapers and $U_k(\omega)$ the DFT of w_k, a least-square fit in the frequency domain yields an estimate $\hat{\mu}$ of the amplitude μ:

$$\hat{\mu}(\hat{\omega}) = \frac{\sum\limits_{k=0}^{K-1} U_k^\star(0)\, y_k(\hat{\omega})}{\sum\limits_{k=0}^{K-1} |U_k(0)|^2}, \tag{18}$$

where the asterisk denotes complex conjugation. A statistical confidence interval can be given for the least-square fit (18) by a Fisher-Snedecor test, or *F-test* (Kendall and Stuart 1979). This test is roughly based on the ratio of the variance captured by the filtered portion of the time series $X(t)$, using K tapers, to the residual variance. By expanding the variance of the model (17), one finds that it is the sum of two terms,

$$\theta = |\hat{\mu}(\hat{\omega})|^2 \sum\limits_{k=0}^{K-1} |U_k(0)|^2 \tag{19}$$

and

$$\psi = \sum\limits_{k=0}^{K-1} |y_k(\hat{\omega}) - \hat{\mu}(\hat{\omega})U_k(0)|^2, \tag{20}$$

that are respectively the "explained" and "unexplained" contributions to the variance.

The random variable

$$F(\hat{\omega}) = (K - 1)\frac{\theta}{\psi} \qquad (21)$$

would obey a Fisher-Snedecor law with 2 and $2K - 2$ degrees of freedom if the time series $X(t)$ were a pure white-noise realization. One can interpret its numerical value for given data by assuming that $\mu = 0$ — i.e., that $X(t)$ is whit e— and trying to reject this hypothesis with the lowest probability of failure.

This harmonic-analysis application of MTM is able to detect low-amplitude oscillations in a relatively short time series with a high degree of statistical significance or to reject a large amplitude if it failed the F-test based on (21), because the F-value $F(\hat{\omega})$ does not depend — to first order — on the magnitude of $\hat{\mu}(\hat{\omega})$. This feature is an important advantage of MTM over the methods of Sec. 3.1, where error bars are essentially proportional to the amplitude of a peak (Jenkins and Watts 1968). It turns out in practice that the F-test above is robust to the white-noise assumption and still gives reasonably good results in the presence of colored noise.

The key assumption of this harmonic-analysis technique is that the time series is produced by a process that consists of a superposition of separate, purely periodic components. If not, a continuous spectrum (in the case of a colored noise or a chaotic system) will be broken down into spurious lines with arbitrary frequencies and possibly high F-values. This is a danger of the method, which can be partially avoided if the power spectrum is also computed by the methods of Secs 3.1 and 3.2, resulting in fairly sharp peaks that might hint at the presence of lines; it is also very important to vary the bandwidth parameter ΩN and the number of tapers K, to ensure the stability of the frequency and amplitude estimates. The application of MTM harmonic analysis to the SOI time series is given in Fig. 11. This analysis estimates correctly the presence of two interannual peaks, which it places at 56 and at 31 months; it breaks, however, the continuous spectrum up into numerous spikes, many of which pass the F-test at 95% and better, thus necessitating other tests in order to be accepted or rejected. Tests for red-noise detection, permitting S/N enhancement with MTM, have been devised by Mann and Lees (1996).

MTM has been applied to interdecadal climate variability (Kuo et al. 1990, Ghil and Vautard 1991, Mann and Park 1995, Mann et al. 1995) and paleoclimatic glaciation cycles (Thomson 1990a, b; Park and Maasch 1993, Yiou et al. 1995), among many other problems. Time-frequency analyses

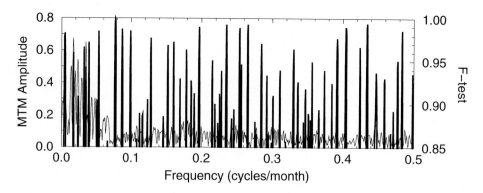

Figure 11: *MTM harmonic analysis of the SOI time series. Light-solid line is amplitude (shown on the left ordinate), while the heavy-solid line is F-test value (on the right ordinate). The bandwidth parameter is $\Omega N = 4$ and $K = 7$ tapers were used.*

(evolutive spectral analyses), using a moving window, were performed with this method by Yiou et al. (1991) and Birchfield and Ghil (1993).

4   M-SSA and Other Examples

Multi-channel SSA (or M-SSA) is a natural extension of SSA to time series of vectors or maps, such as time-varying temperature or pressure distributions over the globe. The use of (M-)SSA for such multivariate time series was proposed theoretically, in the context of nonlinear dynamics, by Broomhead and King (1986*b*) and applied to large-scale atmospheric fields by Kimoto et al. (1991) and Plaut and Vautard (1994). It is formally equivalent to extended EOF (EEOF) analysis (Weare and Nasstrom 1982, Lau and Chan 1985), except that the emphasis in EEOFs was on spatial resolution for a fixed and small window size in time, while in M-SSA spatial resolution is often sacrificed in the interest of a more exhaustive and flexible treatment in the time domain. SSA was applied in the context of ODEs and the reconstruction of (the skeleton of) their attractor; M-SSA is a generalization to partial differential equations (PDEs) and the study of the spatio-temporal structures that characterize the behavior of solutions on their attractor (Témam 1988, Constantin et al. 1989).

Our treatment here follows Plaut and Vautard (1994). Let $\{X_{l,n}, l = 1, \ldots, L, n = 1, \ldots, N\}$ be an L-channel time series with N data points.

We assume that X has (temporal) mean zero and is stationary. The cross-covariance matrix T_X for a chosen temporal lag M has the general block Toeplitz form:

$$
T_X = \begin{pmatrix}
T_{1,1} & T_{1,2} & \cdot & \cdot & & \cdot & & T_{1,L} \\
T_{2,1} & T_{2,2} & \cdot & & & & & \cdot \\
\cdot & \cdot & \cdot & \cdot & & & & \cdot \\
\cdot & & \cdot & \cdot & \cdot & T_{l,l'} & & \cdot \\
& & & & \cdot & & T_{L-1,L} & \\
\cdot & & & & \cdot & & \cdot & \\
T_{L,1} & & \cdot & \cdot & T_{L,L-1} & & T_{L,L}
\end{pmatrix},
\tag{22}
$$

where each $T_{l,l'}$ is an $M \times M$ lag-covariance matrix between channel l and l'. The "local" covariance matrices $T_{l,l'}$ can be estimated using a least-bias formula (Vautard et al. 1992, Plaut and Vautard 1994):

$$
(T_{l,l'})_{jj'} = \frac{1}{N - |j - j'|} \sum_{n=1}^{N-|j-j'|} X_{l,n} X_{l',n+j-j'}.
\tag{23}
$$

Other estimates for the elements of this matrix are also possible (Allen and Robertson 1996), and the spectral results' being independent of the specific formulation is an indication of their robustness.

The next step is to compute the eigenelements of the matrix T_X whose dimension is $LM \times LM$. The LM (distinct) eigenvectors (or EOFs) E^k, $k = 1, \ldots, LM$ are associated with LM (not necessarily distinct) eigenvalues λ_k. The EOF elements E_n^k, $n = 1, \ldots, LM$ can be re-indexed as $E_{l,j}^k$, $l = 1 \ldots L$ and $m = 1, \ldots, M$. The PCs A^k, $k = 1, \ldots, LM$ can be computed by projecting X onto the EOFs:

$$
A_n^k = \sum_{m=1}^{M} \sum_{l=1}^{L} X_{l,n+m} E_{l,m}^k,
\tag{24}
$$

where n varies from 1 to $N - M + 1$. The double summation accounts for the block matrix structure of T_X.

For a given set of indices $\mathcal{K}$, RCs can be obtained by convolving the corresponding PCs with the EOFs. Thus, the kth RC $R_{l,n}^k$, at time n and for channel l, is given by:

$$
R_{l,n}^k = \frac{1}{M_n} \sum_{m=1}^{M} A_{n-m}^k E_{l,m}^k.
\tag{25}
$$

with $M_n \neq M$ when n is smaller than M or larger than $N - M + 1$, as in Eq. (9). This type of reconstruction preserves the phase, as in the single-channel case, and the sum of all RCs gives back the original time series

(Plaut and Vautard 1994). The properties of pairs of near-equal eigenvalues that satisfy suitable significance criteria can also be transposed to M-SSA: RCs based on pairs isolate elementary spatio-temporal oscillations. The ideas of Monte Carlo noise tests can also be applied to M-SSA (Allen and Robertson 1996), by paying additional attention to the spatial structure(s) assumed by the null hypothesis.

An M-SSA application to low-frequency atmospheric variability is shown in Fig. 12, adapted from Plaut and Vautard (1994). M-SSA is applied to a time series of geopotential height anomalies at 70 kPa, covering 45 boreal winters between 1949 and 1994. The analysis is performed on five-day averages (pentads), using ten channels ($L = 10$) obtained as the leading PCs of a spatial EOF analysis (Preisendorfer 1988) of the anomaly fields defined over the Pacific sector (30°–70°N, 140°–260°E). The annual cycle was removed and a time window of 180 days ($M = 36$) is used. The succession of panels in Fig. 12 shows the progression of an oscillating pattern, keyed to the phase index associated with the EOF pair $(11, 12)$; the associated period is 40–45 days. This spatio-temporal evolution could be related to the 50-day Madden-Julian (1971, 1972) oscillation in the tropical Indo-Pacific Ocean or to the —possibly independent — extratropical oscillation reviewed by Ghil et al. (1991) and studied further by Marcus et al. (1994, 1996) and Strong et al. (1993, 1995), whose period is closer to 40 days.

M-SSA has also been applied to both observations and model results on intraseasonal variability by Kimoto et al. (1991) and Keppenne and Ghil (1993). It helped in the study of seasonal-to-interannual variability in the tropical Pacific, for the analysis of data (Jiang et al. 1995) and model simulations (Robertson et al. 1995a, b), as well as to investigate interannual and interdecadal variability in global sea level data (Unal and Ghil 1995) and in United States surface air temperatures (Dettinger et al. 1995b). The channels can be the leading spatial PCs, as in the example of Fig. 12, or a set of low-resolution grid-point time series, as in Jiang et al. (1995), Robertson et al. (1995a, b) or Unal and Ghil (1995).

5   Whereto next?

We have introduced in Sec. 1 the concept of a *ghost limit cycle* (Fig. 1) as the underpinning of the dynamic usefulness of a spectral peak. In a number of instances, from intraseasonal oscillations (Ghil et al. 1991) through paleoclimatic variability (Ghil 1994), the robust identification of an SSA

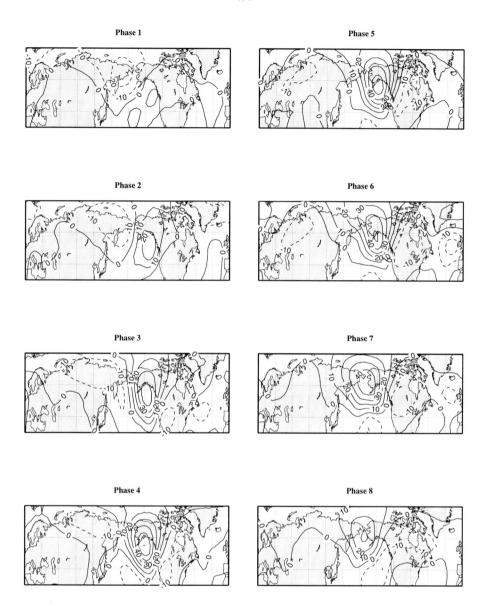

Figure 12: *Composites of 700 mb geopotential height anomalies (in meters) over the Pacific sector; the 8 composites correspond to equidistant phases covering one half of the life cycle of the 40–45 day oscillation isolated as an M-SSA pair (courtesy of R. Vautard).*

oscillatory pair in observations or proxy records has gone hand in hand with the description of a Hopf bifurcation in a (hierarchy of) model(s);

the latter gives rise, via an oscillatory instability, to a limit cycle that eventually becomes destabilized in turn.

The signal-to-noise (S/N) enhancement described in Sec. 2 led in fact, in the case of seasonal-to-interannual variability taken as the Ariadne's thread of methodological exposition, to the robust identification of separate 4/1-, 4/2- and 4/3-year peaks. This separation of the peaks and description of the associated spatio-temporal variability, in data (Rasmusson et al. 1990, Jiang et al. 1995) and coupled general circulation models (Robertson et al. 1995a, b) has helped bolster the case for the Devil's staircase scenario provided by simple and intermediate models of the El Niño/Southern Oscillation (ENSO: Jin et al. 1994, 1996; Tziperman et al. 1994). This scenario for the interaction between the intrinsic ENSO cycle of 2–3 years (depending on models and parameters) and the seasonal cycle could provide the cause for both the warm events occurring in boreal winter and the irregularity of their spacing from year to year.

A similar dialog between spectral studies of (inter)decadal variability and dynamical studies of oscillatory model behavior is taking shape. Interdecadal spectral peaks at 10–11, 14–15 and 25–27 years were identified by combining the application of the various spectral methods described in Sec. 3, in both global (Ghil and Vautard 1991, Vautard et al. 1992) and regional (Plaut et al. 1995) time series. Using (M-)SSA-related but distinct methods, Allen and Smith (1994) described the spatial patterns associated with the peakks at 10–11 and 25–27 years, while the peak at 14–15 years was confirmed by SSA or coral-reef proxy records (Quinn et al. 1993) and its associated spatial pattern described using MTM-related methods of intrumental temperature data (Mann and Park 1995). On the other hand, a plethora of oscillatory mechanisms is emerging in two- (Quon and Ghil 1995) and three-dimensional (Weaver et al. 1993, Chen and Ghil 1995, Rahmstorf 1995) models of the ocean's thermohaline circulation [(THC); see summary table, with further references, in Ghil (1994)], as well as in coupled ocean-atmosphere models (Delworth et al. 1993, Chen and Ghil 1996). As in the case of ENSO variability, a full hierarchy of models will probably be needed to clarify the connections between the (dimly) observed and the (imperfectly) modeled oscillations.

On the methodological side, there is much to be done in terms of refining statistical significance tests for each one of the methods presented here, as well as extending further their multi-channel applications. Beyond this, we showed that SSA can follow well, via RCs, variations in signal amplitude associated with a fairly broad spectral peak (Ghil and Mo 1991a, Plaut and Vautard 1994). MTM-based harmonic analysis can be adapted, by using a sliding window, to follow the slow shift in time of a sharp line's frequency (Yiou et al. 1991, Birchfield and Ghil 1993). When applied to time series (Flandrin 1993), wavelet analysis seems to provide an even more natural way of following a quasi-adiabatic or sudden change in the natural frequency of a climatic oscillator (Meyers et al. 1993, Yiou et al. 1995). Suitable significance criteria for wavelet analysis results need to be developed and a tantalizing question (Yiou 1994) is whether data-adaptive wavelet bases, combining in some sense SSA with wavelets, can be developed, not unlike the way that SSA and MEM turned out to "be made for each other."

Acknowledgments

It is gratifying to thank our SSA, MEM and MTM collaborators, M. R. Allen, P. Billant, M. D. Dettinger, N. Jiang, C. Keppenne, M. Kimoto, M. Mann, J. D. Neelin, C. Penland, G. Plaut, A. R. Robertson, C. M. Strong, R. Vautard, Y. S. Unal and W. Weibel for the pleasure of working together and exchanging ideas; we hope that this review gives credit to the joint work and does not betray their, occasionally different, ideas. Comments and Fig. 1 from K. Ide, as well as Fig. 12 from R. Vautard, are greatly appreciated. M.G.'s SSA work is supported by NSF grant ATM90-13217 and P.Y.'s by the French Commissariat à l'Energie Atomique. M.G. is also indebted to the French Académie des Sciences for the visiting Chair that provided the necessary leisure for the writing. This is IGPP publication no. 4607 and LMCE publication no. 00368.

References

Akaike H. (1969) Fitting autoregressive models for prediction. *Ann. Inst. Statist. Math.* **21**:243–247

Akaike H. (1974) A new look at the statistical model identification. *IEEE Trans. Autom. Control* **19**:716–723

Allen M. (1992) Interactions Between the Atmosphere and Oceans on Time Scales of Weeks to Years. PhD thesis St. John's College Oxford

Allen M. R., Robertson A. W. (1996) Distinguishing modulated oscillations from coloured noise in multivariate datasets. *Clim. Dyn.* (submitted)

Allen M., Smith L. A. (1994) Investigating the origins and significance of low-frequency modes of climate variability. *Geophys. Res. Lett.* **21**:883–886

Allen M. R., Smith L. A. (1996) Monte Carlo SSA: detecting irregular oscillations in the presence of coloured noise. *J. Clim.* (to appear)

Arnold V. I. (1973) *Ordinary Differential Equations.* MIT Press, Cambridge MA London England

Arnold V. I. (1983) *Geometrical Methods in the Theory of Ordinary Differential Equations.* Springer-Verlag, New York Heidelberg Berlin

Benoist J. P. (1986) Analyse Spectrale de Signaux Glaciologiques: Etude des Glaces Sédimentaires Déposées à Dome C, Morphologie du Lit d'un Glacier. Thèse d'Etat USMT Grenoble

Birchfield G. E., Ghil M. (1993) Climate evolution in the Pliocene-Pleistocene as seen in deep sea $\delta^{18}O$ records and in simulations: internal variability versus orbital forcing. *J. Geophys. Res.* **98**(D6):10385–10399

Blackman R. B., Tukey J. W. (1958) *The Measurement of Power Spectra From The Point of View of Communication Engineering.* Dover, New York

Box G. E. P., Jenkins G. M. (1970) *Time Series Analysis, Forecasting and Control.* Holden-Day, San Francisco

Broomhead D. S., King G. P. (1986*a*) Extracting qualitative dynamics from experimental data. *Physica* **D20**:217–236

Broomhead D. S., King G. P. (1986*b*) On the qualitative analysis of experimental dynamical systems. In: Sarkar S. (ed.) *Nonlinear Phenomena and Chaos.* Adam Hilger, Bristol pp 113–144

Burg J. P. (1967) Maximum entropy spectral analysis. In: 37th Ann. Intern. Meeting. Soc. Explor. Geophys., Oklahoma City, Oklahoma

Chatfield C. (1984) *The Analysis of Time Series: An Introduction.* 3 edn Chapman and Hall, New York

Chen F., Ghil M. (1995) Interdecadal variablity of the thermohaline circulation and high-latitude surface fluxes. *J. Phys. Oceanogr.* **25**(11):2547–2568

Chen F., Ghil M. (1996) Interdecadal variability in a hybrid coupled ocean-atmosphere model. *J. Phys. Oceanogr.* (in press)

Childers D. G. (ed.) (1978) *Modern Spectrum Analysis.* IEEE Press, New York

Constantin P., Foias C., Nicolaenko B., Témam R. (1989) *Integral Manifolds and Inertial Manifolds for Dissipative Partial Differential Equations.* Springer-Verlag, New York

Delworth T., Manabe S., Stouffer R. (1993) Interdecadal variations of the thermohaline circulation in a coupled ocean-atmosphere model. *J. Clim.* **6**:1993–2011

Dettinger M. D., Ghil M., Strong C. M., Weibel W., Yiou P. (1995*a*) Software expedites singular-spectrum analysis of noisy time series. *Eos Trans. AGU* **76**(2):12, 20, 21 (available on the World Wide Web at http://www.atmos.ucla.edu/)

Dettinger M. D., Ghil M., Keppenne C. (1995*b*) Interannual and interdecadal variability in United States surface-air temperatures, 1910–87. *Clim. Change* **31**:35–66

Drazin P. G., King G. P. (eds.) (1992) *Interpretation of Time Series from Nonlinear Systems (Proc. of the IUTAM Symposium and NATO Advanced Research Workshop on the Interpretation of Time Series from Nonlinear Mechanical Systems).* University of Warwick, England North-Holland

Eckmann J.-P. (1981) Roads to turbulence in dissipative dynamical systems. *Rev. Mod. Phys.* **53**:643–654

Eckmann J.-P., Ruelle D. (1985) Ergodic theory of chaos and strange attractors. *Rev. Mod. Phys.* **57**(3):617–656 and **57**:1115

Farmer J. D., Ott E., Yorke J. A. (1983) The dimension of chaotic attractors. *Physica D* **7**:153–180

Flandrin P. (1993) *Temps-fréquence.* Hermes, Paris

Ghil M. (1994) Cryothermodynamics: the chaotic dynamics of paleoclimate. *Physica* **D77**:130–159

Ghil M., Childress S. (1987) *Topics in Geophysical Fluid Dynamics: Atmospheric Dynamics, Dynamo Theory and Climate Dynamics.* Springer-Verlag, New York

Ghil M., Mo K. C. (1991*a*) Intraseasonal oscillations in the global atmosphere. Part I: Northern hemisphere and tropics. *J. Atmos. Sci.* **48**(5):752–779

Ghil M., Mo K. C. (1991*b*) Intraseasonal oscillations in the global atmosphere. Part II: Southern hemisphere. *J. Atmos. Sci.* **48**(5):780–790

Ghil M., Vautard R. (1991) Interdecadal oscillations and the warming trend in global temperature time series. *Nature* **350**(6316):324–327

Ghil M., Kimoto M., Neelin J. D. (1991) Nonlinear dynamics and predictability in the atmospheric sciences. *Rev. Geophys.* **36 Supplement** (U.S. National Report IUGG, 1987–91):46–55

Gleick J. (1987) *Chaos: Making a New Science.* Viking, New York

Golub G. H., Van Loan C. F. (1983) *Matrix Computations.* John Hopkins Univ. Press

Grassberger P., Procaccia I. (1983) Measuring the strangeness of strange attractors. *Physica* **D9**:189–208

Guckenheimer J., Holmes P. (1983) *Nonlinear Oscillations, Dynamical Systems and Bifurcations of Vector Fields.* Springer-Verlag, New York

Hannan E. J. (1960) *Time Series Analysis.* Methuen, New York

Hasselmann K. (1976) Stochastic climate models. *Tellus* **6**:473–485

Haykin S., Kesler S. (1983) Prediction-error filtering and maximum-entropy spectral estimation. In: Haykin S. (ed.) *Nonlinear Methods of Spectral Analysis.* Vol 34 of *Topics in Applied Physics.* Springer Verlag, Berlin pp 9–72

Jenkins G. M., Watts D. G. (1968) *Spectral Analysis and its Applications.* Holden-Day, San Francisco

Jiang N., Neelin D., Ghil M. (1995) Quasi-quadrennial and quasi-biennial variability in the equatorial pacific. *Clim. Dyn.* **12**:101–112

Jin F.-F., Neelin J. D., Ghil M. (1994) El Ninõ on the Devil's staircase: Annual subharmonic steps to chaos. *Science* **264**:70–72

Jin F.-F., Neelin J. D., Ghil M. (1996) El Niño/Southern Oscillation and the annual cycle: Subharmonic frequency-locking and aperiodicity. *Physica D* (accepted)

Kaplan J. L., Yorke J. A. (1979) Chaotic behavior of multidimensional difference equations. In: Peitgen H. O., Walther H. O. (eds.) *Functional Differential Equations and Approximation of Fixed Points.* Vol 730 of *Lecture Notes in Mathematics.* Springer-Verlag, New York pp 228–237

Kay S. M. (1988) *Modern Spectral Analysis: Theory and Applications.* Prentice-Hall

Kendall M., Stuart A. (1979) *The Advanced Theory of Statistics.* Vol 2 4 edn Macmillan, New York

Keppenne C. L., Ghil M. (1992) Adaptive filtering and prediction of the Southern Oscillation Index. *J. Geophys. Res.* **97**:20449–20454

Keppenne C. L., Ghil M. (1993) Adaptive filtering and prediction of noisy multivariate signals: An application to subannual variability in atmospheric angular momentum. *Intl. J. Bifurcation and Chaos* **3**:625–634

Kimoto M., Ghil M. (1993) Multiple flow regimes in the Northern Hemisphere winter. Part II: Sectorial regimes and preferred transitions. *J. Atmos. Sci.* **50**:2645–2673

Kimoto M., Ghil M., Mo K. C. (1991) Spatial structure of the extratropical 40-day oscillation. In: Proc. 8th Conf. Atmos. Oceanic Waves and Stability. Amer. Meteorol. Soc., Boston, MA pp 115-116

Kumaresan R., Tufts D. W. (1980) Data-adaptive principal component signal processing. In: Proc. Conf. Decision and Control. IEEE , Albuquerque pp 949–954

Kuo C., Lindberg C., Thomson D. J. (1990) Coherence established between atmospheric carbon dioxide and global temperature. *Nature* **343**:709–713

Lasota A., Mackey M. C. (1994) *Chaos, Fractals, and Noise: Stochastic Aspects of Dynamics*. Vol 97 of *Applied Mathematical Sciences* 2 edn Springer-Verlag, New York

Lau K. M., Chan P. H. (1985) Aspects of the 40-50 day oscillation during the northern winter as inferred from outgoing longwave radiation. *Mon. Wea. Rev.* **113**:1889–1909

Legras B., Ghil M. (1985) Persistent anomalies, blocking and variations in atmospheric predictability. *J. Atmos. Sci.* **42**:433–471

Lichtenberg A. J., Lieberman M. A. (1991) *Regular and Chaotic Dynamics*. Springer-Verlag, New York

Lorenz E. N. (1963) Deterministic nonperiodic flow. *J. Atmos. Sci.* **20**:130–141

Lorenz E. N. (1969) Atmospheric predictability as revealed by naturally occurring analogues. *J. Atmos. Sci.* **26**:636–646

MacDonald G. J. (1989) Spectral analysis of time series generated by non-linear processess *Rev. Geophys.* **27**:449–469

Madden R. A., Julian P. R. (1971) Detection of a 40-50 day oscillation in the zonal wind in the Tropical Pacific. *J. Atmos. Sci.* **28**:702–708

Madden R. A., Julian P. R. (1972) Description of global-scale circulation cells in the tropics with a 40-50 day period. *J. Atmos. Sci.* **29**:1109–1123

Mandelbrot B. (1982) *The Fractal Geometry of Nature*. 2nd edn Freeman, San Francisco

Mañé R. (1981) On the dimension of the compact invariant sets of certain non-linear maps. In: Rand D. A., Young L.-S. (eds.) *Dynamical Systems and Turbulence*. Vol 898 of *Lecture Notes in Mathematics*. Springer, Berlin pp 230–242

Mann M. E., Park J. (1995) Global-scale modes of surface temperature variability on interannual to century timescales. *J. Geophys. Res.* **99**(D12):25819–25833

Mann M. E., Lees J. M. (1996) Robust estimation of background noise and signal detection in climatic time series. *Clim. Change* (in press)

Mann M. E., Park J., Bradley R. S. (1995) Global interdecadal and century-scale climate oscillations during the past five centuries. *Nature* **378**:266–270

Marcus S. L., Ghil M., Dickey J. O. (1994) The extratropical 40-day oscillation in the UCLA general circulation model. Part I: Atmospheric angular momentum. *J. Atmos. Sci.* **51**:1431–1446

Marcus S. L., Ghil M., Dickey J. O. (1996) The extratropical 40-day oscillation in the UCLA general circulation model. Part II: Atmospheric angular momentum. *J. Atmos. Sci.* (accepted)

Meyers S. D., Kelly B. G., O'Brien J. J. (1993) An introduction to wavelet analysis in oceanography and meteorology: with application to the dispersion of Yanai waves. *Mon. Wea. Rev.* **121**:2858–2866

Mitchell J. M. (1976) An overview of climatic variability and its causal mechanisms. *Quatern. Res.* **6**:481–493

Packard N. H., Crutchfield J. P., Farmer J. D., Shaw R. S. (1980) Geometry from a time series. *Phys. Rev. Lett.* **45**:712–716

Park J., Maasch K. A. (1993) Plio-Pleistocene time evolution of the 100-kyr cycle in marine paleoclimate records. *J. Geophys. Res.* **98**:447–461

Penland C., Ghil M., Weickmann K. (1991) Adaptive filtering and maximum entropy spectra, with application to changes in atmospheric angular momentum. *J. Geophys. Res.* **96**(D12):22659–22671

Percival D. B., Walden A. T. (1993) *Spectral Analysis for Physical Applications.* Cambdridge University Press, Cambridge UK

Pike E. R., McWhirter J. G., Bertero M., de Mol C. (1984) Generalized information theory for inverse problems in signal processing. *Proc. IEE* **131**:660–667

Plaut G., Vautard R. (1994) Spells of low-frequency oscillations and weather regimes in the northern hemisphere. *J. Atmos. Sci.* **51**(2):210–236

Plaut G., Ghil M., Vautard R. (1995) Interannual and interdecadal variability in 335 years of Central England temperatures. *Science* **268**:710–713

Preisendorfer R. W. (1988) *Principal Component Analysis in Meteorology and Oceanography.* Elsevier, Amsterdam

Press W. H., Flannery B. P., Teukolski S. A., Vettering W. T. (1988) *Numerical Recipes: The Art of Scientific Computing.* Cambridge University Press

Quinn T. M., Taylor F. W., Crowley T. J. (1993) A 173 year stable isotope record from a tropical South Pacific coral *Quat. Sci. Rev.* **12**:407–418

Quon C., Ghil M. (1995) Multiple equilibria and stable oscillations in thermosolutal convection at small aspect ratio. *J. Fluid Mech.* **291**:33–56

Rahmstorf S. (1995) Bifurcation of the Atlantic thermohaline circulation in response to changes in the hydrological cycle. *Nature* **378**:145–149

Rasmusson E. M., Wang X., Ropelewski C. F. (1990) The biennial component of ENSO variability. *J. Mar. Syst.* **1**:71–96

Robertson A. W., Ma C.-C., Mechoso C. R., Ghil M. (1995*a*) Simulation of the Tropical-Pacific climate with a coupled ocean-atmosphere general circulation model. Part I: The seasonal cycle. *J. Clim.* **8**:1178–1198

Robertson A. W., Ma C.-C., Ghil M., Mechoso C. R. (1995*b*) Simulation of the Tropical-Pacific climate with a coupled ocean-atmosphere general circulation model. Part II: Interannual variability. *J. Clim.* **8**:1199–1216

Rögnvaldsson Ö. E. (1993) Spectral estimation using the multi-taper method. Technical Report RH-13-13 Science Institute, U. of Iceland Reykjavik

Roux J. C., Rossi A., Bachelart S., Vidal C. (1980) Representation of a strange attractor from an experimental study of chemical turbulence. *Phys. Lett. A* **77**:391–393

Ruelle D. (1981) Small random perturbations of dynamical systems and the definition of attractors. *Commun. Math. Phys.* **82**:137–151

Ruelle D. (1990) Deterministic chaos: the science and the fiction. *Proc. Roy Soc. London, Ser. A* **427**:241–248

Ruelle D., Takens F. (1971) On the nature of turbulence. *Commun. Math. Phys.* **20**:167–192 and **23**:343–344

Sauer T., Yorke J. A., Casdagli M. (1991) Embedology. *J. Stat. Phys.* **65**:579–616

Shannon C. E. (1949) Communication in the presence of noise. *Proc. I.R.E.* **37**:10–21

Slepian S. (1978) Prolate spheroidal wave functions, Fourier analysis and uncertainty-V: The discrete case. *Bell. Sys. Tech. J.* **57**:1371–1430

Smale S. (1967) Differentiable dynamical systems. *Bull. Amer. Math. Soc.* **73**:199–206

Strong C. M., Jin F.-F., Ghil M. (1993) Intraseasonal variability in a barotropic model with seasonal forcing. *J. Atmos. Sci.* **50**:2965–2986

Strong C. M., Jin F.-F., Ghil M. (1995) Intraseasonal oscillations in a barotropic model with annual cycle, and their predictability. *J. Atmos. Sci.* **52**:2627–2642

Takens F. (1981) Detecting strange attractors in turbulence. In: Rand D. A., Young L.-S. (eds.) *Dynamical Systems and Turbulence.* Vol 898 of *Lecture Notes in Mathematics.* Springer, Berlin pp 366–381

Témam R. (1988) *Infinite-Dimensional Dynamical Systems in Mechanics and Physics.* Springer-Verlag, New York

Thomson D. J. (1982) Spectrum estimation and harmonic analysis. *IEEE Proc.* **70**(9):1055–1096

Thomson D. J. (1990*a*) Time series analysis of Holocene climate data. *Phil. Trans. R. Soc. Lond. A* **330**:601–616

Thomson D. J. (1990*b*) Quadratic-inverse spectrum estimates: applications to palaeoclimatology. *Phil. Trans. R. Soc. Lond. A* **332**:539–597

Tziperman E., Stone L., Cane M. A., Jarosh H. (1994) El Niño chaos: overlapping of resonances between the seasonal cycle and the Pacific ocean-atmosphere oscillator. *Science* **264**:72–74

Unal Y. S., Ghil M. (1995) Interannual and interdecadal oscillation patterns in sea level. *Clim. Dyn.* **11**:255–278

Vautard R., Ghil M. (1989) Singular spectrum analysis in nonlinear dynamics, with applications to paleoclimatic time series. *Physica* **D35**:395–424

Vautard R., Yiou P., Ghil M. (1992) Singular spectrum analysis: a toolkit for short noisy chaotic signals. *Physica* **D58**:95–126

Weare B. C., Nasstrom J. N. (1982) Examples of extended empirical orthogonal function analyses. *Mon. Wea. Rev.* **110**:784–812

Weaver A. J., Marotzke J., Cummins P. F., Sarachik E. S. (1993) Stability and variability of the thermohaline circulation. *J. Phys. Oceanogr.* **23**:39–60

Whitney B. (1936) Differentiable manifolds. *Ann. Math.* **37**:645–680

Wiggins S. (1988) *Global Bifurcations and Chaos (Analytical Methods).* Springer-Verlag, New York

Yiou P. (1994) Dynamique du Paléoclimat: Des Données et des Modèles. PhD thesis Université Pierre et Marie Curie Paris 6

Yiou P., Genthon C., Jouzel J., Ghil M., Le Treut H., Barnola J. M., Lorius C., Korotkevitch Y. N. (1991) High-frequency paleovariability in climate and in CO_2 levels from Vostok ice-core records. *J. Geophys. Res.* **96**(B12):20365–20378

Yiou P., Ghil M., Jouzel J., Paillard D., Vautard R. (1994) Nonlinear variability of the climatic system, from singular and power spectra of Late Quaternary records. *Clim. Dyn.* **9**:371–389

Yiou P., Jouzel J., Johnsen S., Rögnvaldsson Ö. E. (1995) Rapid oscillations in Vostok and GRIP ice cores. *Geophys. Res. Lett.* **22**(16):2179–2182

Index

List of Participants

Dr. Magdalena ALONSO BALMADSEDA, Universidad de Alcala de Henares, Departamento de Física, Facultad de Ciencias, Campus Universitario, 28871 Alcala de Henares (Madrid), Spain

Mr. Rasmus BENESTAD, University of Oxford, Atmospheric, Oceanic and Planetary Physics, Clarendon Laboratory, Parks Road, Oxford OX1 3PU, U.K.

Prof. Dr. L. BENGTSSON, Director Max Planck Institut für Meteorologie, Bundesstrasse 55, D-20146 Hamburg, Germany

Dr. Roman BEKRYAEV, The Arctic and Antarctic Research Institute, 38 Bering St., 199397 St. Petersburg, Russia

Mr. Arne BIASTOCH, Institut für Meereskunde Kiel, Dept. Theoretische Ozeanographie, Düsternbrooker Weg 20, D-24105 Kiel, Germany

Ms. Ulrike BURKHARDT, Meteorologisches Institut der Universität München, Arbeitsgruppe für Theoretische Meteorologie, Theresienstrasse 37, D-80333 München, Germany

Dr. Maria Antonietta CAPOTONDI, NCAR, Boulder National Center for Atmospheric Research, Advanced Study Program, P.O. Box 3000, Boulder CO 80307-3000, U.S.A.

Mr. Fei CHEN, University of California, Dept of Atmospheric Sciences, 405 Hilgard Avenue, Los Angeles, CA 90024-1565, U.S.A.

Ms. Amy CLEMENT, Lamont-Doherty Observatory of Columbia, University, RT 9W, Palisades, NY 10964, U.S.A.

Dr. Susanna CORTI, Università di Bologna, Dipartimento di Fisica, Gruppo di Dinamica Atmosperica, Via Irnerio 46, I-40126 Bologna, Italy

Dr. Bob DICKSON, MAFF Fisheries Laboratory, Pakefield Road, Lowestoft, Suffolk NR33 OHT, U.K.

Dr. Peter DITLEVSEN, Niels Bohr Institute for Astronomy, Geophysics and Physics, Geophysical Dept, University of Copenhagen, Haraldsgade 6, 2200 Copenhagen N, Denmark

Mr. Gavin ESLER, University of Cambridge, DAMTP, Silver Street, Cambridge CB3 9EW, U.K.

Mr. Augustus FANNING, School of Earth and Ocean Sciences, University of Victoria, P.O. Box 1700, Victoria, B.C., Canada V8W 2Y2

Mr. Laurence FLEURY, Climate Modelling & Global Change, CERFACS (European Centre for Research & Advanced Training in Scientific Computing), 42, av. G. Coriolis, 31057 Toulouse Cedex, France

Prof. M. GHIL, UCLA, Institute for Geophysics and Planetary Physics, Los Angeles, California 90024-1567, U.S.A.

Miss Alessandra GIANNINI, Università di Bologna, Dipartimento di Fisica, Gruppo di Dinamica Atmosperica, Via Irnerio 46, I-40126 Bologna, Italy

Mr. Marco Andrea GIORGETTA, Max-Planck-Institut für Meteorologie, Bundesstrasse 55, D-20146 Hamburg, Germany

Mr. Martin GÖBER, Meteorologisches Institut der Universität Bonn, Auf Dem Hügel 20, D-53121 Bonn, Germany

Dr. Stephen GRIFFIES, Princeton University, Geophysical Fluid Dynamics Laboratory, Route 1, Forrestal Campus, Princeton, NJ 08542, U.S.A.

Mr. Robert HALLBERG, University of Washington, School of Oceanography, WB-10, Seattle, WA 98195, U.S.A.

Mr. Jim HANSEN, University of Oxford, Atmospheric, Oceanic and Planetary Physics, Clarendon Laboratory, Parks Road, Oxford OX1 3PU, U.K.

Mr. Wilco HAZELEGER, Royal Netherlands Meterological Institute, Oceanographic Research Department, P.O. Box 201, 3730 AE De Bilt, The Netherlands

Dr. Mark HOLZER, Canadian Centre for Climate Modeling and Analysis, Atmospheric Environment Service, University of Victoria, P.O. Box 1700, MS 3339, Victoria, B.C., Canada V8W 2Y2

Mr. Thierry HUCK, Laboratoire de Physique des Océans, Unité Mixte de Recherche no. 127, CNRS-IFREMER-UBO, UFR Sciences - B.P. 809, 29285 Brest Cedex, France

Mr. Stephen JEWSON, University of Oxford, Atmospheric, Oceanic and Planetary Physics, Clarendon Laboratory, Parks Road, Oxford OX1 3PU, U.K.

Dr. Ted JOHNSON, University College London, Dept of Mathematics, Gower Street, London WC1E 6BT, U.K.

Dr. Ercan KÖSE, Karadeniz Technical University, Faculty of Marine Sciences TR-61530 Camburnu, Trabzon, Turkey

Dr. M. LATIF, Max Planck Institut für Meteorologie, Bundesstrasse 55, D-20146 Hamburg, Germany

Mr. Gerrit LOHMANN, Alfred-Wegener Institut Bremerhaven, Dept of Theoretical Oceanography, Germany, (NOTE - visiting scholar at University of Gothenburg, Sweden Sept-Dec 1994)

Dr. Jochem MAROTZKE, Dept of Earth, Atmospheric and Planetary Sciences, MIT, Room 54-1514, Cambridge, MA 02139, U.S.A.

Mr. Paul-Antoine MICHELANGLI, Laboratoire de Meteorologie Dynamique, Université Pierre et Marie Curie - C.N.R.S, Tour 15 - 5ème étage - Case courrier 99, 4, place Jussieu, 75252 Paris Cedex, France

Dr. Ron MILLER, Columbia University , Goddard Institute for Space Sciences, Dept of Applied Physics, 2880 Broadway, New York, NY 10025, U.S.A.

Dr. Franco MOLTENI, CINECA, c/o ECMWF, Shinfield Park, Reading, Berkshire RG2 9AX, U.K.

Mr. Kjell Arne MORK, Geophysical Institute, University of Bergen, Allègaten 70, N-5007 Bergen, Norway

Mr. Matthias MÜNNICH, Max-Planck-Institut für Meteorologie, Bundesstrasse 55, D-20146 Hamburg, Germany

Dr. Ragu MURTUGUDDE, NASA Goddard Space Flight Center, Code 971/Bldg. 22, Greenbelt, MD 20771, U.S.A.

Miss Fay NORTLEY, University of Reading, Dept of Meteorology, 2 Earley Gate, Whiteknights Reading, Berks. RG6 2AU, U.K.

Prof. DR. D. OLBERS, Alfred-Wegener-Institut für Polar- und Meeresforschung, Postfach 12 01 61, D-27515 Bremerhaven, Germany

Dr. T. PALMER, ECMWF, Shinfield Park, Reading, Berkshire RG2 9AX, U.K.

Dr. Scott POWER, Bureau of Meteorology, Research Centre, G.P.O. Box 1289K, Lonsdale Street, Melbourne, Victoria, 3001, Australia

Mr. René REDLER, Institut für Meereskunde, Dept Theoretische Ozeanographie, Düsternbrooker Weg 20, D-24105 Kiel, Germany

Prof. Dr. Peter RHINES, WV-10, Dept. of Oceanography, University of Washington, Seattle, WA 98195, U.S.A.

Mr. Norel RIMBU, Bucharest University, Faculty of Physics, Dept of Atmospherics Bucuresti Magurele, P.O. Box 5211, Romania

Mr. Nils RIX, Institut für Meereskunde, Dept Theoretische Ozeanographie, Düsternbrooker Weg 20, 24105 Kiel, Germany

Ms. Carolyn ROBERTS, Meteorological Office, The Hadley Centre, London Road Bracknell, Berks. RG12 2SY, U.K.

Mr. Daniel ROBITAILLE, School of Earth and Ocean Sciences, University of Victoria, P.O. Box 1700, Victoria, B.C., Canada V8W 2Y2

Mr. Tomas N. SAHLEN, University of Stockholm, Department of Meteorology S-106 91 Stockholm, Sweden

Dr. E. SARACHIK, Dept of Atmospheric Sciences, AK-40, University of Washington, Seattle, WA 98195, U.S.A.

Prof. Michael SARNTHEIN-LOTICHIUS, Geologisch-Paläontologisches Institut, Universität Kiel, Ludewig-Meyn-Str. 10, D-24118 Kiel, Germany

Mrs. Nathalie SENNECHAEL, LODYC, Université Pierre et Marie Curie, Tour 14 - 2ème étage - Case 100, 4, place Jussieu, 75252 Paris Cedex 05, France

Dr. Tarmo SOOMERE, Max-Planck-Institut für Meteorologie, Bundesstrasse 55, D-20146 Hamburg, Germany

Dr. Sabrina SPEICH, LODYC, Université Pierre et Marie Curie, Tour 26 - 4ème étage - Case 100, 4, place Jussieu, 75252 Paris Cedex, France

Dr. David STEPHENSON, Climate Modelling & Global Change, CERFACS (European Centre for Research & Advanced Training in Scientific Computing), 42, av. G. Coriolis, 31057 Toulouse Cedex, France

Prof. Th. STOCKER, Physikalisches Institut, Universität Bern, Switzerland

Dr. W. Troy STRIBLING, NASA Goddard Space Flight Center, Code 971, Greenbelt, MD 20771, U.S.A.

Mr. Rowan SUTTON, University of Oxford, Atmospheric, Oceanic and Planetary Physics, Clarendon Laboratory, Parks Road, Oxford OX1 3PU, U.K.

Dr. Amit TANDON, Centre for Earth and Ocean Research, Gordon Head Complex, MS 4015, University of Victoria, P.O. Box 1700, Victoria, B.C., Canada V8W 2Y2

Dr Simon TETT, Meteorological Office, Room H207, The Hadley Centre, London Road, Bracknell, Berks. RG12 2SY, U.K.

Mr. Philippe TULKENS, Institute d'Astronomie et de Geophysique, Georges Lemaitre, Université Catholique de Louvain Chemin du Cyclotron 2, B-1348 Louvain-la-Neuve, Belgium

Prof. M. WALLACE, Dept of Atmospheric Sciences, AK-40, University of Washington, Seattle, WA 98195, U.S.A.

493

Dr. Nanne WEBER, Royal Netherlands Meterological Institute, P.O. Box 201, 3730 AE De Bilt, The Netherlands

Prof. Dr. J. WILLEBRAND, Institut für Meereskunde, an der Universität Kiel, Düsternbrooker Weg 20, D-24105 Kiel, Germany

Mrs. Sophie VALCKE, Institut de Mechanique de Grenoble, Laboratoire des Ecoulements Géophysiques et Industriels, BP 53X, 38041 Grenoble Cedex, France

Dr. Martin VISBECK, Center for Meteorology & Physical Oceanography, Dept of Earth, Atmospheric and Planetary Science, Massachusetts Institute of Technology, 54-1421, Cambridge, MA 02139-4307, U.S.A.

Dr. Michael WINTON, University of Washington, JISAO, GJ-40, Seattle, WA 98195, U.S.A.

The ASI Series Books Published as a Result of
Activities of the Special Programme on Global Environmental Change

This book contains the proceedings of a NATO Advanced Research Workshop held within the activities of the NATO Special Programme on Global Environmental Change, which started in 1991 under the auspices of the NATO Science Committee.

The volumes published as a result of the activities of the Special Programme are: